Wildlife

LAW ENFO

FOURTH EDITION

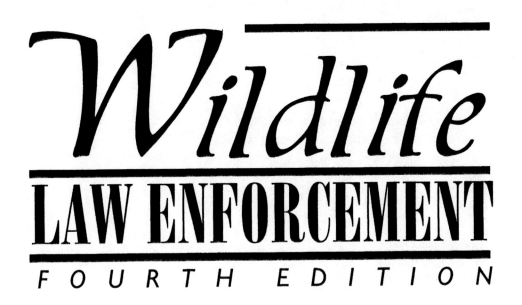

Wildlife
LAW ENFORCEMENT
FOURTH EDITION

WILLIAM F. SIGLER

Utah State University

Boston   Burr Ridge, IL   Dubuque, IA   Madison, WI   New York   San Francisco   St. Louis
Bangkok   Bogotá   Caracas   Lisbon   London   Madrid
Mexico City   Milan   New Delhi   Seoul   Singapore   Sydney   Taipei   Toronto

McGraw-Hill Higher Education

A Division of The McGraw-Hill Companies

Book Team

Editor *Lynne Meyers*
Developmental Editor *Tom Riley*
Production Editor *Audrey A. Reiter/Marla K. Irion*
Photo Editor *Lori Hancock*
Art Editor *Tina Flanagan*
Publishing Services Coordinator/Design *Barb Hodgson*

Vice President and General Manager *Beverly Kolz*
Vice President, Publisher *Earl McPeek*
Vice President, Director of Sales and Marketing *Virginia S. Moffat*
Vice President, Director of Production *Colleen A. Yonda*
National Sales Manager *Douglas J. DiNardo*
Marketing Manager *Amy Halloran*
Advertising Manager *Janelle Keeffer*
Production Editorial Manager *Renée Menne*
Publishing Services Manager *Karen J. Slaght*
Royalty/Permissions Manager *Connie Allendorf*

President and Chief Executive Officer *G. Franklin Lewis*
Senior Vice President, Operations *James H. Higby*
Corporate Senior Vice President, President of WCB Manufacturing *Roger Meyer*
Corporate Senior Vice President and Chief Financial Officer *Robert Chesterman*

Copyedited by Ruth Reeves

Cover/interior design by Schneck-DePippo Graphics
Cover image © R. Sanford/Allstock

Library of Congress Catalog Card Number: 94–070724

ISBN 0–697–20269–0

Printed in the United States of America by Wm. C. Brown Communications, Inc.,
2460 Kerper Boulevard, Dubuque, IA 52001

10 9 8 7 6

To Stacey, Luc, Adam, Marc, and Jim

Contents

Chapter 1

Historic to Present Role of Wildlife Law Enforcement   1

Chapter 2

Administration   17

Chapter 3

The Rights of the Individual   31

Chapter 4

State Jurisdiction over Wildlife   60

Chapter 5

Federal Jurisdiction over Wildlife   68

Index of Cases

Foreword

1994

In every profession there is the commonality of some kind of survey literature which establishes an outline of structure for building of knowledge. In the wildlife law enforcement profession we are fortunate to have a survey text that will readily provide this foundation. The fourth edition of *Wildlife Law Enforcement* reflects the evolution of wildlife law enforcement just as other related texts, for example, provide documentation on the evolution of wildlife management. Beginning students may understand from this edition that the role of wildlife law enforcement has expanded from that of limited fish and wildlife concerns to those of broad environmental matters. Thus, to those contemplating careers in the wildlife law enforcement profession, they will readily understand that the scope of knowledge required is broad indeed. No longer will a simplistic knowledge of fish or wildlife biology suffice for necessary knowledge of principles which must be employed in daily work. Wildlife officers, whether employed in state, federal, or private roles, must have a foundation of knowledge spanning the spectrum from social sciences to ecology. Advanced students, when studying this edition of the textbook, will find new topics also addressed. Administration, covert operations, recreational law enforcement, forensics, and use of aircraft are additions which will stimulate thought and perhaps inspire future research. For "old timers" in the profession, a reading of this edition presents an opportunity to grow in the profession. Information about new legislation, court decisions, and other matters that affect the profession as a whole may fill in gaps of knowledge that have happened as specialized work was pursued.

All who utilize this text will find a valuable source of information that might aid us in our journey to ensure that wildlife law enforcement achieves its full potential as a valuable tool of wildlife management.

A. Eugene Hester
Assistant Regional Director
Law Enforcement
U.S. Fish and Wildlife Service
Fourth Edition

Foreword

1980

Society's first concerns about wildlife in the United States were expressed in the form of proclamations protecting certain animals and encouraging the taking of others held injurious to humans or their property. In 1630, Massachusetts Bay Colony authorized payment of a bounty on wolves, with Virginia following suit shortly thereafter. On February 4, 1646, the Town of Portsmouth, Rhode Island, proclaimed a closed season on deer hunting from May 1 to November 1. Similar closures were practically universal throughout the colonies by 1720.

Much later—in the mid-1800s—the first organized efforts to stem wanton killing of wildlife were taken through citizens' game protective associations. Evolving from this concern came state and federal agencies responsible for protecting and managing fish and wildlife. Strong steps also were taken to curb interstate trafficking and commercialization of wildlife.

As this movement grew, there came demands for hiring individuals to police the lengthening list of laws and regulations on a full-time basis. Resentment of certain segments of society to what was regarded as government's interference with use of fish and wildlife is shown by the names law enforcement officers were called in those early days—catfish cops, brush apes, goose wardens and the like.

Monitoring society's compliance with restrictions on the taking and use of fish and wildlife is a basic element of conservation law enforcement. For the most part, the first fish and game wardens—as most were known—were more remarkable for their personal interest and knowledge of local people and geography than for their technical backgrounds and understanding of law enforcement techniques.

Fish and wildlife enforcement is and will remain an important element of state and federal conservation programs. Substantial sums are spent to train, equip and support law enforcement personnel. Many of today's conservation officers receive police academy training and an increasing number of agencies require applicants to possess a minimum of two years of college preparation. Some require candidates to have a four-year degree. The "catfish cops" of yesterday are today's respected guardians and stewards of increasingly important natural resources.

New laws have vastly broadened conservation officers' responsibilities: threatened and endangered species, marine mammals, international traffic in protected species, littering, watercraft, water pollution and in other areas unimaginable to early conservation officers. The rapid expansion and evolution of conservation law enforcement mandates continuing revision and updating of texts and other instructional materials. This revised edition of Dr. Sigler's *Wildlife Law Enforcement* makes a substantial contribution in that regard.

Daniel A. Poole
Wildlife Management Institute
Third Edition

Foreword

1980

The third edition of *Wildlife Law Enforcement* continues the work of the author in pioneering a book specifically designed for the wildlife law enforcement officer.

All too often the Law Enforcement Officer is referred to by terms which cast a dispersion upon the dignity of the individual and the work which society requires him to do. The public being sovereign, calls a man to perform a public service, underpays him and then permits him to be castigated and belittled by the very element of our society which all of us seek to be protected against—the violator of the laws.

In a sense, we Americans are a perverse people— that is, in the sense that we want a government, but want to be governed as little as possible. This is not a criticism, but a compliment to the nature and temperament of all Americans.

In ordinary talk we refer to professions. We think of our children as engaged in an educational pursuit which will lead them into a profession (i.e. a doctor, lawyer, clergyman, engineer, certified public accountant). Webster confines the definition of a professional person to one of those three learned professions: theology, law and medicine. Time, progress and the intricacies of social life with expanding populations make a broader definition of the professions more to the point. We often hear the expression that so and so is a professional, meaning, of course, that he is trained in the art that he pursues. Within the past ten years the pace of the law with respect to law enforcement agencies in general, and wildlife law enforcement agencies in particular, has increased to such a degree that it is impossible for an untrained individual to effectively become a law enforcement officer. And, therefore, it becomes imperative that the various agencies create a core of professionally trained wildlife law enforcement officers.

In using the word professional or professionalism, there is underscored the necessity of training, retraining, and updating by all means available in order to provide the professional with the tools of his trade, which in the case of the law enforcement officer, are knowledge, procedures and the kind of temperament requisite of any good law enforcement officer.

Enforcement is a harsh word. It implies compulsion. The connotation of the word compliance tends more to soothe the nerves and heart of the outdoor man and woman. You will find much in this volume devoted to public education which seeks to avoid the violation of law, as this is a preference to the capture and the conviction of the violator after the act. It is the prevention of illegal acts through public education, rather than conviction after violation, that results in the greatest conservation of our public resources and the greatest utilization of our wildlife resources. We live in an era where man, through his subdivisions, his industrial development, and his recreational pursuits has infringed upon the natural habitat of our wildlife resources many times. It becomes increasingly important for those wildlife enforcement officers in the field to be professional in their duties and knowledgeable of the tools they have to work with. Science is constantly expanding its fund of knowledge to provide the wildlife officer with evidentiary material. The courts have expanded their acceptance of scientific investigation, all of which aids the officer in the efficient and professional performance of his duties.

This work, now in its third edition, is a step in the right direction. It should instill in the minds of instructors and students the desire to attain a professional status and a more perfect understanding of the role played by those who are engaged in the pursuit of wildlife law enforcement.

George W. Preston
Third Edition

Foreword

1972

This compilation, with revisions, of the text on *Wildlife Law Enforcement* is long overdue. The author of this volume is to be complimented for pioneering this much needed work.

All too often the law enforcement officer is variously referred to as ''cop,'' ''pig,'' ''the fuzz'' or by some other term which casts a dispersion upon the dignity of the work which society requires him to do. The public, being sovereign, calls a man to perform a public service, underpays him and then permits him to be castigated and belittled by the very element of our society which all of us seek to be protected against—the violator of the laws.

In a sense we Americans are a perverse people—that is, in the sense that we want a government, but want to be governed as little as possible. This is not a criticism, but a compliment to the nature and temperament of all Americans.

The man on surveillance and control is usually the first contact we have with ''the law.'' The law enforcement officer who issues the tickets is a person upon whose head is heaped the scorn and abuse of casual and persistent violators of law and order. The officer with high moral standards and a thorough scientific training is able to combat public apathy and condemnation much more effectively than the appointee who is chosen through political influence or is available because of want of other means of earning a livelihood. The first contact with the representative of the law is usually the most impressionable one. From it opinions are formed and these opinions are the formative basis of public response.

In ordinary talk we refer to the professions. We like to think of our sons as engaged in an educational pursuit which will lead them into a profession (i.e., doctor, lawyer, clergyman, engineer, or certified public accountant). Webster confines the definition of a professional man to one of the three learned professions: theology, law and medicine. Time, progress and the intricacies of social life with expanding populations make a broader definition of the professions more to the point. We often hear the expression that so and so is a professional soldier, meaning, of course, that he has been trained in the art of making war. Why not, therefore, create a corps of professionally trained wildlife law enforcement officers?

It is with this thought in mind that this work is presented. It is for those who would so mold their lives that they will not merely be a ''fish cop,'' but one who utilizes all of the tools at his command to make himself efficient, scholarly, sophisticated and worthy of public confidence.

Enforcement is a harsh word. It implies compulsion. The connotation of the word compliance tends more to soothe the nerves and heart of the outdoor man and woman. You will find much in this volume devoted to public education which seeks to avoid the violation of the law, rather than to capture and convict the violator after the act. It is prevention of illegal acts through public education rather than conviction after the violation that results in the greatest conservation of our public resources and the greatest utilization of our wildlife resources.

This work, the first of its kind, is a step in the right direction and should create the urge for subsequent authors to build upon its foundation. It should instill in the minds of instructors and students the desire to obtain a more perfect understanding of the role played by those who are engaged in the pursuit of wildlife law enforcement.

Geo. D. Preston, 1898–1965
George W. Preston
Second Edition

Foreword

1972

The enforcement of laws and regulations has always been an important element in any wildlife conservation program. Historically, regulation of the use and protection of wildlife is one of the oldest conservation concepts. Unless there is a drastic change in the human race, law enforcement will always be necessary to the maintenance of wild creatures.

During the present century, the period in which wildlife management developed into a profession, the emphasis has been on new ideas and new techniques. Relatively little effort was made to provide or train better wildlife officers except in a few states and even fewer schools.

Therefore, when Dr. William F. Sigler of Utah State University instituted classes in wildlife law enforcement it was a long-needed step forward.

When Dr. Sigler wrote *Wildlife Law Enforcement,* an up-to-date text for such training, this help became available. It has been widely used and has helped to emphasize the continued need for trained officers to deal with those elements, in all communities, that persistently refuse to abide by conservation laws.

Now Dr. Sigler is revising and updating this important work and it should be widely used by more universities, as well as by states, in their officer training programs.

This new edition is another distinct contribution to the wildlife management profession and to wildlife itself. I am more than pleased to write a foreword to this book.

Ira N. Gabrielson
Second Edition

Foreword

1956

It has long seemed apparent that schools training wildlife management men were overlooking an opportunity in not including enforcement training in their courses. A growing number of states are raising both the qualification standards and entrance salaries of their law enforcement personnel, but many schools have been reluctant, despite repeated suggestions, to train men for this type of work.

I was therefore much interested when Dr. Sigler started to develop a course in this field some years ago. From this beginning, he has developed a carefully worked-out training program and this well-prepared text.

This text has been prepared as the result of this teaching experience and should provide much help and guidance as similar courses are initiated. It is well documented and should stimulate interest in a field that offers a growing employment opportunity for trained men. Human nature being what it is, there will be a need for good enforcement men as far in the future as anyone can foresee.

This book includes an outline of the basic legal knowledge that officers must have, but more important it provides valuable information on enforcement methods, preparation and presentation of evidence, and the relation of the enforcement staff to the other work of the department.

This book should help materially in providing well-trained men for this phase of management, and I for one am delighted to see it.

Ira N. Gabrielson
First Edition

Foreword

1956

The compilation of a text on wildlife law enforcement is long overdue. The author of this volume is to be complimented for pioneering this much-needed work.

All too often the law enforcement officer is variously referred to as a "flatfoot," "copper," "bull," or by some other term which reflects the dignity of work which society requires be done. The public, being sovereign, calls a man to perform a public service, underpays him, and then permits him to be castigated and belittled by the very element of our society which all of us seek to be protected against—the violator of the laws which we insist must be cast upon our statute books.

In a sense, we Americans are a perverse people—that is, in the sense that we want and must have government, but want to be governed as little as possible. This is not a criticism, but a compliment to the nature and temperament of all Americans.

The man on the "beat" is usually the first contact we have with "the law." The enforcement officer, who issues the tickets, is the person upon whose head is heaped the scorn and abuse of casual and persistent violators of law and order. The officer of high moral standards and thorough, scientific training is able to combat public apathy and condemnation much more efficiently than the appointee who is chosen because of political influence, or because he is available for want of other means of earning a livelihood. This first contact with a representative of the law is usually the most impressionable one. From it, opinions are formed. These opinions are the formative basis of public policy.

In ordinary talk we refer to the professions. We like to think of our sons as engaged in an educational pursuit which will lead them into a profession—likely a doctor, lawyer, engineer, or certified public accountant. Webster confines the definition of a professional man to one of the three learned professions: theology, law, and medicine. Time, progress, and the intricacies of social life with expanding populations make a broader definition of the professions more to the point. We often hear the expression that so and so is a professional soldier, meaning, of course, that he has been trained in the art of making war. Why not, therefore, create a corps of professionally trained wildlife law enforcement officers?

It is with this thought in mind that this work is presented for those who would so mold their lives that they will be not merely the "flatfoot on the beat," but one who utilizes all of the tools at his command to make himself efficient, scholarly, and worthy of public confidence.

Enforcement is a harsh word. It implies compulsion. The connotation of the word compliance tends more to soothe the nerves and heart of the outdoor man and woman. You will find much in this volume devoted to public education which seeks to avoid violation of law, rather than to capture and convict the violator after the act.

This first work of its kind is a step in the right direction and should create the urge in subsequent authors to build upon this foundation. It should instill in the minds of instructors and students a desire to attain a more perfect understanding of the role played by those who will engage in the pursuit of wildlife law enforcement.

Geo. D. Preston
First Edition

Preface

This text presents wildlife law enforcement as an integral part of game and nongame management. Since each of the fifty sovereign states and the federal government have separate and distinct laws and regulations governing wild animal management, it is evident that such a discussion must deal with fundamentals and general principles.

This text addresses three groups: university students who are interested in wildlife law enforcement; line and staff wildlife law enforcement officers; sportsmen and environmentalists who are interested in the role of wildlife law in ecosystem management.

The fourth edition adds five new chapters: Administration, Undercover Investigations in Wildlife Law Enforcement, Forensics, Aircraft in Wildlife Law Enforcement, and Recreational Law Enforcement.

When it comes to gender I would like to quote Norman Cousins (National Forum, Winter, 1991) as my guide: "I realize that the general use of the term 'man' or 'his' when applied to humankind is grossly imperfect and I apologize to anyone who prefers the terms 'his or hers' or 'humankind' or 'persons' in holding to the old convention. I intend no disrespect to half of the human species. As soon as a form is devised that does not sound clanky or that does not call attention to itself, I promise to use it. Language, like all other forms of human relations, has the obligation to reflect increasing awareness of human sensitivity."

Finally, this is not a legal document and should not be accepted as one in North America or elsewhere. Where legal information is needed, the individual should consult several sources, including the code books of the respective states and the United States, Canada, and Mexico, as well as treaties between the United States, sovereign nations, and the several foreign countries.

Acknowledgments

Lt. Colonel Joel M. Brown, Assistant Chief of Law Enforcement, Georgia Department of Natural Resources, Division of Game and Fish.

Tim D. Cosby, Chief, Law Enforcement Section, Alabama Game and Fish Division.

Sergio Elizalde, law student, University of Utah, Salt Lake City, Utah.

Federal Bureau of Investigation, United States Department of Justice, Washington, D.C.

Charles L. Garey, Former Assistant Chief, Law Enforcement, Idaho Fish and Game Department.

Terry Grosz, Assistant Regional Director, Law Enforcement, Rocky Mountain and Prairie States Region of the U.S. Fish and Wildlife Service.

Neill Hartman, Deputy Assistant Regional Director, Law Enforcement, Rocky Mountain and Prairie States Region of the U.S. Fish and Wildlife Service.

Jennifer C. Henry, Sergeant, Office of Law Enforcement, Illinois Department of Conservation.

Lt. W. M. Henry, Training Officer, Department of Conservation, Springfield, Illinois.

A. Eugene Hester, Assistant Regional Director, Law Enforcement, North East Regional Office, U.S. Fish and Wildlife Service.

Bruce Johnson, former Chief, Law Enforcement, Utah Division of Natural Resources.

Edward L. Kozicky, Director of Conservation, Winchester-Western Division of Olin Corporation (retired).

Sumter Moore, Lieutenant, South Carolina Wildlife and Marine Resources Department.

Frank Nesmith, Chief, Law Enforcement, Idaho Fish and Game Department.

David W. Oates, Forensic and Analytical Specialist, Nebraska Game and Parks Commission.

Sydney Peterson, Staff Assistant, Research and Technology Park, Utah State University, word processor and editor.

Conrad G. Seibel, Operations Commander, Alaska Department of Public Safety.

John W. Sigler, former Manager, Environmental Science and Technology Group, Spectrum Sciences and Software, Inc.

Margaret B. Sigler, Editor, Typist, Advisor, Wife.

Wayne H. Watkins, Director, Research and Technology Park, Utah State University.

Larry Wilson, Director, Iowa Department of Natural Resources.

Chapter 1

Historic to Present Role of Wildlife Law Enforcement[1]

HISTORICAL PERSPECTIVE OF WILDLIFE MANAGEMENT

Case 1

A legal system can operate effectively only if most people follow its precepts willingly. Law can operate as the great teacher and civilizer, regulating day-to-day activities peacefully, but only if those whose conduct should be guided by it know what is expected of them (Coughlin 1963).

Since man's beginnings, rules have given order and security to his way of life. The true American contribution to human progress is the notion of law as a check upon power. Struggles over power in other countries that call forth troops, in the United States call forth lawyers. Whenever we buy a car, get a job, hunt or fish, travel, or even go to the grocery store, we are performing an act of legal consequences. Our rights and obligations in all situations are fixed by law and, if need be, determined by the courts. In this sense we are all consumers of the law (Schwartz 1974). The history of American law is said to reflect the history of the United States. If we may apply this to wildlife law enforcement, then the history of those laws reflects the history of wildlife management in North America.

Early Landuser-Hunter Behavior Patterns

To develop a better understanding of enforcement, wildlife management is presented in historical perspective.

Early nomadic family groups lived from uncultivated lands. Their interests were in hunting, fishing, and grazing, but there was no proprietary interest in the land itself. Agricultural settlements soon encroached on this wilderness, leaving the uncultivated part for anyone who was interested in supporting himself. Probably, even in prehistoric times, farmers were encroaching upon the established state of nature (Furst 1946).

Later, as the land became a source of wealth, its control was about the only way of acquiring more wealth than a man could carry in his pockets. It became natural for a person to acquire title to as much land as possible in order to transmit wealth to his descendants. In those days, material possession was probably a matter of force rather than of legal title. It was not until much later that the law recognized and protected personal property.

At first, there was little proprietary interest in any land that was not tillable. However, it soon became apparent that control of the surrounding land was necessary for the cultivation and management of arable land. Depredation by nomadic wilderness dwellers probably brought this about. Large and powerful families developed land holdings around tilled land. These landlords became the rulers and nobles of their time.

According to Toffler, the agricultural revolution swept aside forty-five thousand years of cave dwelling about 8,000 B.C. At that time nomadic peoples shifted from hunting and gathering to domesticating animals for farming and settling on the land. The second wave, according to Toffler, was the Industrial Revolution which began about 1760 and moved the masses from the field to the foundry (Tafoya 1990).

Early English Backgrounds

The following quotation from Nelson (1762) reveals how the idea of privilege became associated in the public mind with the hunting of game:

> To begin with the time of the Britons, when their Princes and great Lords had no Occasion to set apart Places for the Preservation of Game and Beasts of Venary (their Bruery, i.e., Thickets and uncultivated Lands, being such Nurseries and Shelter for them), it was the Interest of both Princes and Lords rather to destroy than preserve them.
>
> During the Wars between the Britons and Saxons, so many of the Britons were killed, and so many fled from the conquering Saxons, that the cultivated Lands were more than sufficient to maintain the Conquerors and the miserable Britons who staid amongst them; for at that Time there were no foreign Markets where the Saxons traded with the Produce of their Lands. When the Saxons found themselves Masters of the British Lands and People, the Saxon Captains, as Conquerors, in Common Council agreed to divide the Lands they had taken amongst themselves, their Friends and Companions in Conquest.
>
> The Woods, Wastes, and Bruery Lands, that were not appropriated to any particular Persons, remained to the Chief Captain, who in Process of Time assumed the Title of King, who, as Occasion offered, granted Parcels of such Woods to whom he thought fit. . . . More and more useless Woods were appropriated and improved; and as Improvements were made, the Game and Beasts of Venary retired from thence for Shelter into the unfrequented Woods, thither the Saxon Kings, that took Delight in Hunting, went for their Diversion, where was such Plenty of Game, that there was no Occasion for restraining Laws to preserve them. . . .
>
> . . . but in Edgar's Time . . . he having an elegant Taste prohibited Hunting his Deer, and appointed Officers to preserve all Game of the Table, in his Woods, who so rigorously put in Execution their Orders, that the Nobility and Gentry were prevented of taking their diversions and their Tenants of their respective Rights; At length this arbitrary Procedure of the Officers grew to so great a Grievance, that Noblemen, Gentlemen, and Farmers, made great Complaints for Want of a Law to ascertain the King's Prerogative and the People's Privilege in this Case, on which King Canute, through his innate Goodness and Justice, in a Parliament holden at Winchester in 1016, brought the Proceedings to a Certainty, that all men might know what they should, and should not do, by publishing Forest Laws, therein setting out the Rounds of his Forests, and limited the Power of the Forest Officers. . . .
>
> William the Conqueror laid waste thirty-six Towns in Hampshire to make a Forest . . . and his Forest Officers . . . exercised such arbitrary Rule, as to abridge even the great Barons of the Privileges they enjoyed under the Saxon and Danish Kings; not at all regarding the liberties given to the Subject by Canute's Forest Laws.
>
> His Son William Rufus is recorded in History for the Severity of his Proceedings against all that hunted in his Forests; inflicting the Punishment of Death upon such as killed a Stag or Buck in his Forests, without any other law than that of his own Will. . . .
>
> In the Reign of King John these and other Oppressions, having exasperated the Barons, they took up Arms . . . and marched to Northampton . . . from whence they sent Letters to the Earls, Barons, and Knights that adhered to the King, and if they would not desert the perjured King, and join them in asserting their Liberties, they would proceed against them as public Enemies.
>
> These Threats drew from the King most of the Barons that had adhered to him, which Defection left the King hopeless and induced him . . . to let the confederated Barons know he would grant them the Laws and Liberties they desired: Upon which a meeting of King and Barons was agreed to be on the fifteenth of June, 1215, at Runnymede between Stains and Windsor, where a conference began between the Barons that adhered to the King and the conferated Barons, who were so superior in Number to the King's Barons, that he seemed

to make no difficulty of granting the Laws and Liberties demanded; which were drawn up as the confederated Lords thought fit, in two Charters, vis. The Great Charter, and the Charter of Liberties and customs of the Forest.

According to Brown (1965), the Forest Charter of Henry III (1225) laid down that no one was to lose life or limb for killing the King's deer. He was, however, to suffer a ''grievous'' fine if he had any money, and if he had none he was to go to prison for a year and a day. But an ''archbishop, earl, or baron,'' on his way to the King via Royal Command, and passing through a royal forest, might kill ''one or two'' deer. This was to be done in the presence of the King's forester if he was present. If he was not, a horn was to be blown for him lest it should seem that the deer were being stolen. There was a like dispensation for those returning home again.

Wildlife Control During the Middle Ages

At the time European land-holding families were acquiring lands, European culture was just emerging from the greatest moral, financial, and spiritual depression in recorded history. People were extremely desperate, superstitious, and confused, resulting in many habit changes, including a loss of interest in hunting and fishing.

When the interest in hunting and fishing diminished, the urge to control them subsided. This was probably tied up with a scarcity of wild game and an inability to harvest it safely away from controlled land holdings. Once landowners gained the right to hunt and fish, they must have quickly seen the advantage of reserving it for themselves, even though this right was not known as such at that time. Landowners declared the game and fish their own and made it a crime for anyone else to hunt or kill. Landowners, then, hunted and fished for either sport or food and controlled the land—prohibiting all others from even trespassing. The peasantry who had lived by fishing and hunting were labeled poachers and outlaws. In Europe, the game laws were then written and enforced for the benefit of ruling classes, and the full weight of the law and social stigma fell upon those who harvested game without permission of the landowner. The general public developed the idea that fish and game laws were for the benefit of the favored few.

European Traditions in North America

Much later, large areas of North America were settled by agricultural people from Europe. They brought with them a tradition of resentment against fish and game laws. There appeared to be an inexhaustible supply of wildlife in early North America, and conservation measures were not considered necessary.

The wilderness was a challenge; something to subdue. How could one abuse an inexhaustible supply of water, fish, game, forests, and soils? Only in the last fifty years have we recognized that there is a limit to our essential resources and that human existence rests upon our ability to equalize use of resources with stewardship of them.

The foundation of American law was the English common law developed over the centuries by English judges. American law treated European law, with its emphasis on codes rather than cases, as instructive rather than authoritative. In early days, the common law was the accepted one because it was the only system that could be taught with the books at hand (Schwartz 1974).

In summary, it is believed that the earliest hunters, thousands of years before recorded history, were nomads who lived off the land. At some point in time, part of these people ceased their wandering and settled down to a primitive type of farming. Conflict arose between the sedentary people and the nomads, and still exists today. Early hunters had no wealth except their primitive weapons and limited personal belongings. The farmers retained title to the land and everything on the land from one generation to the next. They became the stable, wealthy, and influential segment of society. Nomadic hunters must have become a nuisance early, and then a hazard to these tillers of the land. As land holdings increased and hunters were forbidden trespass, the conflict increased. The land was held by force rather than legal title, so the landed gentry armed themselves. In the Dark Ages, the interest in hunting and fishing shifted from nomads to landowners, who claimed title to all game. At this stage the nonlandowner lost both the privilege and the right to hunt and fish.

A rough parallel probably occurred in North America with the increase in population of the American Indian. There was a distinct difference, however. While there were sedentary crop-raising Native Americans for hundreds or even thousands of years, it seems that the warring tribes almost invariably held the upper hand. The title to game was with the one who took it. This was not changed until European immigrants arrived. Unhappy with the way game was managed in their homeland, the immigrants, ironically, fought with the Indians, who wanted only to take game as they needed it.

Before horses were introduced into America by the Spaniards, the dog was the beast of burden and a staple in the Indians' diet (Ewers 1955). Horses made the warring Indian tribes more mobile and, therefore, more powerful. Perhaps if the white man had not interfered, North America would still be controlled by nomadic hunters.

The Public Domain

Many species of wild animals are harbored on the public domain. This includes song birds and nongame mammals such as the Utah prairie dog and black-footed ferret. Much of the prime range for such big game animals as moose, elk, antelope, and deer is public lands. Several species of forest grouse spend part or all of their lives on public lands. A number of America's finest trout streams head in the mountains of Bureau of Land Management and Forest Service controlled areas. The many public marshes are some of the world's finest waterfowl habitat. The vast reservoir of resources harbored on the public domain is for the benefit of the people of the United States of America.

The history of the public domain partially explains the attitudes and behavior of the people who use the lands today. In 1763, England acquired all land in North America east of the Mississippi River by defeating France in the Seven Years War. In 1783, at the Treaty of Paris, the victorious thirteen original American Colonies received British rights to those lands. Thus, a vast new empire—including the future public domain of the United States—passed to a newly independent nation. An empire had been ceded, but ownership remained in doubt. When Maryland refused to ratify the Articles of Confederation until all such claims were ceded through Congress, creations of public domain, owned by the national government, became a prerequisite for forming the nation. The articles were ratified and a new nation with public domain as its patrimony was created (Zimmet 1966).

The subsequent history of the public domain coincides with that of the expansion of the United States. Of the major domestic and foreign problems that have confronted this nation, many have been directly concerned with, or closely related to, matters of governing, acquiring, and disposing of public lands in the expanding West (Zimmet 1966). While the western frontier of the United States was expanding, public lands in the East, then in the Midwest, were quickly passing from public to private ownership. Yet, today nearly one-third of the nation, primarily in the western states, is still public lands (Public Land Law Review Commission 1970).

Now, more than two hundred years after the creation of the original public domain, its management philosophies and practices are far from settled. Public administrators are constantly criticized by vested interest groups and by the general public, which seems each day to behave a little more like other interest groups. In some governmental departments, the public ethic appears to be shifting from production to social consideration, but many critics believe the administrators are slow in their shifts, and trail the public by some distance.

The public lands have consistently generated conflicts between users: whites v. Indians, cattle ranchers v. sheep ranchers, ranchers v. homesteaders, mineral interests and timber operators v. conservationists, and, finally, the most recent and the most powerful in point of numbers, the recreationists v. other landusers. These latter day outdoors people first became interested in the public domain because it appeared to be a vast, unexploited, free recreation resource. They soon discovered that other people had preceded them, and their use, if not their very presence, was resented. Part of the problem hinged on access. Some of the land was accessible only by passing through private land for at least part of the year. Lack of public access to some public domain lands is the focal point of many use conflicts (Munger 1968).

Western ranchers had sought, and won, the use of unappropriated public lands long before the government established a system for administering the acreage. This use was gained to some extent by the ownership of strategic tracts of land acquired by homesteading and other methods. It was this hard-won usage of federal rangelands that ranchers interpreted as an implied proprietary interest in public domain land—an interest they have defended vigorously over the years.

The sharp rise in the demand for outdoor recreation following World War II soon outstripped the supply of readily accessible land. Alarmed landowners began to restrict access to and across private lands. They complained about the irresponsible acts of recreationists, who countered that they were denied the use of public lands (Munger 1968). By the late 1950s the conflict had attracted national attention, and articles appeared in newspapers and sports magazines. Criticism was directed at public land management agencies. People making studies released figures pointing out how millions of acres of public land were blocked to free public access, legally or otherwise. In short, the crunch was on.

The problem was further complicated by the fact that unpatented mining claims were, and still are, not open to the public, although title to these lands rests with the federal government. These claims can block access to other public lands. The legal authority of the federal government to acquire public rights-of-way through condemnation procedures is inherent in the concept of sovereignty and is specifically provided for in the act of August 1, 1888. Nevertheless, condemnation procedures are rarely invoked by the Bureau of Land Management exclusively for acquiring public access. Public rights-of-way usually are acquired along with temporary access roads or general service roads.

Public responsibility and authority, however, are not created solely by laws, executive directives, or court decisions. Other social institutions, less formal but no less powerful, often are important determinants of public land management policies (Munger 1968). This attitude explains why the federal government is into the business of providing recreational services and why it is likely to continue. It also explains why the controversy over public access for recreational use of public land carries with it the threat of major political repercussions (Munger 1968).

HIGHLIGHTS OF WILDLIFE MANAGEMENT

One of the basic concepts of wildlife management is habitat carrying capacity. That is, how many fish will a body of water support, how many deer will a forest safely harbor, how many quail will one square mile of habitat support without unduly exposing them to predators, disease, or adverse weather conditions. Assuming no change in land-use patterns, all lands have a precise carrying capacity, although it may often be difficult to measure or perceive. A given body of water will support a specific poundage of fish, assuming no change in the species composition of the fish or fertility of the water.

Habitat

Density is the number of animals of one or more species carried by a given unit of habitat; for example, twelve deer per square mile. Prime deer habitat in the eastern United States can support about twenty white-tailed deer per square mile. Inferior habitat may carry only two or three. The upper population levels represent carrying capacity, that is, the maximum number of animals any given habitat can support throughout the year. When population densities are low in relation to habitat carrying capacity for a species, females produce high numbers of young in a year. If there is adequate food throughout the year, animals may be maintained in peak physical condition. When animals overwinter with an abundance of stored energy, antler or horn development is accelerated. Young does may reach maturity as soon as they lose their spots and may breed successfully in their first year. Twin fawns are common and the incidence of triplets is not uncommon in median age does. The deer's resistance to parasites and diseases is high and their alertness and agility aid them in avoiding predators (Poole and Trefethen 1977).

According to Poole and Trefethen (1977), when habitat is fully stocked, some form of population control becomes necessary to maintain the animals in health and reproductive vigor. Without control, both the animals and the habitat suffer. For example, in the western states winter habitat is generally the limiting factor for mule deer. If numbers become excessive, the animals die from malnutrition and the habitat may take decades to recover.

Most big game animals prosper in subclimax rather than climax vegetation. When our forebears came to America, they found few deer. But when they cut vast tracts of large trees across the country, it produced an ideal situation for browsing animals. In the more arid regions of the western United States the climax vegetation was grass. Heavy grazing by domestic range animals changed the areas to predominantly browse vegetation, an ideal change for mule deer, which reached peak numbers in the 1940s and 1950s. Deer browsed so heavily that some areas reverted largely to grass, causing some biologists and ranchers to claim mule deer and cattle have a symbiotic relationship. Regulating timber cutting, controlled grazing by domestic animals, and limiting big game populations to carrying capacity of the habitat are all compatible practices.

Unique Problems

Wildlife management problems take some strange twists. Fast-moving vehicles traveling on criss-crossing roads and highways have become a hazard to game, particularly deer. Chain saw enthusiasts, in search of inexpensive fuel, cut snags, or dead trees, which are essential to the welfare of some species of wildlife.

Nongame Wildlife

Any animal defined as nongame by state or federal law may not be killed, harassed, or harmed. Nongame classification includes:

(1) endangered;

(2) of certain migratory species; and

(3) of aesthetic, zoological, or ecological value. Some nongame animals, such as black-footed ferret and snail darter, are protected by the Federal Endangered Species Act and state law.

Others are protected by state law only. Songbirds and shorebirds are protected by the Federal Migratory Bird Treaty Act and state law; and desert tortoise and certain other reptiles are protected by state and federal law. Under the Endangered Species Act, the habitat essential to the species' well-being, as well as the animals, is protected. The laws are administered jointly by state and federal government. The creation of refuges is one method of giving total protection to individual species. The refuges are maintained as long as the species is endangered.

Furbearers

Animals sought for their pelts occupy almost every type of habitat. Martins live in trees in wilderness areas; mink, muskrat, otter, and beaver near water, often close to people; lynx and bobcat in mountainous areas, although they venture far afield; skunks, foxes, raccoons, and opossums in farming areas; and coyotes just about every place, although they appear to not favor deep wilderness. These animals are sought when the fur becomes prime in early winter. A few species are hunted, but most are trapped. The intensity at which a furbearing animal is sought is governed largely by the market, although some are taken for sport.

Forest and Wilderness Game

The forest and the wilderness are havens for deer, elk, moose, black bear, and grizzly bear; they also host forest grouses and some of North America's best trout waters. The grizzly bear shuns humans and human habitation. Elk and deer prefer the wildlands from early spring to fall, but appear to lose their fear of people when winter food in the high country becomes scarce. Moose less often show this tendency. Black bear seek rugged terrain for protection from people, but are not adverse to raiding apple orchards or other food supplies.

Waterfowl

Many North American waterfowl are reared in Canada and winter in Mexico. However, large numbers are produced in the United States and quite a few overwinter in the southern tier of states. There are numerous waterfowl refuges in the United States, particularly in the western states. A part of most refuges is open to hunting on the theory that only the surplus will be harvested.

The U.S. Fish and Wildlife Service and the states have annually managed the waterfowl harvest by differential regulations. This is done primarily through changes in the length of hunting season and daily bag limit.

Waterfowl hunting seasons and bag limits in the Pacific, Central, Mississippi, and Atlantic flyways are determined by the estimated numbers of birds which will move through these areas. Flight patterns from northern breeding grounds are highly predictable. The estimate of fall migrants is based on the number of breeding birds returning to the breeding grounds and their reproductive success. The need for different hunting regulations for various species of ducks is well recognized.

Farm and Ranch Wildlife

Farm and ranch wildlife includes small game animals such as pheasants, quail, and cottontail rabbits, but there may also be waterfowl and deer. Western America ranches play host to many pronghorn antelope, others are on publicly owned rangelands. Farm and ranch game is a product of private land, fed and cared for by the land operator. Hunting is generally by permission. The small animals are short-lived, and each year's harvestable crop is determined by that year's reproductive success, which is determined in part by agricultural practices, some of which favor game and some do not. Farm ponds frequently produce good crops of bass, bluegill, catfish, and trout. Sometimes this asset is ignored by both the owners and anglers.

Predator Control

Predator control for many years meant the removal of undesirable animals, by state or federal agencies and private agricultural interests or individuals, by any effective means. The issue is now so clouded by emotions and political intrigue that neither the problem nor the solution is clear. Ranchers claim the loss of young animals, particularly lambs, is high. Some big game biologists believe the loss of young game animals to predators under adverse conditions represents a serious management problem. Certain people oppose any predator control. Others favor limited control under certain conditions. Eagles, bears, cougars, bobcats, skunks, and foxes all are accused at times, but the coyote, particularly in the western United States, gets most of the blame. And while the partial banning of poisons has given the coyote a reprieve, the high price its pelt demands, plus a bounty, makes it a popular target. The future of predator control is uncertain; the future of certain predators, particularly coyotes, is clear. They will survive.

Feral v. Wild Animals

Feral, or free-running, dogs and wild horses create many common problems. These two animals relate closely to humans, they are capable of reproducing in substantial numbers, and they are hazardous to society if uncontrolled. One of the problems facing enforcement personnel is determining whether the animal in question is feral or wild. Three classic examples appear: the first involves horses and burros; the second, reindeer and caribou in Alaska; the third, the European wild boar and domestic pigs. The solution at times has been pragmatic, if not satisfying. A reindeer in the company of migrating caribou is a caribou. A wild pig, rooting up a garden, in Red Bluff, California, is a feral animal, although it may have come from European wild boar stock. On the other hand, a caribou in a herd of reindeer is a reindeer, and any pig two-thousand-five-hundred feet above Red Bluff, California, in the rugged mountains is a wild pig, and a license is required to hunt it. One way to identify a wild horse as such is geographically. That is, horses appearing in certain areas are wild, assuming they are not branded or otherwise marked as domesticated; in other areas they may be feral.

General Funds

Case 2

Krug[2] does not believe that general funds will ever be adequate to support the kind of wildlife law enforcement programs that will ensure the protection of wildlife populations. This means that hunting and fishing fees and licenses must be continued and commensurate with the rising costs of operating a department.

Fisheries

A popular fishery management theory is that the harvest of the resource should be manipulated to achieve optimum sustained yield. This is accomplished by:

(1) legal restrictions on size of fish, as well as daily and possession limits, type of gear used, and time and place that any given species may be taken;

(2) control of predators and/or undesirable competitors;

(3) on rare occasion sanctuaries;

(4) stocking; and

(5) habitat maintenance or improvement.

Collectively these are the techniques of the administrator and biologist.

State laws and regulations are passed with regard to: biological evidence, attitudes of the public, background, and beliefs of commissioners, who have the final say in passing laws in some states, available personnel and facilities, federal and other laws affecting fishes, endangered and threatened species, and projected revenue. Federal laws and regulations are enacted by Congress and federal agencies in response to the needs of the resource and demands of the people. There are many compromises between biological recommendations, law enforcement capabilities, and political facts of life.

Fishes are divided into warm-water fishes, cool-water fishes, and cold-water fishes. Examples of warm-water fishes are sunfishes, black basses, catfishes, and carp. These fishes are prolific, have rapid growth, many are short-lived, the egg hatching time is short, and there generally is a high percentage of nongame fishes in a given population.

Fly fishing only is
becoming one of the
more efficient
management
techniques.
*Courtesy Wyoming Game
and Fish Department.*

Optimum temperatures to reproduce and grow are 58° to 82° F. Examples of cool-water
fishes are northern pike, muskellunge, and walleye. Their associates are more often
warm-water, rather than cold-water fishes, although they prosper in either habitat.
Optimum temperatures are 48° to 72° F. Examples of cold-water fishes are grayling,
trout, and salmon. The eggs produced by these fishes are low in number, hatching time
may be two to three months, and the total pounds per acre are a small fraction of what
it is in a warm-water fish population. Optimum temperature for them is 38° to 62° F.

Warm-water fish regulations are often open on a year-round basis with no minimum
size limit and liberal creel. Cold-water fish regulations may have a limit on the minimum size and number that may be taken. Cool-water fishes, as expected, fall somewhere between the other two. Popular cool-water game fishes, such as northern pike
and muskellunge, have both a minimum size limit and creel limit, and only one or two
trophy size fish may be taken. Restrictions on legally accepted gear range from small,
barbless, artificial flies to large, multiple ganghooks. The use of live fish as bait is
restricted in many areas because it may introduce undesirable, exotic fishes into a body
of water.

Control of predatory fishes is rarely practiced since they are generally the game fish
in question. The exceptions to this rule include gar and bowfin. The desirable way to
reduce nongame fishes is to introduce highly piscivorous fish, such as northern pike,
lake trout, or brown trout. There is a limit to predator capabilities, and sometimes other
means, such as fish toxicants, are used. To maintain optimum growth of piscivorous
game fishes, ninety to ninety-five percent by weight of the population should be prey
species. Phytoplankton and invertebrate feeders, such as some strains of rainbow trout,
grow rapidly to the size of about two pounds and prosper best in a monoculture.

The stocking of cold-water fishes is an important aspect of management in the
mountainous areas of North America. Rainbow trout are stocked in greater numbers
than perhaps all other trout combined. They are easy to raise and the stocked fish easier
to catch than most other trout. Cutthroat trout are stocked to reestablish populations in
several of its original habitats or to supplement natural reproduction. Brown trout are
prospering in many waters of North America, partly because they are fall spawners

and do not face the fluctuating water levels and high turbidities of spring stream spawners, and also because they are more difficult than most trout to catch. Once a population is established, further stocking is rarely needed. Golden trout and grayling are stocked in a few cold bodies of water. Lake trout, a fall spawner, is most often stocked where it grows large, but may not reproduce. Kokanee (landlocked sockeye) prosper where plankton is abundant. Landlocked silver and chinook salmon have been stocked in the Great Lakes with good success. Warm-water and cool-water fishes are generally stocked to supplement or replace breeding stock, rather than to directly increase the catch.

The U.S. Fish and Wildlife Service policy for stocking catchable trout states that trout will be stocked at densities no higher than necessary to provide up to an average catch rate of 0.5 fish per hour between planting intervals for the length of the stocking season.

Habitat maintenance is the most important, yet most difficult, technique in fisheries management. Desirable water quality characteristics include adequate oxygen and temperature and minimal pollution. Drastic water level fluctuations in either time or place may endanger spawning success or inhibit food production, especially in the littoral zone.

THE ROLE OF LAW ENFORCEMENT IN MANAGEMENT

Neither ingenuous biological plans nor thoughtful and equitable laws can execute themselves. Administrators initiate acts, but the recipients, by their behavior and attitudes, determine what succeeds and what fails. How then can natural resource managers best approach people in order to achieve that fine mix where the resources prosper and people accept what they perceive as fair and equitable.

Laws should be concise and clear. Annual changes in such things as terminology and hunting boundaries should be minimal. Only citations with merit should be made; the officer should not issue borderline citations and let the court decide the merits of the case. Cases should be well researched before they go to court. People who believe they are treated fairly and are kept informed will have faith in the laws and their administration. Individuals and special interest groups should believe their views were considered before the decisions affecting them were made (Sigler 1974).

In states and provinces, the opinion of the local wildlife officers are often more sought after than that of any other agency personnel. They are known locally, they are available, and they project the image of the department. They explain laws, programs, and philosophies of the department. Their work at the grass-roots level is invaluable (Sigler 1975).

Current Issues in Resource Management

The issues facing fish and wildlife managers, or perhaps their intensity and priority, have changed markedly in the last fifteen to twenty years. These issues include:

(1) worldwide environmental problems;

(2) rapidly increasing human populations in many countries;

(3) large and growing numbers of threatened or endangered plants and animals;

(4) the anti-kill movement that strikes mainly at hunting, fishing, and trapping and, to a lesser extent, at domestic operations such as fur farming; and

(5) last, but perhaps most important, a shortage of quality water, especially in the western United States.

Several occurrences are recognized as damaging the environment and its inhabitants: destruction of the ozone layer; depletion of rain forests; enlarging of the greenhouse effect; and increase in acid rain. Two other expanding concerns are the destruction of the riparian (streamside) habitat, especially in the western states, and the loss of wetlands. Only a cursory look at these problems is in order here. However, they all impact wildlife management.

The ozone layer in outer space protects all living things from ultraviolet rays. Rain forests consume carbon dioxide and other air pollutants and give off oxygen. The greenhouse effect is a blanketing of the earth by pollutants that may cause a detrimental level of warming. Acid rain is just that; rain carrying acid that ultimately reaches streams and lakes and lowers or destroys productivity. The world population increases, yet the land becomes less productive. For a detailed discussion see *World on Fire* by George Mitchill, 1991.

Riparian zones, the most fragile part of the ecosystem, are being damaged at an alarming rate. It is estimated that eighty percent of the riparian zone on federal lands in western states has been damaged, much of this by domestic livestock. Riparian zones are important in the arid western states because they help raise water levels, increase the availability of forage, collect sediments, dissipate the energy of floodwaters, and serve as fish and wildlife habitat. Wetlands purify the water, raise the water table, and are host to many living things.

Concern is rising as the number of threatened and endangered plants and animals rapidly increases. Many new species have been added to the list, but very few have been removed. A number of species have gone extinct, while others appear to be headed that way.

The interest and activity of private individuals and groups who believe wild animals are not being managed in a proper way has greatly increased. Many of these private groups are ready with help and advice for management agencies. Some file suits when they want to be heard. Interests range from Ducks Unlimited and The Rocky Mountain Elk Foundation to antihunting and antitrapping groups. The groups and individuals causing great concern for wildlife managers are those that believe we should not kill wild animals. Some carry the philosophy over into medical research, domestic fur breeding, and even rearing livestock.

The industrial era changed the United States to more of an urban society. The outdoors experience of many urban dwellers was far different from that of their ancestors. Waterfowl and shorebirds were seen from a distance, if at all. Big game animals were seen in zoos. The urban populace began to see wild things as a resource to be looked at but not destroyed. Thus, was born the antikill movement. It blossomed slowly at the grass-roots level, but in an organized way. Later, a flurry of activity created a number of organizations with a wide range of views and objectives. Today, some organizations are well funded and have talented personnel.

The antihunting movement is a serious threat, as evidenced in California. In 1970, California passed the California Environmental Quality Act (CEQA), which was similar to the National Environmental Policy Act (NEPA). The courts had exempted federal agencies from adhering to NEPA, and the California Department of Fish and Game assumed that it would be similarly exempted from CEQA requirements. Two groups challenged the bow and arrow season for bears in California, stating that the Commission and Department had failed to prepare an impact report under the provisions of CEQA. The suit failed in a lower court, but the California Supreme Court reversed the decision and stopped the hunt. Justice Augustus Richardson wrote, "California Fish and Game Commission must comply with the appropriate part of the law (CEQA) and

For many, hunting over duck decoys is the epitome of all outdoor experiences.
Courtesy Coleman Marine Products.

may not rely on judicial precedent exempting federal agencies from the National Environmental Policy Act.'' The Department eventually complied and reopened the bear season, but the precedent had been set. Anyone with a lawyer could now ask the Department to prepare an environmental impact report on any of its hunting seasons. It was an opening that would make an Anaheim group drool, but no one took advantage of it for the next ten years (Madson 1991).

In 1985, the California Fish and Game Department decided to open the mountain lion season, which had a long-standing moratorium which expired in 1985. When an initial court challenge failed, opponents of the season sought a law to protect the lions. Their bill died when stockmen raised concern over the control of the lions' depredation

FBI lab experts make careful and detailed analyses of evidence in criminal cases for state and local law enforcement agencies. *Courtesy Federal Bureau of Investigation.*

on livestock. A coalition then started a petition to put the issue directly to the voters and to end mountain lion hunting. California hunters suddenly found themselves faced with an effort that, if passed, would have the force of law. The hunters were poorly organized and financed, and the issue won by a small margin. The mountain lion hunting issue caused the Funds for Animals, a New York-based organization, to reassess the ramifications of previous cases and decide the Fish and Game Department was vulnerable under the Environmental Quality Act. On August 24, 1990, the Department released its Draft Environmental Assessment of the 1990 migratory bird season. On the last day of the thirty-day comment period, a law firm filed some seventy pages of objections, along with one hundred seventy pages of supporting documentation. It was the same approach they had used to gain official standing prior to the bow and arrow bear suit which they had stalled. A rumor immediately went out that the antihunters were planning an assault on California's duck and goose season. At the public meeting, a spokesman for Funds for Animals announced they would not be suing to halt any portion of the migratory bird hunting season. The spokesman then added ''we have decided to challenge the department's ideas about migratory bird management through public discussion and exchange . . . in case there is any confusion, the decision not to pursue litigation is not an endorsement by the Funds for Animals of existing migratory bird hunting regulations.'' And that is where California stands today. The bear season was temporarily interrupted. The Tule elk season was closed and reopened. A law was passed to end mountain lion hunting and the waterfowl season appeared to survive by the barest of margins (Madson 1991).

Deputy Director Paul Jensen of the California Fish and Game Department estimates ''our license sales have dropped from about 800,000 to 400,000 in the last ten years.'' Madson (1991) believes state wildlife departments should recognize that most people do not have any idea how wildlife conservation is funded. Taxpayers assume that they are picking up the tab for the wildlife bureaucracy. Also he believes that a few non-hunters contribute heavily to wildlife conservation, but the overwhelming majority either do not visit wildlife areas or do not feel they have any impact on the wildlife they enjoy and, therefore, should not help fund it. Why do people hunt and fish today? With few exceptions, it is not needed to put food on the table, nor is it that the hunters do not enjoy eating their catch. From a monetary standpoint, it would be much less expensive to go to the nearest supermarket. But that is not the point. Many people feel a need to hunt and fish. Perhaps the genes of our nomadic ancestors lurk deep in some of us, making it necessary for us to hunt and fish, to plan and look forward to a trip with great anticipation, and to feel frustrated and disappointed if we are prevented from going. Some do not feel this need. Perhaps it has been sublimated by generations of people with no opportunity to hunt or fish. Again, not all of our ancestors were nomads. Some were crop growers and were sedentary.

Fish and game managers cannot ignore these groups of nonhunters and nonanglers that probably outnumber hunters, and possibly even anglers. Germany, in a effort to offset nonangler sentiment, has a comprehensive test for anglers before they are issued a three-year license (*The Salt Lake Tribune,* May 30, 1991, page A6). Working with these organizations in developing and implementing fish and game management may lead to a solution.

The above issues are serious and challenging. Perhaps the most serious one facing us today is water management, or a lack thereof. Former Governor of Arizona, and now Secretary of the Interior, Bruce Babbitt believes "the era of surface water development in the West is over." Babbitt continues, the "cost in environmental terms is now higher than most people in the West want to pay and in many places it is not just a question of not taking more water, but of giving water back" (Knickerbocker 1991). Meanwhile, farmers in southern Arizona are stunned by the high cost of water from the CAP (Central Arizona Project) ranging from twenty-five dollars to fifty-two dollars per acre-foot. They are wondering what went wrong with their dream of agricultural abundance. As with many other natural resources, the public interest is supplementing property rights and local political clout in determining water policy across the region. Only two percent of U.S. rivers are free-flowing. There are more than sixty-five thousand large dams, plus another two million small impoundments for flood control, agricultural storage, and recreation. As a result of all this, federal surveys show that fish habitat is suffering in about sixty-eight percent of the nation's streams. We have diverted, dammed or otherwise manipulated changes in most of the flowing waters of the United States (Editorial, West's Water Riddles, the *Christian Science Monitor,* January 21, 1993). Wildlife habitat loss and pollution resulting from agriculture and other commercial water diversions are two key reasons why the EPA is currently taking a hard look at the Clean Water Act. Most western waters are being used for agriculture, much of it provided by projects built during this century by the U.S. Bureau of Reclamation. However, the U.S. Office of Technology Assessment says farmers pay just about seventeen percent of the capital and operating costs of such water. The balance is obviously paid for by the taxpayers. Groundwater, as well as surface water, is being explored and, consequently, there are more opportunities for contamination from dissolved solids, mining, agriculture, storm sewers, and general population growth (Knickerbocker 1991). It is time we take a long hard look at water needs for the next fifty to one hundred years.

Literature Cited

Brown, W. J. 1965. *The field.* London.

Coughlin, G. C. 1963. *Your introduction to law handbook: Practical information about legal problems in everyday life.* New York: Barnes and Noble, Inc.

Ewers, J. C. 1955. *The horse in the Blackfoot Indian culture.* Washington, D.C.: Smithsonian Inst., Bureau American Ethnology, U.S. Govt. Printing Office.

Furst, D. S., Jr. 1946. Outlook in wildlife law enforcement. Paper presented at Intl. Assn. Game, Fish and Forestry Comm., Denver, Colo.

Knickerbocker, B. 1991. The sun is setting on a century of concrete waterways out west. *The Christian Science Monitor,* June 18, 1991, 1–2.

Madson, C. 1991. California, here we come? *Pheasants Forever* 9 (4): 33–37.

Mitchill, G. 1991. *World on fire.* New York: MacMillan Publ. Co.

Munger, J. A. 1968. *Public access to public domain land: Two case studies of landowner-sportsman conflict.* U.S. Dept. of Agriculture, Economic Research Service.

Nelson, W. 1762. *The law concerning game. Of hunting, hawking, fishing and fowling, etc. and of forest, chases, parks, warrens, deer, doves, dove-cotes, conies.* 6th ed. London: E. Richardson and C. Lintot for T. Waller.

Poole, D. A., and J. P. Trefethen. 1977. Habitat's essential role in big game management. In *North American Big Game,* ed. W. H. Nesbitt and J. S. Parker. Washington, D.C.: The Boone and Crockett Club and The Natl. Rifle Assn. of America.

Public Land Law Review Commission. 1970. *One-third of the nation's land.* Washington, D.C.: U.S. Govt. Printing Office.

Schwartz, B. 1974. *The American Heritage history of the law in America.* Ed. A. M. Josephy. New York: American Heritage Publ. Co., Inc.

Sigler, W. F. 1974. Professionalism in wildlife law enforcement. Paper presented at 30th Annual Conference Assn. Midwest Fish Game Law Enforcement Officers, Helena, Mont.

———. 1975. Recommended. B.S. Degree for state wildlife law enforcement officers. *Wildlife Society Bull.* 3 (4): 173–175.

Tafoya, W. L. 1990. The future of policing. *FBI Law Enforcement Bull.* (Federal Bureau of Investigation, U.S. Dept. of Justice, Washington, D.C.) 59 (1): 13–17.

Zimmet, S. 1966. *The public domain.* Bureau of Land Management, U.S. Dept. of Interior, Washington, D.C.

Notes

1. Also see Environmental Law Institute. The Evolution of National Wildlife Law. Council on Environmental Quality. 485 pages. 1977.

2. Krug, Alan S., Midregional Executive, Natl. Wildlife Federation. Letter to author, 31 Aug. 1978.

Recommended Reading

Bean, M. J. 1983. *The evolution of national wildlife law.* 2d ed. New York: Praeger Publishers.

Calkins, F. 1970. *Rocky Mountain warden.* New York: Alfred A. Knopf, Inc.

Dent, A. 1975. Game and poachers in Shakespeare's England. *History Today* 25: 782–786.

Freidmann, W. 1967. *Legal theory.* 5th ed. New York: Columbia Univ. Press.

Hershey, W. S. 1971. Impact of federal treaties on fish and game resources. In *Proceedings Western Assn. of Game and Fish Comms.,* 216–220.

Jones, D. J. V. 1979. The poacher: A study in Victorian crime and protest. *Historical Journal* 22 (4): 825–860.

Kirby, C. 1933. The English game law system. *American Historical Review* 38 (2): 240–262.

Lund, T. A. 1980. *American wildlife law.* Berkeley: Univ. of California Press.

MacDowell, D. M. 1978. *The law in classical Athens.* Ithaca, New York: Cornell Univ.

Palmer, T. S. 1904. *Hunting licenses—their history, objects, and limitations.* USDA Biological Survey Bull., No. 19.

Smith, B. 1991. The ecological wisdom of hunting. *Pheasants Forever* 9 (4): 44–46.

Tober, J. A. 1981. *Who owns the wildlife?* Westport, Conn: Greenwood Press.

Trefethen, J. B. 1975. *An American crusade for wildlife.* New York: Winchester Press.

Williams, R. W. Jr. 1907. *Game commissions and wardens— their appointment, powers, and duties.* USDA Biological Survey Bull., No. 28.

Williams, T. 1992. Incite: Overkill. *Audubon* 94 (4).

Multiple-Choice Questions

Mark only one answer.

1. What is probably the most serious problem facing wildlife managers and the general public today?
   a. the antihunter, anti-kill groups
   b. the loss of wildlife habitat
   c. water management or lack thereof
   d. well-intentioned but ill-advised sportsmen
   e. wilderness habitats

2. According to Nelson, which of the following rulers exacted the most vicious penalty for unauthorized killing of stags in his forest?
   a. William the Conqueror
   b. King John
   c. Edgar
   d. William Rufus
   e. King Arthur

3. According to Brown, the forest charter was under the reign of:
   a. Henry III
   b. King John
   c. King Arthur
   d. William the Conqueror
   e. Edgar

4. According to Toffler:
   a. the cave-dwelling period lasted about 45,000 years
   b. the agricultural revolution started about eight-thousand years ago
   c. the Industrial Revolution began about 1760
   d. nomads were largely replaced by farmers with the progression in time
   e. the Industrial Revolution began about 1950

5. Early nomadic people:
   a. had an interest in hunting and fishing
   b. had no proprietary interest in the land
   c. periodically settled down for a few years to raise crops
   d. found themselves, at times, in conflict with the more sedentary people
   e. had very little personal property

6. People perform an act of legal consequence when they:
   a. go fishing
   b. drive the car to work
   c. assume a new employment position
   d. go to church
   e. all of the above

7. What historical event made the Native American (Indians) nomads far more powerful than ever before?
   a. the introduction of the dog
   b. the introduction of the horse
   c. weather changing to a gradually warming climate
   d. drought that destroyed nonnomadic peoples
   e. changing grazing patterns of the buffalo

8. One of the following statements is not true:
   a. when population densities are low, female deer produce a higher number of young per year
   b. when population diversities are low, female deer produce a lower number of young per year
   c. when deer overwinter where there is an abundance of food, aftergrowth is better
   d. twin fawns are more common when does have an abundance of food in winter
   e. deer avoidance of predators is best when there is abundant food

9. Which of the following is not a true wilderness animal?
   a. black bear
   b. grizzly bear
   c. timber wolf
   d. coyote
   e. moose

10. In what year did England acquire all North American lands east of the Mississippi river:
    a. 1673
    b. 1763
    c. 1777
    d. 1783
    e. 1761

11. Which is not typically a warm-water fish?
    a. sunfishes
    b. catfishes
    c. carp
    d. black basses
    e. walleye

12. In what year did the Treaty of Paris pass to the thirteen original colonies the land that was to become the public domain?
    a. 1763
    b. 1776
    c. 1774
    d. 1792
    e. 1783

13. Public lands of the West have consistently generated conflicts between:
    a. recreationists v. other landusers
    b. whites v. Indians
    c. livestock owners v. homesteaders
    d. loggers v. conservationists
    e. all of the above

Chapter 2

Administration

LEADERSHIP

The director of a natural resources department should be a conceptionalist, that is, more than just someone with ideas. The director should be a leader with entrepreneurial vision and the time to spend thinking about the forces that will affect the future of the resource and the department. Leaders must create for their organizations clear-cut and measurable goals based on advice from all elements of the community. They must be allowed to proceed toward these goals without being crippled by bureaucratic machinery. Leaders must be allowed to take risks, to embrace error, and to use their creativity to the maximum as well as encourage people who work with them, or for them, to use theirs. Leaders are people who do the right thing; managers are people who do things right. Both roles are crucial, but they differ profoundly (Bennis 1991). One of the most important aspects of leadership is surrounding oneself with the most capable, most intelligent help possible. Leaders should not be intimidated by people who may be more capable than they, in given areas, but should utilize those people to their fullest.

Leaders must lead and not manage, and they must make sure that managers make their own decisions. Bennis (1991) believes that American organizations are underled and overmanaged. He believes that this is also true of most of the industrialized world. Bennis believes there are four competencies in every great leader: management of attention, management of meaning, management of trust, and management of self. The first leadership competency is the management of attention through a set of intentions or visions which give the direction. The second leadership competency is management of meaning, which implies that the leaders must be able to communicate their visions to other people. The third competency is management of trust. Trust is essential to all organizations. The main component of trust is reliability or consistency. People would rather follow individuals they can count on, even when they disagree with their viewpoints, than people who are constantly shifting their stands on issues. Bennis points out that, when Margaret Thatcher was reelected Prime Minister of Great Britain, her critics predicted she would revert to defunct labor policies. She did not. She was constantly focused and "all of a piece." The fourth leadership competency is management of self. Management of self is knowing one's skills and deploying them effectively. Management of self is critical. Without it leaders can do more harm than good. Some employees have said that a leader who does not waste their time, but knows precisely what he wants of them, is the type they are looking for.

Perhaps failure is a word that leaders should not become acquainted with. They should use synonyms such as mistake, foulup, error, or false start. Leaders and followers alike should realize that everyone makes mistakes, many some days. Napoleon is given credit for the statement, "I make so many mistakes that I no longer blush." Leaders should make people feel significant and important. This helps the individual and the organization. Administrators should also tell employees, that in learning and competence matters, there are no failures, only mistakes. They should give feedback on what they want done next. People like to be a part of the community; they want to belong. Where there is leadership there is a team, a family or unit. There is a sense of loyalty and pride in the organization. Work should be exciting, stimulating, and fascinating.

One of the important concepts in the workplace is quality. Response to quality is a feeling that is often connected with our experiences of such things as beauty and value. Closely connected to quality is the dedication or even the love of work. This dedication brings forth quality and is the force that energizes high performance systems. If we love our work we need not be afraid of punishment for making mistakes. Ultimately, in great leaders, and the organizations surrounding them, there is a fusion of work and play (Bennis 1991).

Former U.S. Senator from Iowa, John C. Culver (1987), points out that leadership today depends heavily on pollsters, consultants, image makers, and the media. He states that the traditional theory of education in classical Athens recognized that certain technical skills were needed for leadership. But without training in logic and rhetoric, for example, one could never climb the ladder of success in government. Also needed was commitment to pursue the truth, a sense of public service, good judgment about the ends as well as the means, and moral qualities. These were essential to reach the level of balance which was, hopefully, the goal of education.

This classical tradition was challenged by thinkers who called themselves sophists. The sophists maintained that technical skills were an end in and of themselves. They believed that one's talent for arguing a point was more important than the judgment used in deciding which position to defend. They were clever rather than wise. They pursued success instead of excellence and they honored intelligence above character. Culver believes that much of what passes for political education today is using the sophists' approach. People are more obsessed with the mechanics of reaching the goal rather than the goal itself. He believes that we are training technicians where we should be preparing leaders and stressing the fundamentals of value (Culver, J. C. 1987. Leadership. Editorial in *The Salt Lake Tribune*, June 1).

Hansen (1991) believes that the authoritarian military style of management and management practices used during and after the Industrial Revolution to control unskilled factory laborers is passé today. This is primarily true because today's personnel are better educated and more technically competent than in the past. Officers today are not willing to accept autocratic leadership that requires them to follow orders without question. Leaders must master skills such as patience, understanding, fairness, and good judgment. Such practices as public criticism, tactlessness, and unfairness in dealing with officers are destructive to the organization. Instead, today's leaders should stress the importance of considering, caring, and loyalty. Some chiefs of law enforcement, when they receive citizens' complaints about an officer, may assume the officer is guilty. This can demoralize the officer and undermine employee confidence in leadership. A leader will assume an employee is innocent of any wrongdoing until the facts prove otherwise.

Supervisors should study various leadership models and select one that best suits their personality and problems. One model assumes that there are two basic styles of leadership—job orientation and employee orientation. Job-oriented leaders are primarily concerned with the problems at hand and rely on the formal power structure and close supervision to accomplish the task. Employee-oriented leaders are concerned with maintaining good subordinate relations, and tasks are delegated according to the employee's ability.

Another model considers five management styles: task management, country club management, impoverished management, management of the middle-of-the-road type,

Wood ducks are
arguably the most
beautiful of all
waterfowl.
Courtesy U.S. Fish and
Wildlife Service, Dave
Menke.

and team management. Task management is concerned with achieving results by planning, directing, and controlling a subordinate's work. Country club management stresses the importance of good employee relationships. The impoverished management supervisor attempts to maintain organizational membership. The middle-of-the-road manager tries to maintain both good employee relationships and production. The team manager maintains a high degree of production through delegation of tasks to subordinates, and uses their input in the decision-making process. Team management is considered the most effective of the five styles.

Consider these four basic leadership styles—telling, selling, participating, and delegating. The telling style is highly task oriented and low on relations. This works best on new employees and employees with limited abilities or no experience. With experienced, motivated, and willing employees, the best results often come from team management, which relies on delegating tasks. Obviously, the management style and approach must take into consideration the experience level of the employees, their job maturity, and their ability to learn and perform new tasks. Hansen (1991) believes that the three leadership traits associated with top-level effectiveness are intelligence, personality, and ability. There are times when some employees seem to not respond to any type of leadership. Sound management practices mandate the use of discipline only when other reasonable courses fail. Disciplinary action should be quick and fair. Failure to dismiss or discipline an employee who obviously should be dismissed or disciplined will lose the respect of the experienced personnel. Last, but far from least, promotion policies should be fair, a system which considers the merits of the person in relation to the proposed job description.

WRITTEN POLICY

Written policies and procedures are valuable to both the administrator and the employee. Experienced employees are guided and, to a large degree, protected when written policy and procedure is understood and followed. There is an important difference between policies that must be followed as written and procedures that suggest ways of handling situations but leave areas for officer discretion based on individual

Cases 1–7

circumstances. Procedures are preferable in all but high liability areas such as use of firearms and vehicle pursuit. Administrators are better able to defend themselves and employees who follow instructions in these situations.

When writing policies, the law enforcement administrator should keep several considerations in mind. The first is to remember that no guideline covers all aspects of law enforcement. Attempting to cover all possible contingencies is not only impossible, but undesirable. Writing such policies would lose the element of officer discretion, which is needed to do a good job. Secondly, the tone of the policy should be positive. Instructions written primarily to protect administrators from litigation invariably take on a negative tone, which is translated to the officer. Thirdly, avoid glittering generalities—phrases that sound nice but have little meaning in a practical sense. One example is ''if in the best judgment of the officer,'' which means virtually nothing. Guidelines, on the other hand, may be the critical element that leads to effective and efficient operation of the organization (Auten 1988). How many facets of a program are covered in writing is determined by the administration.

REGULATING EMPLOYEE SPEECH

Law enforcement organizations may impose reasonable restriction on work-related speech of agency personnel. The primary objective of written speech policies is to adopt legally defensible rules and procedures that accommodate the interests of the manager, employee, and public. As in many cases, the best guidelines are formed by court decisions. In *Broaderick v. Oklahoma,*[1] the U.S. Supreme Court set forth a rigorous standard for the facial invalidation of catchall regulations. In *Broaderick,* the Court ruled that a regulation which affects both speech and conduct is unconstitutional only if its overbreadth is substantial, in other words, if it is a catchall type statement. For example, there is a broad difference between a speech which criticizes the department and conduct unbecoming an officer, inasmuch as the latter may be speech, action, or other conduct. *Parker v. Levy*[2] upheld two sections of the Uniform Code of Military Justice which stated ''conduct unbecoming an officer and a gentlemen'' and ''all disorders and neglects to the prejudice of the good order and discipline of the armed forces.'' In this case an army physician was convicted by a general court martial.[3] In *Barrett v. Thomas,*[4] regulations in the Dallas County sheriff's office were declared officially invalid because they attempted to limit unprotected speech through a prophylactic approach.[5] The courts were particularly troubled by the impermissible legal breadth of these rules (Schofield 1985). In *Barrett* this represents a clear application of the *Broaderick* pure speech v. speech plus conduct rule. In *Bence v. Breier,*[6] the U.S. Circuit Court of Appeals for the Seventh Circuit ruled unconstitutional on vague grounds a catchall rule in the Milwaukee Police Department which prohibited conduct unbecoming a member and detrimental to the service. It should be pointed out, that in *Bence,* the ruling was directed at the homogenous unit where members perform more or less similar functions. *Snepp v. United States*[7] made national news when the Central Intelligence Agency refused to let Frank Snepp, a former employee, publish memoirs that were considered detrimental to the agency. Snepp had executed, as a condition of employment, a secrecy agreement that expressly stipulated that he would not publish any material relating to the agency during or after his employment without specific prior approval by the agency. In his tour in Vietnam during the period of employment, Snepp was granted frequent access to classified information. Later he became dissatisfied with the manner in which the CIA was conducting its affairs in Vietnam and resigned in 1976. He subsequently published a book entitled *Decent Interval,* which contains serious allegations of misconduct in regard to CIA participation in Vietnam. The Court ruled in favor of the CIA. The Court did not address the scope of Snepp's First Amendment rights to publish or to speak about the CIA. The more narrow question before the Court concerned Snepp's legal obligation to submit to a prepublication review (Schofield 1985). Law enforcement employees are entrusted with a special responsibility, and they must conduct themselves in a professional manner. Employees are subject to disciplinary action for speeches constituting treason, libel, slander, perjury, incitement to riot, or knowingly making false statements regarding departmental operations. When employees are off duty and out of uniform they enjoy the same speech rights as other citizens, except for restrictions on partisan political speech imposed by law or specific instruction under departmental policy. Regulations of employee speeches within clearly established boundaries mitigate many of the constitutional concerns which are often associated with unpredictable standards. A vague or overbroad regulation can do considerably more harm than good. Organizational policy that encourages responsible employee criticism may heighten the confidence of the public and the law enforcement agency and improve employee morale. In the final analysis, a well-written and organized employee speech policy will benefit everyone involved, including the general public (Schofield 1985).

Deadly Force

Probably no police activity is in more need of a firm written policy than that of the use of deadly force. Most departments prohibit or strongly discourage firing warning shots. Many departments discourage or prohibit firing at moving vehicles unless the occupant of the vehicle is shooting. All departments recognize the right of an officer to use deadly force in self-defense or in the defense of others. But many differ, however, on standards to be used when a person is in jeopardy. Questions that may be asked are: Have defensive measures other than shooting failed? Is the threat imminent to life of the officer or someone nearby? Does the officer have reason to believe that death or serious injury will result if deadly force is not used? The FBI, for example, does not allow the use of deadly force with a fleeing suspect. Other questions asked are: What constitutes flight, and what distinction, if any, should be made between adults and juveniles or between violent and nonviolent crimes (Wilson 1980)?

Intradepartmental Communication

One of the difficulties law enforcement administrators have is maintaining a flow of two-way information. It is all too easy for field officers not to discuss problems or even communicate that they have a problem. One way of avoiding this is to form a group of employees who recommend a plan for bettering two-way communication. This employee council should not include administrators. Field officers should be chosen from each region or district and from the central office. Minutes should be taken of all meetings and distributed to departmental employees so they are kept informed of council activities and discussions. Copies should also be sent to all administrative staff. The burden can be lessened for council members by limiting terms to one year or less and by rotating positions, such as chair (Skidmore 1987).

Public Complaints

One of the common complaints by the public is that enforcement officers never listen to them. Often, the public does not know how to file a formal complaint. This is easily solved with a written complaint procedure which is made known to the general public, as well as field and staff employees. Citizens' negative perceptions are serious because they threaten the community trust which is essential to all enforcement agencies. When citizens believe that law enforcement officers are accountable to no one, they become alienated from and distrustful of those officers. One indication of this concern is the increased public demand for external control of law enforcement agencies, most notably civilian review boards, which may or may not be taken care of by fish and game

Mallards often land in small potholes.

councils or boards. Both the rights of the officer and the citizen should be considered, and there should be an avenue of approach for each individual or group (Duffy 1983).

Accident and Incident Reports

Accidents or incidents that involve injury to a person or a property are a continuing problem for wildlife law enforcement personnel. These can often result in insurance claims and lawsuits. It is important that all pertinent information available at the time of the accident or incident be documented for future reference. There should be a person to report to within a specified time and forms for recording extent of injury or damage, if medical help was needed, if the scene was secured to avoid any further damage or any possibility of destroying evidence. The person or persons involved should be instructed that their comments may provide evidence, but may be used against them if it is an admission of guilt. Names and addresses of witnesses should be obtained whenever possible. Questions by reporters or attorneys should be referred to a designated higher authority.

Officer Stress

Most police administrators have come to realize that officers suffering from emotional, physical, and behavioral symptoms may have critical stress problems. Employee turnover, sick leave abuse, increase in alcoholism, extreme aggressiveness, or substance abuse are a few of the signs that officers may need help from the administration and their peers. Because of these problems, it is imperative that administrators establish a relief process within their command. Officers should watch for signs of stress within themselves and in peers, and should balance job demands accordingly (Bratz 1986).

Administrative Liability

Agencies are responsible for the training of officers, including the use of firearms. In the past, officers were required only to show a proficiency on the shooting range. This has been demonstrated to be inadequate. The firearms training course should be able to establish a valid and defensible goal that will withstand the scrutiny of the courts. Some courses might create situations based on actual officer incidents. A training

An officer in summer uniform.
Courtesy Idaho Fish and Game Department, Rick Gilchrist.

Cases 8–9

course with both friendly and unfriendly targets appearing at random may be used to evaluate an officer's stress. A computerized record should be kept of each officer's range score. These records should be able to withstand the scrutiny of the court if there is a shooting incident (Schraeder 1988).

The U.S. Supreme Court in 1961 ruled that state and local police officers can be sued in their individual capacities in federal court for damages pursuant 42 USC Section 1983, *Monroe v. Pape*.[8] In 1978, the Court ruled for the first time that a municipal corporation (read department) may be held liable under section 1983 when it implicates or executes a formal policy ordinance or regulation officially adopted by its officers and which results in a constitutional deprivation (*Monell v. New York City Department of Social Services*).[9] Both of these cases revolved around municipal (read department) liability for inadequate training and supervision. In *Monell,* the Court made it clear that municipal liability was predicated solely upon the unconstitutional conduct of municipalities, avoiding the idea that liability can be visited upon a city on a respondent superior theory. This theory imposes liability upon an employer for the wrongful action of an employee regardless of the absence of fault by the employer. Plaintiffs attempting to sue a municipality under section 1983 for injuries caused by unconstitutional conduct of police officers are required, as in *Monell,* to establish that their injuries were caused either by the municipal policy or one of its customs. If plaintiffs are unable to find a policy which is unconstitutional on its face, then they must establish that the department adopted a custom of inadequate training or supervision. However, the exact level of

Cases 10–14

this inadequate training and supervision is a matter of considerable dispute among the justices of the courts (Callahan 1989). For example, in *City of Oklahoma City v. Tuttle,*[10] Supreme Court Chief Justice Rehnquist's views appear to be that liability should not exist when the only proof offered is a single, unconstitutional act of a subordinate police officer. Rather, in order to prove that a custom of inadequate training caused an injury, the plaintiff will likely be required to establish a history or a pattern of similar incidents which were unaddressed or unremedied by policy-making personnel (Callahan 1989). In *Spell v. McDaniel,*[11] the Fourth Circuit affirmed a verdict for the plaintiff and observed that two theories of municipal liability have emerged when no special policy can be found which is per se unconstitutional. First, there can be fault in deficient officer training programs which cause unconstitutional violation by untrained or mistrained officers. In the second case of ''in the failure of police managers to correct a widespread pattern of unconstitutional conduct by subordinates,'' the court stated that in appropriate circumstances either of these theories may be invoked. Department policymakers, acting with deliberate indifference to the constitutional rights of the citizens, may be successfully litigated against. Departments should review training and supervisory procedures relating to all high-risk activities. This includes the use of firearms, deadly nonlethal force, high-speed pursuit, constitutional criminal law, and civil rights and prisoner safety in detention facilities. No policy should fall below state standards. With respect to firearms training, officers should train and qualify with the guns they carry on the job, and they should not be permitted to carry a weapon unless they have qualified with that particular weapon. Departments should ensure that hiring policies of new officers are carefully reviewed and evaluated. Background investigations and appropriate testing of potential applicants should be required (Callahan 1989).

Following the *Monell* decision, there were many suits directed against cities that were based on the claim that the city had adopted a custom or policy of inadequate training of officers. During this time there was considerable judicial disagreement concerning the standard by which municipalities should be judged in these suits. The Supreme Court resolved much of this uncertainty in its 1989 decision in *City of Canton, Ohio v. Harris.*[12] In *Canton,* the lower courts held for the plaintiff, but in a landmark decision, the U.S. Supreme Court reversed lower court ruling. It held that inadequate police training can serve as a basis for liability only where the failure to train amounts to deliberate indifference by the city policymakers in regard to the constitutional rights of the person contacted by police officers. By adopting the higher deliberate indifference standard, the Court rejected the post-negligence standard that had been adopted by many lower federal courts. The Court explained that inadequate training meets the deliberate indifference standard only when the need for more or different training is necessary and the failure to implement is likely to result in constitutional variation. The Court offered two examples of deliberate indifference. First, when the city policymakers know that officers are required to arrest fleeing felons and are armed to accomplish that goal, then the need to train officers in constitutional limitations regarding the use of deadly force in apprehending felons is obvious. Second, deliberate indifference could be based on a pattern of officer misconduct which should have been observed by police officials, and the necessary remedial training provided. Two cases, *Gutierrez-Rodriguez v. Cartagena*[13] and *Dobos v. Driscoll,*[14] upheld the deliberate indifference standard. No aspect of policing brings forth more passionate concern or more divided opinions than the use of deadly force. It should be noted that police administrators are only personally liable for their unconstitutional action or inaction and are not liable for the misconduct of subordinates, unless their actions as a police supervisor are the cause of the constitutional injury. The Court has also noted that police supervisors are not liable simply because a subordinate employee violated someone's rights (Callahan 1990).

Retirement

Retirement is looked forward to by many people with mixed emotions. It may offer stress-free days with a peaceful and happy future. Or it may be a time when the retiree wakes up in the morning and wonders, ''What do I do now?'' The answer is to plan ahead—for many months or even years. Retirement may be a fishing vacation, a loafing vacation, the taking on of another job, or perhaps going back to school to pick up things that were missed along the way. Officers should consider the effects of retirement not only on themselves, but on their families and even close friends. Officers who are used to giving orders and being respected may find, upon retirement, that they miss this authority.

There are many reasons why officers leave enforcement. Some leave to find new and better opportunities, and others leave because of job stress or dissatisfaction. Perhaps there are administrative problems or even disciplinary ones. Then there are people who stay because they like the work and cannot see themselves doing anything else. Still others stay because of the financial burdens that cannot be handled on a lower paying job. Others have a fear of starting something new. Administrators can educate officers about retirement problems through information and consultation with retirees. (Violanti 1990).

Pursuit Driving

Departmental Responsibility for Liability Reduction

Organizations must carefully evaluate pursuit policies, training, supervision, and post-incident evaluation.

Policy development. A policy should be tailored to the department's operational needs, geographic peculiarities, and training capabilities. A written policy also provides a basis for holding officers accountable for pursuit-related conduct and for giving them general guidelines on how to pursue each step. Commentators continue to debate the relative merits and disadvantages of written departmental policy concerning vehicle pursuit. Most experts, however, recommend that law enforcement organizations adopt a written policy that imposes specific controls on the operation of vehicles and officers.

Training. Lack of adequate training can easily contribute to a pursuit-related accident. It is easy to lose control of a vehicle that is driven beyond the capability of either the driver or the vehicle. Whatever the circumstances, officers should not drive beyond their capabilities. If an officer becomes too emotionally involved in a pursuit, it should be discontinued.

Supervision. In most cases, an officer not in the pursuit vehicle should be in charge of the pursuit. This person is better qualified to objectively oversee the pursuit and decide whether it should continue and how each step should be taken.

Evaluation and documentation. Law enforcement officers should provide an ongoing process of evaluation and documentation of pursuit-related incidents regardless of the outcome.

Conclusion. Drivers of police vehicles have a statutory duty to drive with due regard for the safety of others. Liability must be based on proof that police conduct in breaching a duty owed was the proximate cause of a pursuit-related accident. It is not a valid defense against departmental liability to argue that a breach of a duty trained officer in pursuit driving was due to inadequate resources or lack of training facilities (Schofield 1988).

The FBI National Crime Information Center (NCIC) maintains records available to law enforcement agencies. *Courtesy Federal Bureau of Investigation.*

ADMINISTRATORS BEWARE OF THE TYRANNY OF SMALL DECISIONS

Administrators were warned in the above heading. It was first developed by Kahn (1966) in relation to market economics and further refined by Odum (1982) in regard to environmental degradation. Ideally, society's problems are resolved through a system of nested levels of public decision. Low-level decisions are made by an individual or a small group of individuals. Theoretically, the more important the decision, the higher the level of decision making. At the highest level, decision makers are experts whose joint decisions will form the rules for decisions which are to be made later, at a lower level. Unfortunately, Odum says, important decisions are often reached in an entirely different manner. They are comprised of a series of small, apparently independent decisions often made by groups of individuals or just one individual. A big decision occurs (*post hoc*) as an accretion of these small decisions. The central question is never really addressed at the higher level. Much of the current confusion and problems surrounding environmental issues today can be traced to decisions that were never consciously made, but simply resulted from a series of small decisions. Odum points out that, in the loss of coastal wetland on the U.S. east coast between 1950 and 1970, fifty percent of the marshland was destroyed without any conscious effort. It amounted to a series of small decisions, such as the conversion of small tracts of marshlands to housing and various other uses, that in themselves did not seem particularly important at the time.

Odum (1982) states that the Florida Everglades have suffered a serious ecological shock, not from a single adverse decision, but from a multitude of small decisions which include: one more drainage canal, one more roadway, one more retirement village, and one more well to provide Miami with drinking water. No one made the choice to reduce the annual flow of water into the Everglades, but a series of small decisions did, in effect, do just that. Many threatened or endangered species of wildlife have been subject to this type of decision making. The green turtle, for example, has lost most of its nesting beaches through a series of developments and human intervention. The decline was furthered by anglers harvesting 'just one more turtle' despite its threatened status. Many of today's eutrophic lakes have suffered from the process of small decisions, such as adding one more sawmill, one more area of nonpoint pollution, another water diversion that returned agricultural fertilizer to the stream, and the use of pesticides on the watershed.

Odum (1982) and Edmondson (1991) believe that we must have at least a few scientists who study whole systems and help us avoid the consequence of many small decisions. Environmental science teachers should include in their courses some examples of large scale processes that result in human-induced problems. Does Odum believe that the solution to this tyranny of small decisions will come about in the near future? The answer is, tragically, no.

Preemployment Background Investigation

Case 15

The preemployment background investigation may be the most important investigation that a law enforcement agency has to conduct. A proper and thorough investigation conducted by an agency can eliminate undesirable applicants and hire qualified, dedicated employees. If the preemployment investigation is not conducted, the agency may expose itself to a vast array of libelous situations, occupational problems, or, at the very least, a nonproductive employee (Wright 1991).

More law enforcement agencies are being held accountable for the actions of their employees, and an increasing number of agencies are being sued for negligent hiring and negligent retention. A negligent hiring suit is generally based on the legal concept of ''let the master answer.'' This type of suit alleges that the employer is negligent by placing the employee in a position where the employer knows or should have known that the individual was incompetent. Negligent retention is a breach of an employer's duty to monitor an employee's unsatisfactory performance and take corrective action through training, reassignment, or discharge. A department may be held liable if it knowingly allows an officer who cannot successfully qualify with a handgun to continue carrying the weapon. In a U.S. Supreme Court decision in 1986, in *Daniels v. Williams*,[15] it was held that negligence is not actionable as a constitutional violation. The viability of negligent hiring or retention suits are now dependent on each state's tort law. Each department should be familiar with the relevant statute governing negligent hiring and retention suits in their state.

Aside from avoiding legal damages, an investigation should give some indication as to the competency, motivation, and personal ethics of the applicant. These are important factors that should be made known to the department before an individual is hired. If derogatory information becomes known after hiring, it can even jeopardize criminal cases involving the officer (Wright 1991).

During the prehiring discussion with the prospective employee, applicants should be advised of information on salary, benefits, working conditions, vacations, days off, probation status and duration, civil service and union rules, overtime pay policies, and retirement. Prospective officers should be asked why they want a career in this particular type of law enforcement, and how their spouses will feel about them working in the law enforcement field. The applicant should be allowed to ask questions. The investigating officer should contact applicant's previous employers if possible, see a birth certificate, driver's license, social security card, proof of whatever education the person claims, and proof of military service. A credit check and a driving record should also be recorded. Many administrators also prefer to interview an applicant's spouse (Wright 1991).

Literature Cited

Auten, J. H. 1988. Preparing written guidelines. *FBI Law Enforcement Bull.* (Federal Bureau of Investigation, U.S. Dept. of Justice, Washington, D.C.) 57 (5): 1–7.

Bennis, W. 1991. Learning some basic truisms about leadership. *National Forum* 71 (Winter) (1): 12–15.

Bratz, L. L. 1986. Combating police stress. *FBI Law Enforcement Bull.* (Federal Bureau of Investigation, U.S. Dept. of Justice, Washington, D.C.) 55 (1): 1–7.

Callahan, M. 1989. Municipal liability for inadequate training and supervision—divergent views. *FBI Law Enforcement Bull.* (Federal Bureau of Investigation, U.S. Dept. of Justice, Washington, D.C.) 58 (3): 24–30.

———. 1990. Deliberate indifference: The standard for municipal and supervisory liability. *FBI Law Enforcement Bull.* (Federal Bureau of Investigation, U.S. Dept. of Justice, Washington, D.C.) 59 (10): 27–32.

Duffy, J. F. 1983. PERF acts to improve citizen complaint procedures. *FBI Law Enforcement Bull.* (Federal Bureau of Investigation, U.S. Dept. of Justice, Washington, D.C.) 52 (5): 11–14

Edmondson, W. T. 1991. *The uses of ecology—Lake Washington and beyond.* Seattle and London: Univ. of Washington Press.

Hansen, P. 1991. Developing police leadership. *FBI Law Enforcement Bull.* (Federal Bureau of Investigation, U.S. Dept. of Justice, Washington, D.C.) 60 (10): 4–8.

Kahn, A. E. 1966. The tyranny of small decisions: Market failures, imperfections, and the limits of economics. *Kyklos* 19: 23–47.

Odum, W. E. 1982. Environmental degradation and the tyranny of small decisions. *Bioscience* 32 (9): 728–729.

Schofield, D. L. 1985. The constitutionality of organizational policies regarding employee speech. *FBI Law Enforcement Bull.* (Federal Bureau of Investigation, U.S. Dept. of Justice, Washington, D.C.) 54 (9): 21–31.

———. 1988. Legal issues of pursuit driving. *FBI Law Enforcement Bull.* (Federal Bureau of Investigation, U.S. Dept. of Justice, Washington, D.C.) 57 (5): 23–30.

Schraeder, D. E. 1988. Firearms training/civil liability: Is your training documentation sufficient? *FBI Law Enforcement Bull.* (Federal Bureau of Investigation, U.S. Dept. of Justice, Washington, D.C.) 57 (6): 1–3.

Skidmore, J. W. 1987. The employee council. *FBI Law Enforcement Bull.* (Federal Bureau of Investigation, U.S. Dept. of Justice, Washington, D.C.) 56 (3): 13–15.

Violanti, J. M. 1990. Police retirement: The impact of change. *FBI Law Enforcement Bull.* (Federal Bureau of Investigation, U.S. Dept. of Justice, Washington, D.C.) 59 (3): 12–15.

Wilson, J. Q. 1980. Police use of deadly force. *FBI Law Enforcement Bull.* (Federal Bureau of Investigation, U.S. Dept. of Justice, Washington, D.C.) 49 (8): 16–21.

Wright, T. H. 1991. Preemployment background investigations. *FBI Law Enforcement Bull.* (Federal Bureau of Investigation, U.S. Dept. of Justice, Washington, D.C.) 60 (11): 16–21.

Notes

1. 413 U.S. 601 (1973).
2. 417 U.S. 733 (1974).
3. The case involved an army physician who was convicted by a general court martial after making public statements urging black soldiers to not obey orders to go to Vietnam.
4. 649 F. 2d 1193 (5th Cir. Unit A 1981), cert. denied, 102 S. Ct. 1969.
5. The regulations provided that gossip about affairs of the department, making unauthorized statements, or unauthorized reviewing of confidential information are prohibited.
6. 510 F. 2d 1185 (7th Cir. 1974), cert. denied, 95 S. Ct. 804.
7. 444 U.S. 507 (1980).
8. 365 U.S. 167 (1961).
9. 436 U.S. 658 (1978).
10. 471 U.S. 808 (1985).
11. 824 F. 2d 1380 (4th Cir. 1987).
12. 109 S. Ct. 1197 (1989).
13. 882 F. 2d 553 (1st Cir. 1989).
14. 537 N.E. 2d 558 (1989), cert. denied, 110 S. Ct. 149.
15. 106 S. Ct. 662 (1986).

Recommended Reading

Bennis, W., and B. Nanus. 1985. *Leaders*. New York: Harper and Row.

Fischer, J. 1971. *How our laws are made*. Washington, D.C.: U.S. Govt. Printing Office.

Fleck, R. 1982. Physically fit employers, more efficient workers. *Bureau of National Affairs Communicator* (Summer): 16.

Garrison, V. J. 1980. Motivation as it applies to conservation officers. In *Proceedings Southeast Assn. of Fish and Wildlife Agencies*, 649–653.

Harris, J. H. 1972. An examination of law enforcement recognition by the United States Forest Service in the Intermountain Region. Ph.D. diss., Utah State Univ. Logan, Utah.

Hickman, C. R. 1990. *Mind of a manager, soul of a leader*. New York: John Wiley and Son.

Kotter, J. P. 1990. *A force for change*. New York: The Free Press.

Service, J. G. 1987. Let the master answer. *Security Management Magazine* (May), 100–102.

———. 1988. Negligent hiring: A liability trap. *Security Management Magazine* (Jan.), 65–68.

Spaniol, J. F. Jr. 1959. *The United States Courts*. Washington, D.C.: U.S. Govt. Printing Office.

Spomer, R. 1992. Winning the conservation battle; losing the PR war. *Pheasants Forever* 10 (2).

Swanson, C. R., L. Territo, and R. W. Taylor. 1988. *Police administration: Structure, processes, and behavior*. New York: MacMillan.

Vance, D. C. 1982. Regulations and the wildlife resource. In *Proceedings Western Assn. of Fish and Wildlife Agencies*, 84–87.

Zaleznik, A. 1977. Managers and leaders: Are they different. *Harvard Business Review*.

Multiple-Choice Questions

Mark only one answer.

1. The sophists of ancient Greece believed that:
   a. intelligence was more important than character
   b. technical skills are an end in and of themselves
   c. a talent used in arguing a point is more important than the point
   d. it is better to be clever than to be wise
   e. all of the above
2. The Supreme Court ruled that the guidelines for employee speech policy are unconstitutional if catchall types of statements are used as in:
   a. *Parker v. Levy*
   b. *Barrett v. Thomas*
   c. *Jones v. Altman*
   d. *Broaderick v. Oklahoma*
   e. *Snepp v. United States*
3. Officer stress may show up as:
   a. substance abuse
   b. sick leave abuse
   c. extreme aggressiveness
   d. alcoholism
   e. all of the above
4. In a preemployment background investigation, which is correct?
   a. can be casual since it is rarely used later
   b. is very important and it should be thorough
   c. is required by law but can be routine
   d. is rarely important
   e. is not followed by most agencies

5. Why are small decisions and their sometimes long-range effects so often overlooked by administrators?
   a. they appear to relate only to a small segment of society
   b. they seem innocuous and, therefore, harmless
   c. they frequently never come to the attention of the administration
   d. they are unguided because they are never addressed by high-level administrators
   e. all of the above
6. Chief Justice Rehnquist ruled in what case that a single inconsistent act by an officer does not constitute a custom?
   a. *Monroe v. Pope*
   b. *Dobos v. Driscoll*
   c. *Barrett v. Thomas*
   d. *Parker v. Levy*
   e. *City of Oklahoma City v. Tuttle*

7. The qualifications of a director include:
   a. being a conceptionalist
   b. doing the right thing
   c. leading rather than managing
   d. managing one's own attention span
   e. all of the above
8. What word would Bennis like to see dropped from the vocabulary of administrators?
   a. mistakes
   b. errors
   c. foulups
   d. false start
   e. failure

Chapter 3

The Rights of the Individual

MAN AS VIEWED BY THE FOUNDING FATHERS

The founding fathers insisted that the individual did not belong to the state, but had a personal and immortal soul that should be beyond the dictates of any totalitarian regime. On that principle they established a constitutional government that was under and not above the law. This was, and still is, the most powerful and popular political idea in the world (Smith 1987).

Paige Smith, in the First Annual Tanner Academy Lecture, offers some interesting comments on the Federal Constitution and its framers (Smith 1987). The Federal Constitution, we often say, is one of the greatest documents in all history, but we seldom pay any attention as to how a small group of provincial politicians were able to compose such an enduring work. Certainly the framers were remarkable men. What is most notable about them is that they lived at a junction of two great currents of intellectual history, one as old as Greece and Rome, and the other as recent as the so-called Enlightenment. As Smith says, the framers were representatives of what he calls the classical-Christian conscience. Co-joining the words classical and Christian may have an odd ring to the modern ear, but the two traditions shared a common skeptical view of human nature. They believed that individuals and classes hungered for wealth and power and sought it often by dubious means, and when they obtained it they almost invariably used it to victimize and exploit those without wealth and power. This was primarily the poor. Smith cites three specific instances. In the Federal Debates, Gouverneur Mouris, one of the more conservative delegates, urged that provisions be written into the Constitution to protect the poor against the rich. Mouris argued, ''The rich will strive to establish their domain and to enslave the rest. They always did. They always will'' (Smith 1987). He believed that the best protection against the rich was to form them into a separate interest where they could be controlled or balanced by other interests. If they were mixed with the poor they would establish an oligarchy.

Madison began with the assumption that a faction, the pursuit of their own interest at the expense of other legitimate interests, is a characteristic of all society. ''So strong is this propensity of mankind to fall into mutual animosities, . . . that where no substantial occasion presents itself, the most frivolous and fanciful distinctions have been sufficient to kindle their unfriendly passions and excite their most violent conflicts'' (Smith 1987).

Smith (1987) says that the most trenchant analysis of human beings as ''fallen'' comes from the pen of a semiliterate Massachusetts farmer, William Manning. Manning said, ''Men are born & grown up in this world with a vast variety of capacityes, strength & abilityes both of Body & Mind, & have strongly implanted within them numerous pashons & lusts continually urging them to fraud violence & action of injustis toards

Hunter safety training is
generally mandatory for
first-time hunters.
Courtesy Crossman Guns.

one another.'' He continued that man knew right from wrong, yet he is sustained by
the degree of heaven to hard labor for living in this world. He added that man has a
desire for self-support, self-defense, self-love, and self-defeat, self-importance and
self-aggrandizement. Smith says that Manning's views of human nature are rather grim
but were shared by his contemporaries who were delegates to the Federal Convention.
The delegates did not set out to devise a government for enlightened, rational, human
beings. They set out to fashion one for very fallible human beings. They were nothing
if not realistic in their assessment of their fellow countrymen.

The spirit of enlightenment was expressed in the Declaration of Independence. The
Declaration was the manifesto of what Smith (1987) calls, in its American form, the
Secular-Democratic Consciousness. It shared the French enthusiasm for reason, science, and progress. It subscribed to the notion so plainly rejected by the framers of the
Constitution that ''the voice of the people is the voice of God.'' It was, Smith says, as
much as anything else a declaration against history. There was no historical evidence
for its proposition that men were created equal. All evidence of history was against
such a proposition. Men and women were nowhere observed to be equal in talents,
fortunes, or rights, and the vast majority of the inhabitants of this earth did not enjoy
liberty, in fact, they hardly knew the meaning of the word. As for the pursuit of hap-
piness, the struggle to survive was the fate of most humans. By contrast, the Constitution was an almost painfully practical, down-to-earth document, as befitted the
temper of its framers. There was nowhere in it any reference to the Almighty, and when
a delegate to the convention suggested each session be opened with a prayer for divine
guidance, Alexander Hamilton replied that he saw no need for outside assistance. Smith
believes the Federal Constitution could not have been drafted even a decade later. The
consciousness that shaped it, the profound respect for the lessons of history that un-
dergirded it, the view of man as fallible, fallen creatures whose manifold weaknesses
must be taken into account in any effort to frame a popular government, was already

Cases 1–4

in decline. Smith adds that the Federal Constitution is a remarkably enduring document that was, in a sense, a gift of the preceding two thousand years to the people of the United States.

SEARCH AND SEIZURE

''The right of the people to be secure in their persons, houses, papers, and effects, against unreasonable searches and seizures, shall not be violated, and no Warrants shall issue, but upon probable cause, supported by Oath or affirmation, and particularly describing the place to be searched, and the persons or things to be seized.'' (The Fourth Amendment)

A fundamental principle of search and seizure law which has been emphasized in decisions of the Supreme Court, is that the police, whenever practicable, must obtain advance judicial approval of search and seizure by way of a warrant. However, it is well settled that a search incident to a lawful arrest is a traditional exception to the warrant requirement of the Fourth Amendment (Rissler 1978). The reasons for this exception are quite basic and were identified by the Supreme Court in its landmark decision in *Chimel v. California*.[1] The permissible scope of search incident to arrest has two levels: search of the actual person of the arrestee, and the other, a search of possessions within the area of the arrestee's immediate control (*United States v. Robinson*).[2]

There seems to be little disagreement about the limits of search at the subject's arrest. It may extend to subject's body, clothing, and personal items located on or in subject's clothing, such as wallets and cigarette packages, but the area of permissible search beyond the person of the arrestee has been subject to different interpretations.

Since *Chimel*, it has been generally understood that an officer may search the area within his immediate control, otherwise known as the ''grabbing distance.'' However, *United States v. Chadwick*[3] contains language requiring officers and departments to reexamine existing policies concerning searches incidental to arrest. Chadwick was arrested, and his footlocker and its keys seized. The defendant and the footlocker were then transported to the Boston Federal Building where an hour and a half later the footlocker was searched without the benefit of a search warrant. The court ruled that once the law enforcement officers ''have reduced luggage or other personal property not immediately associated with the person of the arrestee to their exclusive control and there is no longer any danger that the arrestee might gain access to the property to seize a weapon or destroy evidence,'' a search of that property is no longer an incident of the arrest. The essentials were the same in *United States v. Ester*.[4] The court said, in brief, that having reduced luggage or other personal property not immediately associated with the person of the arrestee to their exclusive control and there being no danger, a warrantless search cannot be justified as being incident to an arrest (Rissler 1978).

It appears that the full impact of *Chadwick* is not known. However, it seems wise to assume that searches incident to arrest are acceptable; when an individual is arrested, his person and personal property immediately associated with him may be searched completely. On the other hand, the *Chadwick* decision indicates that ''area'' searches may no longer be conducted once officers have reduced possessions located in the immediate vicinity of the arrestee to their exclusive control and they are no longer accessible to the arrestee. Possible exceptions to this might be if the officers had reason to believe that the items seized contain a dangerous explosive or evidence that would be destroyed or altered with the passage of time (Rissler 1978).

Cases 5–10

The Fourth Amendment affords an individual protection against unreasonable search and seizure. And an arrest is a seizure in a constitutional sense. Until 1914, the Supreme Court had not provided a constitutionally required remedy for violations of the Fourth Amendment. However, in *Weeks v. United States,*[5] the Court adopted an exclusionary rule for the federal courts. The penalty for violation of personal rights against unreasonable search and seizure was loss of evidence. The primary purpose of the exclusionary rule is to deter unlawful police conduct. An officer will be discouraged from making illegal arrests and unreasonable searches if the evidence obtained will be rendered inadmissible. The second purpose is to protect judicial integrity; that is, by preventing the use of evidence illegally seized, the doctrine avoids making a court party to the unlawful act. The exclusionary rule concludes that, on balance, it is preferable to protect the Constitution by the exclusion remedy, rather than to make certain all criminals are convicted. Thus, the traditional test of evidence admissibility—relevancy, materiality, and competency—no longer solely control use of evidence in a criminal prosecution (McLaughlin 1976).

One of the most publicized Supreme Court decisions, *Mapp v. Ohio,*[6] made the exclusionary evidence rule applicable to state criminal proceedings. In *Mapp,* the Court held that the Fourth Amendment rights of privacy are enforceable against the states through the due process clause of the Fourteenth Amendment by the sanction of exclusion.

In 1963, the decision of *Wong Sun v. United States*[7] introduced the phrase ''primarily illegality,'' which may be equated with unlawful arrest or unlawful search. *Wong Sun* held that where evidence is derived from an exploitation of the illegality, such evidence, physical or testimonial, may be excluded. In confession cases where there is an apparent causation factor between an unlawful arrest and an incriminating statement taken thereafter from a suspect in custody, both federal and state courts are prepared to exclude the statement (McLaughlin 1976).

The Supreme Court in *Miranda v. Arizona*[8] announced specific guidelines for police to follow prior to conducting custodial interrogation. Ironically, the *Miranda* requirement also afforded prosecutors an opportunity to preserve confessions which were obtained following illegal arrest. It was argued by some, that notwithstanding an improper arrest, the conferral of Miranda warnings and the obtaining of a waiver were intervening circumstances which severed the chain of causation between arrest and confession. In short, Miranda compliance alone was said to ''dissipate the taint'' between the primary illegality and the challenged evidence, thus bringing the case within a recognized exception to the derivative evidence rules[9] (McLaughlin 1976).

Nine years after the Supreme Court's *Miranda* decision, a lingering issue of importance to law enforcement officers was resolved. The issue was whether Miranda warnings alone insulate a confession taken from a person in custody following an illegal arrest. The answer to this is, it does not (McLaughlin 1976). The decision that provided this answer was *Brown v. Illinois.*[10] A week after a crime was committed three officers went to Brown's apartment where two broke in and searched it. Later, when the defendant returned, he was arrested outside his apartment. The officers had no warrant authorizing either the entry or the arrest. The defendant was taken to the police station where he was warned of his Miranda rights. About an hour after the arrest, an interrogation of the defendant began, which resulted in his signing an incriminating statement. Later, an Assistant State's Attorney interrogated the defendant after conferring the Miranda warnings. The defendant was indicted, tried, and

Cases 11–22

convicted of murder, both incriminating statements being admitted into evidence at trial. On appeal, the Supreme Court of Illinois reaffirmed the conviction. The U.S. Supreme Court reversed with no dissents, all justices agreeing that Miranda warnings alone did not break the causal chain between an unlawful arrest and inculpatory statement subsequently obtained. The Court said, in part, that the exclusionary rule, when used to effectuate the Fourth Amendment, serves interests and policies distinct from those served under the Fifth Amendment. The Fourth Amendment rule of exclusion is aimed at deterring lawless police conduct and preserving judicial integrity, that is, barring the use in court of evidence unconstitutionally obtained. If officers, knowing that the Fourth Amendment has been violated, can overcome this by the simple expediency of giving a Miranda warning, the intent of the Fourth Amendment is defeated. The Court noted that the Fifth Amendment is a device used to protect an individual's right against compulsatory self-incrimination. The warnings have never been regarded as a means either of remedying or determining violations of the Fourth Amendment rights (McLaughlin 1976).

COVERAGE OF THE FOURTH AMENDMENT

This amendment denounces only such searches and seizures as are "unreasonable," and is to be construed in the light of what was deemed an unreasonable search and seizure when it was adopted and so as to conserve public interests as well as the rights of individuals.[11] It applied only to governmental action, not to the unlawful acts of individuals in which the government has no part.[12] . . . but the amendment is applicable to search warrants issued under any statute, including revenue and tariff laws.[13]

Security "in their persons, houses, papers and effects" is assured to the people by this article. Not only the search of a dwelling, but also of a place of business,[14] a garage,[15] or a vehicle,[16] is limited by its provisions. But open fields are not covered by the term "house"; they may be searched without a warrant.[17]

The right to search the person upon arrest has long been recognized.[18]

The Fourth Amendment has been construed

. . . as recognizing a necessary difference between a search of a store, dwelling house, or other structure in respect of which a proper warrant readily may be obtained, and a search of a ship, motor boat, wagon, or automobile for contraband goods, where it is not practicable to secure a warrant because the vehicle can be quickly moved out of the locality or jurisdiction in which the warrant must be sought. . . . The measure of legality of such a seizure is, therefore, that the seizing officer shall have probable cause for believing that the automobile which he stops and seizes has contraband therein which is being illegally transported.[19]

Where officers have reasonable grounds for searching an automobile which they are following, a search of the vehicle, immediately after it has been driven into an open garage, is valid.[20] The existence of reasonable cause for searching an automobile does not, however, warrant the search of an occupant thereof although the contraband sought is of a character which might be concealed on the person.[21]

". . . nor shall any person be subject for the same offense to be twice put in jeopardy of life or limb; nor shall be compelled in any criminal case to be a witness against himself. . . ." It has held, however, that where the same act is an offense against both the state and federal governments, its prosecution and punishment by both governments is not double jeopardy.[22]

Cases 23–43

By common law, not only was a second punishment for the same offense prohibited, but a second trial was forbidden whether or not the accused had suffered punishment, or had been acquitted or convicted.[23] This clause embraces all cases wherein a second prosecution is attempted for the same violation of law, whether felony or misdemeanor.[24] Seventy-five years earlier a closely divided Court held that the protection against double jeopardy prevented an appeal by the government after a verdict of acquittal.[25] A judgment of acquittal on the grounds of the statute of limitations is a protection against a second trial,[26] as is also a general verdict of acquittal upon an issue of ''not guilty'' to an indictment which was not challenged as insufficient before the verdict.[27] When a federal trial judge, even when there is evidence of improper conduct on the part of the prosecutor and lack of credibility of government witnesses, refuses to hear additional testimony, and directs a judgment of acquittal, the double jeopardy clause precludes a second trial of the defendants.[28] Similarly, when the accused, by appeal, obtains a reversal of a conviction of second degree murder resulting from a trial for first degree murder, the clause prevents a second trial for the greater offense; for the jury's finding at the first trial is to be viewed as an implied acquittal of first degree murder.[29] On the other hand, a petitioner, whose conviction for conspiracy to evade income taxes was set aside on appeal, was not subjected to double jeopardy by the action of a Federal Court of Appeals. In this instance the court, after having originally directed a verdict of acquittal, ordered a new trial upon rehearing. Nor was he exposed to double jeopardy by that court's conclusion that he might be tried on the alternative theory that the original conspiracy had continued long enough to preclude application of the statute of limitations, and that his conviction may have been based on the ''impermissible theory'' that a subsidiary conspiracy had extended the statute of limitations which had run against the main conspiracy.[30]

Whether or not the discontinuance of a trial without a verdict bars a second trial depends upon the circumstances of each case.[31] Discharge of a jury because it is unable to reach agreement,[32] or because of the disqualification of a juror,[33] does not preclude a second trial. Where, after a demurrer to the indictment was overruled, a jury was impaneled and witnesses sworn, the discharge of the jury to permit the defendant to be arraigned did not bar a trial before a new jury.[34] When a judge, on his motion and without the approval or objection of defense counsel, withdrew a juror and declared a mistrial in the interest of the defendant, conviction of the latter at a second trial did not give rise to double jeopardy; for when substantial justice cannot be obtained without a mistrial, the latter may be ordered, even over the objection of the defendant, without foreclosing a second trial.[35] Nor did the withdrawal of charges, after a trial by a general court-martial had begun, because the tactical situation brought about by the rapid advance of the army made continuance of the trial impracticable, bar a trial before a second court-martial.[36] On the other hand, when a prosecutor, having been apprised of possible absence of a witness before the case was called, requested discharge of the jury because his key witness on two of seven counts had failed to appear, and the court rejected petitioner's motion that the trial continue on all but these two counts, the subjection of petitioner to second jury trial exposed him to double jeopardy.[37] An accused is not put in jeopardy by preliminary examination and discharge by the examining magistrate,[38] by an indictment which is quashed,[39] nor by arraignment and pleading to the indictment.[40] In order to bar prosecution, a former conviction must be pleaded.[41]

A plea of former jeopardy must be based upon a prosecution for the same identical offense.[42] The test of identity of offenses is whether the same evidence is required to sustain them; if not, the fact that both charges relate to one transaction does not make a single offense where two are defined by the statutes.[43] Where a person is convicted

Cases 44–48

of a crime which includes several incidents, a second trial for one of those incidents puts him twice in jeopardy.[44] Congress may impose both criminal and civil sanctions with respect to the same act or omission,[45] and may separate a conspiracy to commit substantive offense from the commission of the offense and affix to each a different penalty.[46]

RECENT COURT RULINGS AFFECTING THE FOURTH AMENDMENT

During its 1990–1991 term, the U. S. Supreme Court ruled on eight cases that are of importance to law enforcement officers. These cases, affecting the Fourth and Fifth Amendments, follow (McCormack 1991). In *Bostick,*[47] the Court ruled that law enforcement officers who approach a seated bus passenger and request consent to search the passenger's luggage do not necessarily seize the passenger under the Fourth Amendment. The test applied in these situations is whether or not a reasonable passenger would feel free to decline the request or otherwise to terminate the encounter. The defendant in this case was on a bus traveling from Miami, Florida, to Athens, Georgia. When the bus stopped in Ft. Lauderdale, two police officers involved in drug interdiction efforts boarded the bus and, without reasonable suspicion, approached the defendant. They asked to see his ticket and identification and asked consent to search luggage for drugs. The defendant agreed and during the search the officers found cocaine.

The Florida Supreme Court ruled that cocaine had been seized in violation of the Fourth Amendment. The court noted that the defendant had been illegally seized without reasonable suspicion, and that an impermissible seizure necessarily resulted any time police board a bus, approach a passenger without reasonable suspicion, and ask to search luggage. The U.S. Supreme Court reversed the decision and held that this type of drug interdiction effort may be permissible so long as the officers do not convey any message that compliance to their request is a required one. The Court noted that previous cases have permitted police to approach individuals in airports for the purpose of asking questions to verify identification and seek consent to search luggage.

The Supreme Court recognized that the defendant who was seated on a bus may not have felt free to leave. However, the Court rejected a "free-to-leave" test for determining whether a Fourth Amendment seizure occurs in such cases in which defendants who are in the middle of an ongoing trip are free to leave whether the police are present or not. Instead, the Court ruled that the proper question to determine whether or not an impermissible seizure occurred is whether a reasonable person would be free to decline the officer's request or to otherwise terminate the encounter.

In *Hodari,*[48] the Court ruled that a Fourth Amendment seizure does not occur when law enforcement officers are chasing a suspect, unless the officers apply physical force or the suspect submits to an officer's show of authority. In this case, police encountered four or five youths huddled around a sports car in a high crime area in Oakland, California. The youths scattered and one officer gave chase on foot. The defendant, who was apparently looking over his shoulder, emerged from an alley and unknowingly ran toward the pursuing officer. When he saw that the officer was ten to twenty feet away and was approaching him, the defendant discarded some crack on the ground and was arrested. The California Court of Appeals concluded that the defendant was seized without reasonable suspicion, and that the crack that he discarded was, therefore, the fruit of an illegal search.

The U.S. Supreme Court reversed the ruling that a Fourth Amendment search occurs only when a fleeing person yields to a show of authority or is physically grasped by

Cases 49–51

the officer. The Court noted that a "show of authority" is defined in terms of whether a reasonable person would have believed that he or she was, or was not, free to leave. Even assuming that the officer's act of running toward the defendant was a sufficient show of authority for seizure, the Court concluded that, since the defendant did not comply with or submit to any show of authority, he was not seized until he was actually tackled. Therefore, the drugs that the defendant discarded before being tackled were not seized under the Fourth Amendment and should not be excluded from evidence.

In the *County of Riverside,*[49] the Court ruled that a person arrested without a warrant must generally be provided with a judicial determination of probable cause within forty-eight hours after arrest, including intervening weekends or holidays. In this case, an arrestee allegedly did not receive a prompt judicial probable cause determination following his warrantless arrest as required by the Fourth Amendment. A federal district court issued an injunction requiring probable cause determination within thirty-six hours of arrest, which was upheld on appeal by the U.S. Court of Appeals for the Ninth Circuit.

The U.S. Supreme Court vacated that judgment and held that a judicial determination of probable cause within forty-eight hours of arrest will, as a general matter, be constitutional, unless an arrestee can prove the probable cause determination was delayed unreasonably. The Court analyzed the competing interests that exist between the need for flexibility on the part of the state judicial systems and the unfair burden that prolonged detention places on persons who are arrested based on incorrect or unfounded suspicion. The Court concluded that states should be allowed flexibility to experiment with combining a judicial probable cause determination as well as other judicial proceedings, such as bail hearings and so forth (McCormack 1991).

In *Acevedo,*[50] the U.S. Supreme Court overruled its prior decision in *Arkansas v. Sanders*[51] and upheld under the automobile exception to the warrant requirement, the warrantless search of a container placed into a vehicle, even though the probable cause to search was focused on that container. In this case, police observed the defendant leaving an apartment and carrying a brown paper bag. They believed they had probable cause that it contained marijuana. They saw the defendant place the paper bag in the trunk of his car. As he started to drive away police officers stopped him, opened the trunk, and searched the bag, which contained marijuana. The California Court of Appeals ruled that the marijuana should be suppressed in the light of the *Sanders* rule since the probable cause to search was directed specifically at the bag, and the warrantless search of the bag exceeded the scope of automobile search exceptions.

The U.S. Supreme Court reversed and held that containers placed in vehicles may be searched without a warrant, even when there is no probable cause to search and it focuses solely on these containers. The Court offered the following reasons in support of its decision to overturn *Sanders,* which would have required a warrant to search the bag:

(1) The *Sanders* rule afforded, at most, minimal protection to privacy interests and has confused courts and police officers;

(2) the *Sanders* rule may have encouraged some law officers to articulate that probable cause existed to search for evidence in a vehicle, resulting in searches of an entire vehicle without warrant; and

(3) even when the *Sanders* rule applied, officers could still seize packages found in a vehicle and wait for a search warrant, which could be obtained in a vast majority of cases.

The Court emphasized that since the police did not have probable cause to believe that contraband was hidden in the car other than the paper bag, a search of the entire car

Cases 52–53

would have been without probable cause and violation of the Fourth Amendment (McCormack 1991).

In *Jimeno*,[52] the Court held that a person's general consent to search the interior of a car includes, unless otherwise specified by the consent order, all containers in a car that might reasonably hold the object of the search. In this case, an officer followed a defendant's car after overhearing what he thought might be a drug transaction. After observing the car make an illegal turn, the officer stopped the car and told the defendant that he suspected him of carrying drugs in the car and asked for permission to search. The defendant consented, and on the car's floorboard the officer found an open paper bag containing a kilogram of cocaine.

The Supreme Court held it was objectively reasonable for the officer to conclude that a general consent to search the defendant's car included consent to search a paper bag on the floor of the car. The Court stated that the objective reasonableness test, used to determine the scope of consent search, assesses what the typical reasonable person would understand, based on the conversation between the officer and the suspect. The Court concluded that, when an officer has obtained a consent to search for drugs, it is objectively reasonable to search containers that might hold drugs.

It is important to note, however, that the Court distinguished this case from a case in which police are given consent to search the trunk of a car for drugs and encounter a locked brief case. The Court cautioned that a consent to search the trunk for drugs would not allow officers to pry open a locked brief case that was present in the trunk (McCormack 1991).

RECENT RULINGS AFFECTING THE FIFTH AMENDMENT

In *Minnick*,[53] the Court ruled that once a custodial suspect requests counsel in response to a Miranda warning, law enforcement officers may not attempt to reinterrogate the suspect unless the suspect's counsel is present or if the suspect initiates the contact with law enforcement.

In this case, the defendant escaped from jail in Mississippi and was, thereafter, involved in two murders. The defendant was eventually arrested in California and interviewed the next day by two FBI agents. After the agents gave the defendant Miranda warnings, he provided information but told them to come back Monday when he had a lawyer. After the interview, the defendant met several times with his appointed attorney. On Monday, after the defendant talked to his attorney, a deputy sheriff from Mississippi interviewed him. After again being advised of his Miranda rights, the defendant described in detail his escape and participation in the murders. Trial court did not suppress the defendant's statements to the deputy sheriff, and the Mississippi Supreme Court upheld the trial court's ruling.

The U.S. Supreme Court reversed the Mississippi Supreme Court, which admitted the defendant's statement to the deputy. The Court held that after an in-custody accused invokes the right to counsel, *Miranda* bars law enforcement officers from initiating interrogation of the accused person unless he has counsel at the time. Since the defendant's attorney was not present, the Court ruled that the subsequent waiver was invalid, and the confession to the deputy sheriff was in violation of *Miranda*.

The Court stated that in light of the purpose of the *Miranda* decision and to provide clear and unequivocal guidelines to law enforcement, recontact with an in-custody suspect is not permissible, unless the suspect has counsel with him at the time. It also noted that a valid waiver of Miranda may be obtained after counsel has been requested if the accused initiates a conversation or contact with enforcement officers (McCormack 1991).

Cases 54–55

In *McNeil*,[54] the Court held that an in-custody suspect who requests counsel at judicial proceedings, such as an arraignment or initial appearance, is only invoking the Sixth Amendment rights to counsel as to the charged offense, and is not invoking the Miranda Fifth Amendment-based right to have counsel present during interrogation when he is in custody. Thus, officers are not prohibited from later approaching an in-custody suspect for interrogation about uncharged crimes. In *McNeil*, the defendant was arrested for armed robbery committed in West Allis, Wisconsin, and was represented by counsel at his initial appearance. Later in the day, a detective visited the defendant in jail to question him about a separate incident involving a murder and armed robbery. After the detective had advised the defendant of his Miranda rights, the defendant waived those rights and provided accounts of his involvement in the Caladonia murder and armed burglary.

The Wisconsin Supreme Court refused to suppress the defendant's incriminating statements. The court found that his appearance with counsel at the initial appearance hearing concerning the West Allis armed robbery did not constitute an invocation of his Fifth Amendment Miranda right to counsel so as to prevent police questioning on an unrelated offense committed elsewhere.

The U.S. Supreme Court agreed with the Wisconsin Supreme Court and ruled that a defendant who appears at a formal judicial proceeding with counsel or requests counsel at such proceedings, is, in fact, invoking solely the Sixth Amendment right to counsel. This prohibits a police-initiated interrogation without the accused's counsel present, only concerning the offense. The Court reviewed the purpose and nature of an invocation of counsel under *Miranda*. It stated that a request for counsel in response to a *Miranda* finding prohibits police-initiated recontact for the purpose of obtaining confessions concerning other criminal matters, unless the suspect's counsel is present. The Court concluded that if Sixth Amendment right to counsel—invoked by the defendant in this case—was defined to be nonoffense specific, effective law enforcement would be seriously impeded. This is true since most suspects in pretrial custody suspected of involvement in other crimes would not be approachable by police.

In *Fulminante*,[55] a divided Court decided that a confession between prison inmates was involuntary and inadmissible in this case. However, the Court also noted that, in certain cases, the admission into evidence of an involuntary confession may be a harmless error, if it is harmless beyond reasonable doubt.

In this case, the defendant, who was incarcerated after being convicted for possession of a firearm by a felon, was also a suspect in the murder of his daughter. However, no charges had been filed against him concerning the murder. While in prison, the defendant befriended a fellow cellmate who was an FBI informant masquerading as an organized crime person. The informant's cellmate questioned the defendant about rumors that he was suspected of killing the child, but the defendant denied it.

Later, the defendant admitted to the informant that he had choked, sexually assaulted, and shot his daughter. The Arizona Supreme Court ruled that the confession to the informant should have been suppressed because it was involuntary, and that the admission of an involuntary confession can never be a harmless error.

The U.S. Supreme Court upheld the Arizona Supreme Court in that the confession was involuntary and also that his admission was not harmless error. However, the U.S. Supreme Court overruled the Arizona Court finding that the admission of an involuntary confession is always error and ruled that the admission into evidence of an involuntary confession may, in certain circumstances, be harmless error (McCormack 1991).

Recording and
reviewing officer-
sportsman contacts may
save misunderstandings
later.
*Courtesy Idaho Fish and
Game Department.*

THE SIXTH AMENDMENT

Case 56–57

In *Massiah v. United States,*[56] Massiah and Colson were seamen on a ship that customs agents searched. They found 3½ pounds of cocaine. Both seamen were subsequently arrested and indicted for federal drug violations. Following a plea of not guilty, they were released. Later, a customs agent met with Colson and, as a part of his cooperation, Colson consented to have a portable transmitter placed under the front seat of his car. When Massiah was in Colson's car, Colson induced him to talk freely about the shipment of cocaine. The entire conversation was overheard by agent Murphy, who was secreted in a nearby car. At the Massiah trial, defense counsel objected to Murphy's testimony concerning Massiah's statement on the grounds that the government had obtained these statements through illegal eavesdropping. The trial judge rejected this argument and allowed Murphy's testimony, which convicted Massiah. Massiah was sentenced to nine years imprisonment. Massiah appealed his conviction to the Court of Appeals for the Second Circuit Court restating his argument that the agent had obtained his statement in Colson's car as a result of illegal eavesdropping. Massiah also added a new argument to his appeal by alleging that the government had violated his Sixth Amendment right to the assistance of counsel by having one of its agents approach him in order to obtain incriminating information after he had been indicted and had retained a lawyer (Riley 1983a). In support of this, Massiah cited the 1959 Supreme Court decision of *Spano v. New York,*[57] noting that Spano's conviction was overturned by the Supreme Court because his will was "overborne." Therefore, his confession was involuntary. The Second Circuit Court rejected Massiah's arguments, ruling that the constitutional standard for the admissibility of confession in a criminal case is involuntariness, and there was no evidence that Massiah's statement to Colson had been coerced. Later Massiah's case was argued before the Supreme Court, which reversed Massiah's conviction and, in the process, created a new constitutional standard for the admissibility of confessions based on an accused's Sixth Amendment right in all criminal prosecutions . . . "to have the assistance of counsel for his defense." In brief, the Court held that Massiah's Sixth Amendment right to counsel was denied when his own incriminatory words were used against him at his trial and in the absence of counsel (Riley 1983a).

Cases 58–59

In *Kirby v. Illinois,*[58] the Supreme Court reviewed a number of prior decisions concerning the Sixth Amendment right to the assistance of counsel. Kirby's case did not include a confession. The attachment of the right to counsel was a major issue and the Court ruled that the right attaches when "adversary judicial criminal proceedings" have been initiated against the defendant "whether by way of formal charges, preliminary hearing, indictment, information, or arraignment." After *Kirby,* a defendant who finds himself in a *Massiah*-type situation no longer has to prove that he had been indicted at the time of his statements, which were deliberately elicited in order to make a Sixth Amendment argument. Instead, he need only show that, at the time, the statements were deliberately obtained by the government in the absence of counsel (Riley 1983a).

The point at which a defendant's right to counsel attaches in a criminal case is a crucial factor in the Sixth Amendment analysis. The attachment of right in a Sixth Amendment is meaningless, of course, unless the right is violated. In *Massiah,* there would have been no violation had not the officers, through Colson, deliberately solicited incriminating statements from Massiah and in the absence of his counsel. The fundamental principle of constitutional criminal procedure is that the exclusionary rule operates only to exclude evidence that has been obtained illegally, that is, as a result of unconstitutional conduct on the part of federal or state officials or their agents. Incriminating statements deliberately elicited by private persons who are not following government instructions should not be subject to exclusion even though the defendant's Sixth Amendment right to counsel was attached at the time of the statements (Riley 1983b).

The question of when a defendant has waived his or her rights to counsel is not easily answered. In 1938, *Johnson v. Zerbst,*[59] the Supreme Court held that the Sixth Amendment right to counsel can be waived as long as the officers prove that the defendant understood what his right was and evidenced his intention to waive it. Based on these requirements for a valid waiver, it is not surprising that the government did not argue waiver in *Massiah.* The *Massiah* rule has been expanded substantially in recent years. For example, the Supreme Court in 1972 made it clear that a defendant no longer must show that he was indicted at the time his statements were deliberately elicited in order to make a Sixth Amendment argument. Instead, he need only show that, at the time the statements were obtained by the officers, in the absence of counsel, there had been an arraignment or a preliminary hearing, and information had been filed, or he had otherwise formally been charged. Furthermore, some courts have ruled that filing of a complaint in issuance on arrest warrant is a formal charge. That triggers a defendant's Sixth Amendment right to counsel. The most common method of obtaining a waiver of a defendant's right to counsel is to advise the defendant of his Miranda rights and then obtain a knowing, intelligent, and voluntary waiver of those rights. However, at least one Federal Circuit Court of Appeals has ruled that where a defendant has been charged by indictment a waiver of Miranda rights is not sufficient to waive Sixth Amendment rights (Riley 1983b).

COVERAGE OF THE FOURTEENTH AMENDMENT

The key phrases of the Fourteenth Amendment are: privileges and immunities, due process of law, and equal protection of the laws. The guarantee against unusual searches and seizures contained in the Fourth Amendment is applicable to state officers by reason of the due process language of the Fourteenth Amendment. Practically all routine law enforcement work has a potential of becoming the subject of complaint by

Cases 60–70

an irate citizen who is willing to go to court. Therefore, one of the responsibilities of each law enforcement officer is to recognize and protect the rights, privileges, and immunities of persons within the jurisdiction the officer serves.

A part of the Fourth Amendment has long been interpreted to embrace security from the arbitrary intrusion by the police. Plaintiffs have also contended that the use of deadly force against a nonviolent, fleeing felon is cruel and unusual punishment, in violation of the Eighth Amendment. The Fifth Amendment provides, in part, that ''no person shall be . . . deprived of life, liberty, or property, without due process of law. . . .'' The Fourteenth Amendment applies the same limitation on the states.

SELF-INCRIMINATION

Source of the Clause

''Nor shall be compelled in any criminal case to be a witness against himself.'' The source of this clause was the maxim that ''no man is bound to accuse himself'' (*nemo tenetur prodere*—or *accusareseipsum*), which was brought forward in England late in the sixteenth century in protest against the inquisitorial methods of the ecclesiastical courts. At that time, the common law itself permitted accused defendants to be questioned. What the advocates of the maxim meant was merely that a person ought not to be put on trial and compelled to answer questions to his detriment, unless he had first been properly accused, i.e., by the grand jury. But the idea, once introduced, gained headway rapidly, especially after 1660, when it came to have attached to it most of its present corollaries.

Scope of the Privilege

Under the previously mentioned clause, a witness, in any proceeding whatsoever, in which testimony is legally required, may refuse to answer any questions, his answer to which might uncover further evidence against him.[60] The witness must explicitly claim his constitutional immunity or he will be considered to have waived it,[61] but he is not the final judge of the validity of his claim.[62] . . . A witness presumably may refuse to answer questions on the grounds that he thereby would expose himself to prosecution by a state.[63] Conversely, the admission in federal court of testimony given by the defendant in a state court under a grant of immunity may be void.[64] . . . If the accused takes the stand in his own behalf, he must submit to cross-examination,[65] whereas if he does not, it is by no means certain the trial judge in a federal court may not, without violation of the clause, draw the jury's attention to the fact.[66]

Neither does the amendment preclude the admission in evidence against the accused of a confession made while in the custody of officers, if the confession was made freely, voluntarily, and without compulsion or inducement of any sort.[67]

In *McNabb v. United States,*[68] the Court reversed a conviction in a federal court, based on a confession obtained by questioning the defendants for prolonged periods in the absence of friends and counsel, and without their being brought before a commissioner or judicial officer, as required by law. Without purporting to decide the constitutional issue, Justice Frankfurter's opinion urged that the duty of the Court, in supervising the conduct of the lower federal courts, is to establish and maintain ''civilized standards of procedure and evidence.''[69] And in *Mallory v. United States*[70] where petitioner was found guilty on evidence which included a confession obtained during a ten-hour police interrogation following apprehension and prior to arraignment, and without his having been warned that any statement he might make could be used against him or of his right to counsel, the conviction was reversed for failure to comply with the requirement of Fed. R. Crim. P. 5(a) that ''arraignment be without unnecessary delay'' and the *McNabb* rule that the delay ''must not be of a nature to give opportunity for the extraction of a confession.''

Cast of a shoe imprint
may help prove a
person was at the scene
of a crime.
*Courtesy Federal Bureau
of Investigation.*

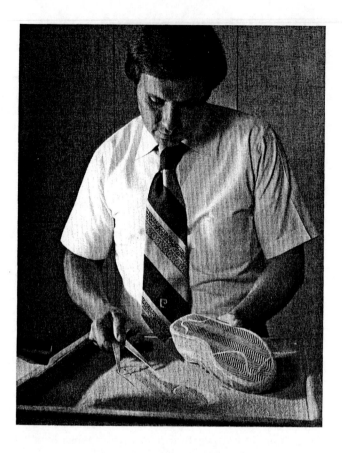

Cases 71–73

The introduction into evidence against one who was being prosecuted by a state for
illegal possession of morphine, two capsules of which he had swallowed and had been
forced by the police to disgorge, was held to violate due process of law.[71]

Finally, this amendment, in connection with the interdiction of the Fourth Amendment against unreasonable searches and seizures, protects an individual from the compulsory production of private papers which would incriminate him. The scope of this
latter privilege was, however, greatly narrowed by the decision in *Shapiro v. United
States*.[72] There, by a five-to-four majority, the Court held that the privilege against
self-incrimination does not extend to books and records which an individual is required
to keep to evidence his compliance with lawful regulations. ". . . the privilege which
exists as to private papers cannot be maintained in relation to 'records required by law
to be kept in order that there may be suitable information of transactions which are the
appropriate subjects of governmental regulation and the enforcement of restrictions
validly established.' "[73]

Search by Consent

Search by consent is an investigative technique frequently used by enforcement officers
where premises are protected against unreasonable search as defined by the Fourth
Amendment of the Constitution. Properly made, such searches are deemed reasonable.
The law consistently approves this method, when legal prerequisites are satisfied, but
does not favor it. The law prefers those searches be made with a warrant. The warrant,
lawfully issued only upon a finding of probable cause by a neutral and detached magistrate, describes the premises to be searched, shows when the premises may be entered
and then specifies the things that may be sought during the search. Such limitations

A Serological Unit examiner can identify blood stains by chemical analysis. *Courtesy Federal Bureau of Investigation.*

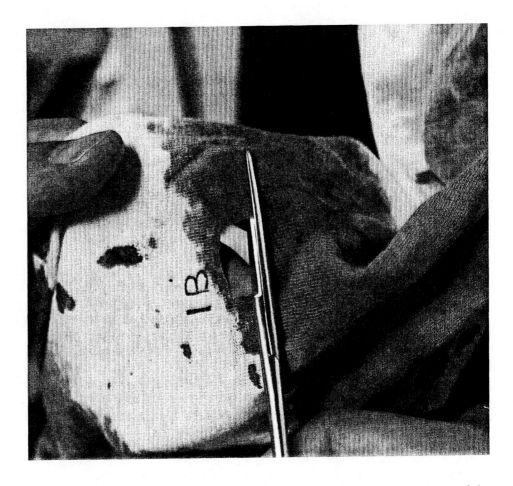

Cases 74–75

obviously are not contained in a search by consent; consequently the tendency of the court is to require the searching officer to present convincing proof that his conduct was reasonable throughout the search (McLaughlin 1977).

As indicated in *Schneckloth,*[74] a consent to search is not a waiver of a constitutional right as that phrase has been frequently used by the Court. The Court has held that prosecution must demonstrate an "intentional relinquishment or abandonment of a known right or privilege" in order to prove a waiver. The Supreme Court in *Chimel v. California*[75] reduced the permissible scope of search made incidental to an arrest inside premises by holding that incidental search must be restricted to the person of the arrestee and areas under his immediate control, and defining that area as one from which he could seize a weapon or destroy evidence. The immediate effect of *Chimel* is that it definitely narrows the zone of search following arrest. The indirect effect of the decision may be more important in that an arrest can no longer rely on the broad power enjoyed prior to 1969. Any search pursuant to an arrest made within premises and beyond the immediate vicinity of the arrestee, absent an emergency, must be made under the authority of a search warrant or with the consent of the party empowered to give consent. That is, *Chimel* bars search incidental to arrest (McLaughlin 1977).

Consent search is upheld frequently where there is an inquiry into all the circumstances and that effort is made to mediate fairly between the interest of society, effective law enforcement, and the right of the defendant to privacy. An officer preparing for a consent search should take four steps: (1) determine whether the premises are

Cases 76–84

protected by the Fourth Amendment; (2) identify the person lawfully entitled to possession; (3) obtain from that person a voluntary relinquishment of constitutional rights declared in the Fourth Amendment; and (4) conduct the search within the limitations expressed or implied in the consent.

In *Katz v. United States*,[76] the Supreme Court held that the Fourth Amendment protects ''people, not places.'' However, while Katz sought to eliminate places as the principal Fourth Amendment consideration, it appears that subsequent court decisions either have returned to or never abandoned the analysis of ''protected areas,'' where an individual has a reasonable expectation of privacy. He may expect no such consideration in open fields (*Patler v. Slayton*).[77]

An officer who equates curtilage with the area wherein an individual possesses a reasonable expectation of privacy will be correct most of the time. Curtilage is easily identified in many urban communities, but is more difficult in rural, sparsely populated areas. The questions of where the curtilage ends and the open fields begin is difficult to answer.

The doctrine of search and seizure permits the taking of property which has been abandoned, so long as the abandonment has not been caused by prior illegal police conduct. Such things as automobiles, garbage, narcotics, gambling records, weapons, etc., are easily recognized as abandoned, but premises such as houses, apartments, hotel rooms, places of business, etc., can also be abandoned. When this happens, the former possessor of the right to assert that his rights were violated by police entry, has no standing to object.

An officer seeking permission to search must obtain this authority from the person in lawful possession of the premises. This does not mean ownership nor mere lawful presence. A legal co-occupant may speak for himself as well as the other partner; and any evidence that is found may be used against either one of them (*United States v. Matlock*).[78] Authority must be real rather than apparent. For example, a store clerk or someone visiting the premises may not give permission (*Moffet v. Wainwright*).[79] On the other hand, it is proper for officers to walk into a commercial establishment open to the public and to look around (*United States v. Berrett*).[80] High school students are protected from unreasonable searches and seizures, even in school, by employees of the state whether they be police officers or schoolteachers (*People v. D*)[81] (McLaughlin 1978).

Electronic Tracking Devices and the Fourth Amendment

The use of electronic tracking devices (beepers) raises the question of whether or not the subject's Fourth Amendment rights have been violated. In the case of *United States v. Knotts*,[82] the Court held that monitoring a beeper in public places or places open to visual observation is not a Fourth Amendment search. On the other hand in *United States v. Karo*,[83] the Court held that monitoring a beeper inside a private premise, that is, a place not open to visual surveillance, is a search which, in the absence of an emergency, requires a warrant. In *Karo*, the government contended that requiring a warrant to monitor a beeper which had been removed from public view could have the practical effect of requiring a warrant for every case. However, the courts, recognizing the need for flexibility in applying the warrant requirement, have declined to impose the same strict standards ordinarily associated with the traditional search warrant. The apparent object is to establish some degree of judicial control over electronic surveillance without unreasonably hindering legitimate law enforcement activities.

In *Katz v. United States*,[84] the Supreme Court held that the Fourth Amendment protects people but not places. It is, however, true that the nature of the place or

Cases 85–89

property into which enforcement officers intrude can be highly significant in determining the extent of the Fourth Amendment protection and application. For example, a residence, because of the traditionally high expectation of privacy associated with private dwellings, is accorded the highest level of Fourth Amendment protection, whereas an open field has no protection at all.

In between these two extremes are found the three general types of property to which beepers are most frequently applied. They are movable containers, vehicles, and aircraft. The Supreme Court has held that because of the high level of privacy associated with personal luggage and other movable containers whose contents are concealed, searches of such containers must be authorized by a warrant.[85] It seems that the installation of a beeper inside a personal container to which the level of protection attaches, will likewise, in the absence of an emergency, require a warrant. However, there may be special exceptions (Hall 1985).

Beepers in Vehicles

The Supreme Court has not yet decided whether the installation of a beeper in or on a vehicle constitutes a Fourth Amendment search, and lower court holdings have been inconclusive. In *United States v. Moore,*[86] a Federal Appellate Court held that the mere installation of a beeper on the exterior of a vehicle is not a search. However, in *United States v. Neet,*[87] at least one Federal District Court has taken the opposite view.

The Supreme Court has traditionally viewed vehicles as being distinct from other kinds of property. The very nature of vehicles and their use serves to reduce the level of privacy normally associated with other property. This creates a reduction in Fourth Amendment rights.[88] It should be noted that the installation of a beeper on the interior of a vehicle is more likely to cause the courts to consider the installation to be a Fourth Amendment search (Hall 1985).

Beepers in Aircraft

Aircraft, in general, have been treated by the courts in about the same manner as automobiles and other vehicles in regard to the Fourth Amendment, and the installation of a beeper inside the aircraft has been treated as a search. Beeper installation inside aircraft has usually been accomplished under court order.[89] Certain considerations suggest the wisdom of assuming that installation of beepers are a Fourth Amendment rule and, therefore, necessitate a warrant. This is for two reasons. First, some courts consider any evidence acquired as a result of using an improperly installed beeper as having been tainted by initial illegality and, therefore, subject to exclusion. Second, as the Supreme Court noted in *Karo,* even when a warrantless installation is permissible, it cannot be anticipated when the vehicle or other property to which it is affixed will move into a private area. The government argued that a warrant to monitor a beeper once it had been withdrawn from public view would have the practical effect of requiring a warrant in every case. The Court's answer was that this was hardly a compelling argument against the requirement, although the Fourth Amendment does not specifically impose a time limit on the life span of a search warrant. However, in its interpretation of the Fourth Amendment, the Supreme Court has viewed the imposition of time constraints by search warrant as additional protection. The courts have not suggested what might be a reasonable length of time. Therefore, some reference to the lower federal courts becomes necessary (Hall 1985). The nature of the case should suggest a reasonable time limit for the warrant to be in effect. It is a generally accepted rule that search warrants are to be executed within the territorial jurisdiction of the issuing court. In the light of special problems associated with beeper surveillance, the courts have declined to hold the authorizing court orders to be subject to all the same procedural requirements as a standard search warrant (Hall 1985).

Video Surveillance

Cases 90–94

Video, or television, surveillance can provide necessary and proper security in undercover operations. It also allows officers to record their observations on tape. It preserves evidence of the observed criminal action, which may constitute overwhelming evidence of the defendant's guilt when replayed in court. Video surveillance may also help insure the safety of undercover officers and informants. There are three distinct situations in which audio or television surveillance can be used. First, the officer may record activities that occur in an area that is either viewable by the public or commonly accessible to the public, such as public streets or hallways of a public building. Second, the officer may record activities with the consent of a participant, such as an undercover officer's purchase of drugs in the officer's rented hotel room. Third, officers can nonconsentually enter a private area protected by the Fourth Amendment and install a video recording device (Fiatal 1989a).

The Fourth Amendment prohibits unreasonable search and seizure and requires that warrants for both arrests and search be issued upon a probable cause determined by the individual who is neutral from law enforcement.[90] Video surveillance obtained in violation of the Fourth Amendment requirements may fall prey to the exclusionary rule and may lead to civil liability for officers or departments. In the benchmark case of *Katz v. United States,*[91] now over twenty-five years old, the Supreme Court interpreted and defined that portion of the Fourth Amendment which prohibits unreasonable search. The Supreme Court ruled that Katz had a constitutionally protected expectation of privacy in his telephone conversation made from within a telephone booth. His conversation was intercepted by the use of a hidden microphone. The Supreme Court stated that a search must generally be conducted pursuant to search warrants unless it fits into one of the traditionally recognized exceptions to the warrant requirement, such as a search incident to arrest, consent or emergency searches (Fiatal 1989a). The Court ruled that the search was conducted without a warrant and did not fall within any exceptions to the requirements. Therefore, it was unreasonable and the conversation was excluded from evidence.

In public areas, an officer is not considered to be conducting a search when he watches activities which are publicly viewable. If people in general can observe areas from a place they are lawfully entitled to be, like the street or sidewalks or open field areas, an officer can likewise conduct a legal surveillance from a vantage point. In *United States v. Felder,*[92] investigators observed the defendant by use of a concealed television camera in an area at his place of employment. This area was both accessible and viewable by other employees during working hours. The court determined that the surveillance did not intrude into any reasonable expectation of privacy and that a search warrant was not required. In *United States v. Cuevas-Sanchez,*[93] the U.S. Court of Appeals for the Fifth Circuit found that investigators had executed a search when they conducted an extended video surveillance of the defendant's back yard. The court ruled that this was within the curtilage of his residence. The investigators had placed a television camera atop a power pole overlooking the defendant's ten-foot privacy fence. The camera was there for twenty-four hours a day for thirty days. The court ruled that this was not a one-time overhead flight or glance over the fence of a passerby, and, therefore, concluded that the defendant's expectations of privacy from this type of video surveillance was reasonable (Fiatal 1989a).

Video surveillance, when the consenting party is present, presents a different set of values. In *United States v. White,*[94] enforcement officers were able to overhear the conversation between the defendant and a government informant through a microphone and transmitter worn by the informant. These officers later related the overheard contents at the trial. The Supreme Court determined that, under the circumstances, the

Cases 95–101

defendant forfeited any expectancy of privacy in his conversation when he volunteered to the informant, this despite the defendant's misplaced belief that the informant would not repeat what he said. The Court said that the government did not search or intrude into the defendant's reasonable expectations of privacy (Fiatal 1989a). Although federal and state wiretapping standards require tapes of court-ordered audio surveillance be judicially sealed to guarantee their integrity, courts have ruled that this same sealing provision is not applicable to video surveillance.

However, officers should undertake proper chain of custody procedures and security measures so that they can later vouch for the integrity and the authenticity of the video tapes. Finally, the U.S. Supreme Court has acknowledged that the Fourth Amendment does not require police officers to delay in the course of an investigation if to do so would gravely endanger their lives or the lives of someone else.[95] In such emergencies, a police officer can conduct a warrantless search necessary to control any threat to life if the action is no greater than necessary to alleviate or eliminate the threat (Fiatal 1989b).

Confessions

The constitutional standard for the admissibility of a confession in a criminal prosecution is voluntariness.[96] This appears to be easily defined. It is not, however, because it has been modified and molded by courts over the years to accommodate applicable rules of evidence and the development of constitutional principles. Some of the factors underlying the voluntariness doctrine include:

(1) rules of evidence which require confessions to be reliable;

(2) the Fifth Amendment requirement that no person ''shall be compelled in any criminal case to be a witness against himself''; and

(3) the Fourteenth Amendment, which provides *inter alia,* that is, no ''state [shall] deprive any person of life, liberty, or property without due process of law'' (Riley 1982).

It is understood that a certain amount of deception is used by enforcement officers in obtaining a confession. The question is, does this deception violate the paramount principles underlying the voluntariness doctrine. It is recognized as a fundamental requirement of due process that for a confession to be admissible it must be reliable. In deciding the issue of reliability, the court reviews all the facts and circumstances surrounding the confession, paying special attention to the interrogation methods employed by the officers. The court looks at the approach used by the officers because of its inherent capabilities of producing untrue statements. For instance, in *Payne,* the defendant was deprived of food. In *Ashcraft,* there was an extended interrogation, and in *Brown,* there was police brutality.[97] Deception, however, has not been placed in this category. The reasons for this are that few, if any, interrogators use deception for the purpose of obtaining a true confession, and more importantly, the courts generally refuse to accept the argument that the types of deception litigated are inherently coercive and that their use is likely to result in an innocent man confessing to crimes he did not commit. Some examples of cases that did not render an otherwise voluntary confession inadmissible are:

(1) the weapon used in a crime has been recovered, *Moore v. Hopper;*[98]

(2) the suspect has been identified by a witness, *United States ex rel. Caminito v. Murphy;*[99]

(3) his partner has given a confession, *Frazier v. Cupp;*[100]

(4) his fingerprints or other evidence were found at the scene of the crime, *Roe v. People of State of New York;*[101] and

Cases 102–104

(5) claiming that polygraph tests showed he was lying, when in fact it did not. The courts have found that as a general rule, the use of deception does not by itself render a confession inadmissible on the grounds that it is unreliable.

The officer, however, should not discount the probability that extreme deception will be viewed in a negative manner by the court. For example, in *Miranda v. Arizona*,[102] the Supreme Court described an interrogation practice known as "reverse lineup." In this technique, the defendant is placed in a lineup and identified by fictitious witnesses as the perpetrator of numerous offenses. It is hoped that the defendant will, in desperation, confess to the crime under investigation in order to escape prosecution of additional offenses. In *Miranda,* the Court did not formally rule on the lawfulness of the reverse witness technique because it was not directly at issue in the case, but it was highly critical because the technique is such that an innocent man could presumably be induced to confess a crime he did not commit (Riley 1982).

Certain interrogation techniques and methods, regardless of whether they produce a confession, are so repugnant to constitutional doctrine of "fundamental fairness," that they should be judicially condemned through implementation of *per se* rules of inadmissibility. The argument for *per se* rules of exclusion generally has been advanced on the theory that it is unfair and a violation of due process to allow law enforcement officers to benefit from illegal acts or from official misconduct. The courts, however, have not been sympathetic to the argument that the use of artifice, stratagem, and deception is similarly inconsistent with a basic notion of fairness. For example, in *Moore v. Hopper,*[103] the defendant was arrested for murdering his girlfriend's husband. The girlfriend, when interviewed, advised authorities that she had been with the defendant after the murder and had observed him throw the murder weapon, a gun, off a bridge. During questioning, the defendant was advised that the girlfriend had given such a statement. Additionally, he was told that the murder weapon had been recovered when, in fact, it had not. The court, finding no problem with this ruse, noted that because of the nature of criminal activity, artifice and stratagem are necessary tools of the law enforcement officer. However, in *Spano v. New York,*[104] the Court reviewed a case where the defendant, after having been first indicted and then arrested for first degree murder, was subjected to a continuous questioning to obtain a confession. When the questioning produced only repeated requests by the defendant to see his attorney, the interrogators resorted to what is commonly known as the "false friend" technique. This technique involved a newly commissioned peace officer named Bruno, who was close to the defendant. Bruno was instructed to tell the defendant falsely that he, Bruno, would be in a lot of trouble if the defendant did not confess. On the fourth attempt this ploy worked, and Spano made a confession. The Supreme Court reversed Spano's conviction on grounds that his will was overborne by official pressure, fatigue, and sympathy falsely aroused. Furthermore, in addressing the interrogation methods that were used, the Court stated: "the abhorrence of society to the use of involuntary confessions does not turn alone on their inherent untrustworthiness. It also turns on the deep-rooted feeling that the police must obey the law while enforcing the law; that in the end, life and liberty can be as much endangered from illegal methods used to convict those thought to be criminals as for the criminals themselves" (Riley 1982).

A confession is not admissible even though reliable unless it is the product of a defendant's free and rational choice. This is a restatement of the constitutional right of the defendant to decide, in an atmosphere free of undue pressure, whether he is going to provide the authority with the confession or remain silent. The Supreme Court has articulated this principle in the following way: "Is the confession the product of an essentially free and unconstrained choice by its maker? If it is, if he has will to

Cases 105–107

confess, it may be used against him. If it is not, if his will has been overborne and his capacity for self-determination critically impaired, the use of his confession offends due process,'' *Culombe v. Connecticut.*[105] In *Frazier v. Cupp,*[106] the defendant was arrested for homicide. During the questioning, the defendant admitted that at the time of the crime he had been with his cousin Jerry Lee Rawls. However, he denied that he and Rawls had been with the victim or any other third party. At this point, the officers questioning Frazier advised him that Rawls had been arrested and had confessed. Following some additional discussion, the defendant, Frazier, provided a full written confession. In upholding the confession the Supreme Court found ''the questioning was of short duration, the petitioner was a mature individual of normal intelligence. The fact that the police misrepresented the statement that Rawls had confessed, while relevant, was insufficient in our view to make this otherwise voluntary confession inadmissible.'' However, in *United States ex rel. Everett v. Murphy,*[107] the defendant was convicted of murder in the first degree in state court. In his petition in federal court the defendant argued that the confession used to convict him was involuntary because of the circumstances under which it was obtained. The court found that on July 3, 1959, one Finocchiaro was found unconscious and bleeding on the ground of a lot and died the following day. Eleven days later Everett was arrested. During questioning, Everett was advised that he had been arrested for assault and robbery. Everett was then told that Finocchiaro was presently at police headquarters and, while he had not been injured, he was angry over what had happened. The officers then told Everett if he confessed they would speak with Finocchiaro and attempt to calm him down. Everett then confessed, and the statement was admitted at trial. The court, after reviewing Everett's confession, found it to be involuntary because Everett had been arrested illegally, subjected to extended incommunicado interrogation, and not advised of his rights. Further, the confession followed the use of deception and a false promise of assistance by the police. However, the court noted that it might have ignored the deception of Everett as to Finocchiaro's survival of the attack if that had been the only basis upon which the confession was challenged. In finding the confession involuntary the court placed heavy emphasis on the fact that the effect of telling the defendant that the victim had survived the attack was to make more plausible the police officer's promise that if Everett confessed he would be protected from the angry Finocchiaro. This the court did not find acceptable (Riley 1982).

As a general rule, the courts have found that the use of artifice, stratagem, and deception by law enforcement officers does not, standing alone, render an otherwise involuntary confession inadmissible. The use of extreme deception, for example, the reverse lineup technique in *Miranda,* could result in finding evidence inadmissable by the courts on the grounds that it renders a confession unreliable, violating the constitutional doctrine of fundamental fairness, or overbearing a defendant's right of free and rational choice (Riley 1982). For a further discussion of individual rights see United States Senate [bulletin] 1972.

RIGHTS OF PRIVATE CITIZENS UNDER WILDLIFE LAW

Ownership of Wildlife a Privilege

Since the state and federal governments are custodians of game in its wild state and in their sovereign capacity, it follows that an individual cannot obtain absolute property rights to game except upon such conditions, restrictions, and limitations as the state or federal government permits. Further, the individual acquires absolute property rights to wildlife only as a matter of privilege and not as a right. The government can and does impose conditions under which an individual may acquire property rights to wildlife. Either the legislature alone, or the legislature and a state fish and game department

Cases 108–114

(as authorized by the legislature) may impose conditions deemed necessary and expedient, so long as these do not contravene any principle of the Constitution.

It was held in Indiana that when a citizen accepts the state's grant, he also accepts it impressioned with all the restrictions and limitations. When he acquires property under a game license, he does so with the full notice of his qualified rights, and if he loses that which he has taken or held possession of upon forbidden terms, he has lost nothing in its forfeiture that belonged to him.[108]

In the case of *Magner v. People*[109] it was held that to hunt and kill game is a boon or privilege granted expressly or implicitly by sovereign authority, but not a right inherited by each individual. Consequently, nothing is taken away from the individual when he is denied a privilege, such as limited hunting seasons. The individual may not acquire ownership of game by capturing, killing, or reclaiming, except as permitted by law.

Acquiring Ownership of Wildlife

There are two ways that individuals can gain legal title to wildlife. They may provide their own brood stock and have property set up as a game farm or fish hatchery under proper permit, or they may take wildlife under legal regulations and reduce it to their possession. The following discussion deals with the latter provision.

The pros and cons of reducing game to possession are discussed in the case of *Dapson v. Daly.*[110] In this case the plaintiff sued for the carcass of a deer killed by the defendant. The plaintiff wounded the deer and was in the act of pursuing it when the defendant fired at the deer. The deer immediately dropped and was dead when the defendant reached it. The defendant carried the carcass away, after which he was sued by the plaintiff for its recovery. The court refused to order the carcass returned to the plaintiff for the following reasons:

1. The plaintiff failed to show that he had taken possession of the deer even though it had been wounded by him.

2. Evidence presented in court indicated that the fatal shot had been fired by the defendant, and thus the animal was taken into possession by him from a wild state.

Other cases that clarify the rights, title, and remedies of hunters in respect to game which is being pursued, or which has been killed or wounded, follow in annotated form.

Game

1. The pursuit of a wild animal gives no title thereto; and an action will not lie against a man for killing and taking an animal *ferae naturae,* though the killing was done in view of another who started it and was on the point of taking it. But, on the other hand, title may accrue to one though he has not actually seized the body of the animal, as where he has ensnared the animal or has actual dominion over it by such means as will prevent its escape.[111]

2. The instant a wild animal is brought under the control of a person, so that actual possession is practically inevitable, a vested property interest in it accrues, which cannot be divested by another intervening and killing it.[112]

3. A partridge shot by one hunter and picked up by a second hunter was alive but in a dying condition. The court ruled that the partridge was in possession of the first hunter since it had been sufficiently reduced to possession.[113]

Fish

1. Where fish have been caught in a net which has not been drawn in and from which it is not absolutely impossible, but practically so, for them to escape, the owners of the net have acquired . . . a property in them. . . .[114]

Cases 115–121

Rights of the Property Owner

Final Authority

2. But in *State v. Thomas*[115] it was held that the owner of fish traps containing an opening about 3½ feet square through which the fish entered, did not have a property in the fish while in such traps, since the court ruled that the avenue of escape must be closed, or the chance of escape at least reduced to a minimum.

The mere fact that a landowner posts a property against hunting, fishing, or trespassing in no way gives the owner title to the game. The state holds ownership of game within its boundaries, but the individual owner of real estate has an interest in the game on the premises. This is not an absolute interest, but rather a qualified one. That is, no one has the right to go on the landowner's premises, without permission, to hunt or fish.

In Minnesota, the court held that the individual has the right to exercise exclusive and absolute domain over his property, and the unqualified right to further restrict taking game.[116] In other words, the restriction does not modify the act but prohibits it.

A private owner has the exclusive right to hunt on his land and may prohibit others.[117] In the case of *L. Realty Co. v. Johnson,*[118] the Supreme Court of Minnesota said: "While true that the title to all wild animals is in the State and the owner of premises whereon it is located has only a qualified property interest therein, yet he has the right to exercise exclusive and absolute dominion over his property, and incidentally the unqualified right to control and protect the wild game thereon."

A state license does not authorize hunting on the lands of a private owner. A Wisconsin court maintained:

> No person has a right to go upon the land of another against the latter's will, or to so intrude upon the right of such other to the exclusive use of lands for any purpose merely because he possesses a state license to hunt. Such a license does not affect the relations of the licensee with such other in the slightest degree. A violation of the latter's rights by such person, which would be actionably wrong if he were not armed with a license to hunt, would be such wrong if he were so armed. It is a mistaken notion that such a license gives the holder thereof any right whatever to trespass upon the property rights of others.[119]

The Wisconsin court also ruled: "The exclusive use of one's own property is a right protected by the Constitution. The Legislature cannot authorize another to enter the premises for the purpose of taking game."[120] In fact, the owner's rights include the right to prohibit a hunter from shooting over the premises of an adjoining owner, or from going on the premises to retrieve game which has fallen there.[121]

A point that should be kept in mind is that fish and game laws are passed by legislators and administered by state and federal enforcement agencies, but they are interpreted by judges and juries. It has been said that a law is nothing more than a prediction of what a judge or a jury will do under a particular circumstance. In borderline cases, one may find that it is often impossible to predict what a verdict will be. In such cases, great stress is placed on the interpretation of the law. It is true that two courts may interpret a ruling in such diverse ways that one court might convict an individual for certain offenses and another court might acquit the individual. Reversals by higher courts demonstrate this point.

Literature Cited

Fiatal, R. A. 1989a. Lights, camera, action: Video surveillance and the Fourth Amendment (part 1). *FBI Law Enforcement Bull.* (Federal Bureau of Investigation, U.S. Dept. of Justice, Washington, D.C.) 58 (1): 23–30.

———. 1989b. Lights, camera, action: Video surveillance and the Fourth Amendment (conclusion). *FBI Law Enforcement Bull.* (Federal Bureau of Investigation, U.S. Dept. of Justice, Washington, D.C.) 58 (2): 22–31.

Hall, J. C. 1985. Electronic tracking devices following the Fourth Amendment (conclusion). *FBI Law Enforcement Bull.* (Federal Bureau of Investigation, U.S. Dept. of Justice, Washington, D.C.) 54 (3): 26–31.

McCormack, W. U. 1991. Selected Supreme Court cases: 1990–1991 term. *FBI Law Enforcement Bull.* (Federal Bureau of Investigation, U.S. Dept. of Justice, Washington, D.C.) 60 (11): 26–32.

McLaughlin, D. J. 1976. Miranda and the derivative evidence rule—Brown v. Illinois. *FBI Law Enforcement Bull.* (Federal Bureau of Investigation, U.S. Dept. of Justice, Washington, D.C.) 45 (1): 12–14.

———. 1977. Search by consent. *FBI Law Enforcement Bull.* (Federal Bureau of Investigation, U.S. Dept. of Justice, Washington, D.C.) 46 (12).

———. 1978. Search by consent. *FBI Law Enforcement Bull.* (Federal Bureau of Investigation, U.S. Dept. of Justice, Washington, D.C.) 47 (1–2).

Riley, C. E. 1982. Confessions and interrogation: The uses of artifice stratagem and deception. *FBI Law Enforcement Bull.* (Federal Bureau of Investigation, U.S. Dept. of Justice, Washington, D.C.) 51 (4): 26–31.

———. 1983a. Confessions and the Sixth Amendment right to counsel (part 1). *FBI Law Enforcement Bull.* (Federal Bureau of Investigation, U.S. Dept. of Justice, Washington, D.C.) 52 (8): 24–31.

———. 1983b. Confessions and the Sixth Amendment right to counsel (conclusion). *FBI Law Enforcement Bull.* (Federal Bureau of Investigation, U.S. Dept. of Justice, Washington, D.C.) 52 (9): 24–31.

Rissler, L. E. 1978. Search incident to arrest—new restrictions on an old doctrine. *FBI Law Enforcement Bull.* (Federal Bureau of Investigation, U.S. Dept. of Justice, Washington, D.C.) 47 (9).

Smith, Paige. 1987. The origins of the Federal Constitution. First Annual Tanner Academy Lecture, Utah Academy of Sciences, Arts, and Letters.

U.S. Congress. Senate. 1972. *Layman's guide to individual rights under the United States Constitution.* Washington, D.C.: U.S. Govt. Printing Office.

Notes

1. 395 U.S. 752 (1969).
2. 414 U.S. 218 (1972).
3. 53 L.Ed. 2d 538 (1977).
4. 442 F. Supp. 736 (S.D.N.Y. 1977).
5. 232 U.S. 383 (1914).
6. 367 U.S. 643 (1961).
7. 371 U.S. 471 (1963).
8. 384 U.S. 436 (1966).
9. Ibid.
10. 45 L. Ed. 2d 416 (1975).
11. *Carroll v. United States,* 267 U.S. 132 (1925).
12. *Burdeau v. McDowell,* 256 U.S. 465 (1921).
13. *Nathanson v. United States,* 290 U.S. 41 (1933).
14. *Gouled v. United States,* 255 U.S. 298 (1921).
15. *Taylor v. United States,* 286 U.S. 1 (1932).
16. *Carroll v. United States,* 267 U.S. 132 (1925).
17. *Hester v. United States,* 265 U.S. 57 (1924).
18. *Weeks v. United States,* 232 U.S. 383 (1914).
19. *Carroll v. United States,* 267 U.S. 132 (1925); *Husty v. United States,* 282 U.S. 694 (1931); *Brinegar v. United States,* 338 U.S. 160 (1949).
20. *Scher v. United States,* 305 U.S. 251 (1938).
21. *United States v. Di Re,* 332 U.S. 581 (1948). See also *Henry v. United States,* 361 U.S. 98 (1959), wherein federal officers, who had been investigating a theft from an interstate shipment of whiskey, obtained a tip from the employer of one of the defendants implicating the latter in certain unexplained, undefined interstate shipments. Thereafter, the officers twice observed cartons being placed in a motor car in a residential area, whereupon they followed, stopped the car, and found therein and seized cartons of radios stolen from an interstate shipment. Probable cause was held to be wanting for the reason that stopping the car in an alley in a residential district and the picking up of packages at this point, rather than a terminal, were outwardly innocent and the movement of the defendants, none of whom had ever been suspected of any crime, did not have the appearance of fleeing men acting furtively.
22. *United States v. Lanza,* 260 U.S. 377 (1922); *Jerome v. United States,* 318 U.S. 101 (1943); *Bartkus v. Illinois,* 359 U.S. 121 (1959); *Abbate v. United States,* 359 U.S. 187 (1959). In the last two cases,

three Justices said prosecutions for the same offense are so grossly unfair as to be violative of due process and that under our Constitution powers are allocated between federal and local governments in such a manner that the basic rights of each can be protected without double trials. They would overrule the principle set forth in the Lanza case. Justice Brennan, with Justice Douglas and Chief Justice Warren, in dissenting in the Illinois case, contended that the latter in actuality was a second federal prosecution originated by federal officers, who, defeated in their attempt to convict Bartkus in a federal court, induced state authorities to institute the second prosecution and assisted the latter by gathering evidence and supplying witnesses.

Cf. *Petite v. United States,* 361 U.S. 529, 531 (1960), referring to a Department of Justice press release, April 6, 1959, announcing a policy against ''duplicating federal-state prosecutions.''

23. *Ex parte Lange,* 18 Wall. 163, 169 (1874).
24. Ibid. 172, 173.
25. *Kepner v. United States,* 195 U.S. 100 (1904). This case arose under the Act of Congress of July 1, 1902 (32 STAT. 691) for the temporary civil government of the Philippine Islands. To the same effect are *United States v. Sanges,* 144 U.S. 310, 323 (1892) and *United States v. Evans,* 213 U.S. 297 (1909), both cases arising within the United States. *Cf. United States v. Shotwell Mfg. Co.,* 355 U.S. 233, 243, 250 (1957), in which the Court ruled that where conviction upon a jury verdict of guilty was reversed by a Court of Appeals for failure to grant a motion to suppress certain evidence, its own order granting the government's petition for certiorari, vacating the judgment of the Court of Appeals, and remanding the case to the District Court for further proceedings to determine, on the basis of new evidence, the proper disposition of the motion to suppress did not give rise to double jeopardy. Since the jury found defendants guilty, and the questions of the admissibility of the evidence was a preliminary one, not submitted to the jury, the remand created no question of double jeopardy. Three dissenting Justices (Black, Douglas and Warren, C. J.) contended that the Court should have denied certiorari in order that the defendants, in the new trial ordered by the Court of Appeals, would not be subjected to piecemeal prosecution, or a ''partial new trial.'' Although ''not technically [infringing] . . . the protection against double jeopardy,'' the opportunity afforded the government ''to introduce new evidence in an attempt to save a conviction it has lost in the Court of Appeals'' seemed to them violative of the spirit of the double jeopardy clause.
26. *United States v. Oppenheimer,* 242 U.S. 85 (1916).

27. *United States v. Ball,* 163 U.S. 662, 669 (1896).
28. *Fong Foo v. United States,* 369 U.S. 141 (1962). Justice Clark dissented.
29. *Green V. United States,* 355 U.S. 184 (1957). The 1905 case of *Trono v. United States,* 199 U.S. 521, which held that the accused, by his appeal, had waived his right to avail himself of a prior acquittal of a greater offense, was limited to its own peculiar facts and setting (trial in a newly acquired Philippine Islands with alien legal customs). Dissenting, four justices (Frankfurter, Harlan, Burton, and Clark) accused the Court of having overruled the Trono case inasmuch as the latter had been rested, not upon the criminal procedure of the Philippines, but upon a construction of the Fifth Amendment. Citing *Stroud v. United States,* 251 U.S. 15 (1919), they contended that historical precedent abundantly discloses that the double jeopardy clause is inapplicable to a judgment annulled at the request of the accused.
30. *Foreman v. United States,* 361 U.S. 416 (1960). A person can be tried a second time when his conviction is set aside by his own appeal, notwithstanding that the petitioner, in the instant case, did not request a new trial with respect to that portion of the charge to the jury dealing with the statute of limitations. Since the action of the Court of Appeals on rehearing was not based on new evidence, the Court declared inapplicable its prior holding in *Sapir v. United States,* 348 U.S. 373 (1955), wherein it vacated the second judgment of a Court of Appeals directing a new trial on the government's motion, founded upon a newly discovered evidence and reinstated the latter Court's initial judgment reversing and remanding the case with instruction to dismiss the indictment. In *Sapir,* however, Justice Douglas maintained that an acquittal based on lack of evidence concludes a prosecution and that the grant of a new trial violated ''the command of the Fifth Amendment.''
31. *Wade v. Hunter,* 336 U.S.l 684, 689 (1949).
32. *United States v. Perez,* 9 Wheat. 579 (1824); *Logan v. United States,* 144 U.S. 263, 298 (1892).
33. *Simmons v. United States,* 142 U.S. 148 (1891); *Thompson v. United States,* 155 U.S. 271 (1894).
34. *Lovato v. New Mexico,* 242 U.S. 199 (1916).
35. *Gori v. United States,* 367 U.S. 364, 368, 370 (1961). To warrant the conclusion reached, according to four dissenting Justices (Douglas, Black, Brennan, and Warren, C. J.), the mistrial must have been ordered at the request of the defendant or with his consent.
36. *Wade v. Hunter,* 336 U.S.l 684, 89 (1949).
37. *Downum v. United States,* 372 U.S. 734, 739–743 (1963). Noting that the prosecutor's failure to ascertain whether the key witness would be present when the jury was assembled was an ''excusable oversight'' attributable to his preoccupation with

another trial conducted earlier on the same day, that
the subpoena for the witness had never been served,
that petitioner having been arraigned in the presence
of the first jury, suffered no proven disadvantage in
being subjected to trial by a second jury, and that
conduct of the trial under alternatives suggested by
petitioner might have exposed him to multiple
prosecutions, four dissenting Justices (Clark, Harlan,
Stewart, and White) would have applied the doctrine
of *Wade v. Hunter,* namely that the ends of justice
required discontinuance of the first trial. *See United
States v. Tateo,* 377 U.S. 463 (1964).

38. *Collins v. Loisel,* 262 U.S. 426 (1923).
39. *Taylor v. United States,* 207 U.S. 120, 127 (1907).
40. *Bassing v. Cady,* 208 U.S. 386, 391–392 (1908).
41. *United States v. Wilson,* 7 Pet. 150 (1833).
42. *Burton v. United States,* 202 U.S. 344 (1906); *United
    States v. Randenbush,* 8 Pet 288, 289 (1834).
43. *Morgan v. Devine,* 237 U.S. 632 (1915). *See also
    Carter v. McClaughry,* 183 U.S. 365 (1902); *Albrecht
    v. United States,* 273 U.S. 1 (1927).
44. *Hans Nielsen, Petitioner,* 131 U.S. 176, 188 (1889).
45. *Helvering v. Mitchell,* 303 U.S. 391 (1938).
46. *Pinkerton v. United States,* 328 U.S. 640 (1946);
    United States v. Bayer, 331 U.S. 532 (1947); *Pereira
    v. United States,* 347 U.S. 1 (1954).
47. *Florida v. Bostick,* 111 S. Ct. 2382 (1991).
48. *California v. Hodari D.,* 111 S. Ct. 1547 (1991).
49. *County of Riverside v. McLaughlin,* 111 S. Ct. 1661
    (1991).
50. *California v. Acevedo,* 111 S. Ct. 1982 (1991).
51. 442 U. S. 753 (1979).
52. *Florida v. Jimeno,* 111 S. Ct. 1801 (1991).
53. *Minnick v. Mississippi,* 111 S. Ct. 486 (1990).
54. *McNeil v. Wisconsin,* 111 S. Ct. 2204 (1991).
55. *Arizona v. Fulminante,* 111 S. Ct. 1246 (1991).
56. 377 U.S. 201 (1964).
57. 360 U.S. 315 (1959).
58. 406 U.S. 682 (1972) (Plurality opinion).
59. 304 U.S. 458 (1938).
60. *McCarthy v. Arndstein,* 266 U.S. 34, 40 (1924); *Boyd
    v. United States,* 116 U.S. 616 (1886); *Counselman v.
    Hitchcock,* 142 U.S. 547 (1892); *Brown v. Walker,*
    161 U.S. 591 (1896).
61. *Rogers v. United States,* 340 U.S. 367, 370 (1951);
    United States v. Monia, 317 U.S. 424, 427 (1943).
62. *Hoffman v. United States,* 341 U.S. 479, 486 (1951);
    Mason v. United States, 244 U.S. 362, 365 (1917).
63. *Murphy v. N.Y. Waterfront Comm.,* 378 U.S. 52
    (1964), overruling *United States v. Murdock,* 382,
    U.S. 141, 149 (1931). In *Hutcheson v. United States,*
    369 U.S. 599, 604–611, 622, 641 (1962), the Court
    refused to consider a request that *United States v.
    Murdock* be overruled for the reason that the
    petitioner, appearing as a witness before a Senate
    committee, expressly refused to invoke the privilege
    against self-incrimination. However, the committee
    was recorded as respecting the invocation of that
    privilege by colleagues, present with petitioner, who
    clearly disclosed that the sole justification of their
    plea was fear of exposure to state, rather than federal,
    prosecution. Justices Douglas and Brennan were in
    agreement in Hutcheson as to the desirability of such
    reversal, the former on the grounds that, under
    Murdock and related precedents, a witness is
    "whipsawed between state and federal agencies,
    having no way to escape federal prison unless he
    confesses himself into a state prison." Under the
    necessary and proper clause (Art. I, 8, cl. 18)
    Congress was acknowledged to possess the power to
    enact 18 U.S.C. 3486 (later repealed) providing that
    testimony given by a witness in congressional
    inquiries shall not be used in evidence in any court;
    and a witness's failure to claim his privilege was held
    not to deprive him of the protection afforded by this
    stipulation in state, no less than federal, courts.
    Adams v. Maryland, 347 U.S. 179 (1954).
64. *Murphy v. N.Y. Waterfront Comm.,* overruling
    Feldman v. United States, 322 U.S. 487 (1944) and
    Knapp v. Schweitzer, 357 U.S. 371 (1958). "Once a
    defendant demonstrates that he has testified under a
    state grant of immunity, to matters related to the
    federal prosecution, the federal authorities have the
    burden of showing that its evidence is not tainted by
    establishing that it had an independent, legitimate
    source for the disputed evidence." Yet in *Knapp v.
    Schweitzer* the Court had contended that it was too
    great a price for our federalism to submit to the
    "sterilizing . . . power of both governments by not
    recognizing the autonomy of each within its proper
    sphere." (357 U.S. at 381.)
65. *Brown v. Walker,* 161 U.S. 591 (1896); *Johnson v.
    United States,* 318 U.S. 189 (1943). *See also Brown
    v. United States,* 356, U.S. 148, 158–159, 161 (1958),
    wherein it was held that when a witness, who is not
    compelled to testify, voluntarily takes the stand and
    testifies at length in his own defense, he thereafter, on
    cross-examination, cannot refuse to answer questions
    relevant to his testimony on direct examination. Thus,
    if a witness in a denationalization proceeding, on
    direct examination, answers questions as to his
    loyalty and nonparticipation in subversive activities,
    all of which testimony is favorable to his case, he
    cannot invoke the privilege upon cross-examination
    and refuse to answer a query as to Communist Party
    membership. On the ground that witnesses in civil
    proceedings do not enjoy the same privileges as to
    refusal to testify accorded criminal defendants and,
    accordingly, that the waiver rule, applied in criminal
    cases, ought not to be extended to civil suits, Justice

Black together with Justices Douglas and Brennan and Warren, C. J., dissented and cited *Arndstein v. McCarthy,* 254 U.S. 71; 262 U.S. 355; 266 U.S. 34, in support of their conclusion that a witness in a civil case, being unable to anticipate the range of inquiry upon cross-examination, is not to be viewed as determining the area of disclosure and inquiry by his testimony on direct examination and therefore does not forfeit the right to claim the privilege unless he makes disclosures amount to an admission of guilt or incriminating facts. *Cf. Grunewald v. United States,* 353 U.S. 391, 415–425 (1957), wherein it was held that when defendant, on direct examination, answers certain questions in a manner favorable to his innocence, it is prejudicial error for the trial court, over defendant's objections, to permit the government on cross-examination and with a view to impugning the defendant's credibility, to extract from defendant an admission that, during a prior appearance before a grand jury in response to a subpoena, he had invoked the privilege to avoid answering the same questions. There is grave danger, that Court acknowledged, that the trial jury might make impermissible use of testimony by equating invocation of the privilege with guilt. Concurring, four Justices (Black, Brennan, Douglas, and Warren, C. J.) were of the opinion that the value of the constitutional privilege would be destroyed "if persons can be penalized for relying on it."

66. *Cf. Twining v. New Jersey,* 211 U.S. 78 (1908). However, a defendant in a federal prosecution enjoys a statutory right, upon request, to have the jury instructed that his failure to testify creates no presumption against him—18 U.S.C. 3481; *Bruno v. United States,* 308 U.S. 287 (1939). *See also Johnson v. United States,* 318 U.S. at 196.

67. *Pierce v. United States* 160 U.S. 355 (1896); *Wilson v. United States,* 162 U.S. 613 (1896); *United States v. Mitchell,* 322 U.S. 65 (1944).

68. 318 U.S. 332 (1943). *Massiah v. United States,* 377 U.S. 201, 1010 (1964).

69. In *Upshaw v. United States,* 335 U.S. 410 (1948), a sharply divided Court found the McNabb case inapplicable to a case in which respondent, while under arrest for assault with intent to rape, was brought, by extended questioning, to confess to having previously committed murder in an attempt to rape.

70. 354 U.S. 449 (1957).

71. *Rochin v. California,* 342 U.S. 165 (1952).

72. 335 U.S. 1 (1948).

73. In a dissenting opinion Justice Frankfurter argued: "The underlying assumption of the Court's opinion is that all records which Congress in the exercise of its constitutional powers may require individuals to keep in the conduct of their affairs, because those affairs also have aspects of public interest, become 'public' records in the sense that they fall outside the constitutional protection of the Fifth Amendment. The validity of such a doctrine lies in the scope of its implications. The claim touches records that may be required to be kept by federal regulatory laws, revenue measures, labor and census legislation in the conduct of business which the understanding and feeling of our people still treat as private enterprise, even though its relations to the public may call for governmental regulation, including the duty to keep designated records. . . . If Congress by the easy device of requiring a man to keep the private papers that he has customarily kept can render such papers 'public' and nonprivileged, there is little left to either the right of privacy or the constitutional privilege."

74. *Schneckloth v. Bustamunte,* 412 U.S. 218 (1973).

75. 395 U.S. 752 (1969).

76. 389 U.S. 347 (1967).

77. 503 F. 2d 472, 478 (4th Cir. 1974).

78. 415 U.S. 164 (1974).

79. 512 F. 2d 496 (5th Cir. 1975).

80. 513 F. 2d 154 (1st Cir. 1975).

81. 315 N.E. 2d 466, 467 (N.Y. 1974).

82. 75 L.Ed. 2d 55 (1983).

83. 82 L. Ed. 2d 530 (1984).

84. 389 U.S. 347 (1967).

85. *United States v. Chadwick,* 433 U.S. 1 (1977); *Arkansas v. Sanders,* 442 U.S. 753 (1979).

86. 562 F. 2d 106 (1st Cir. 1977, cert. denied, 435 U.S. 926 (1978).

87. 504 F. Supp. 1220 (D.Col.1981).

88. *Carroll v. United States,* 267 U.S. 132 (1925); *Chambers v. Maroney,* 399 U.S. 42 (1970); *United States v. Ross,* 456 U.S. 798 (1982).

89. *United States v. Erickson,* 732 F. 2d 788 (10th Cir. 1984); *United States v. Long,* 674 F. 2d 848 (11th Cir.), cert. denied, 455 U.S. 919 (1982); *United States v. Cady,* 651 F. 2d 290 (5th Cir.), cert. denied, 455 U.S. 919 (1981); *United States v. Bruneau,* 594 F. 2d 1190 (8th Cir.), cert. denied, 444 U.S. 847 (1979); *United States v. Miroyan,* 577 F. 2d 489 (9th Cir., cert. denied, 439, U.S. 896, (1978).

90. *Coolidge v. New Hampshire,* 403 U.S. 443 (1971); *Connally v. Georgia,* 429 U.S. 245 (1977).

91. 389 U.S. 347 (1967).

92. 572 F. Supp. 17 (E.D. Pa. 1983).

93. 821 F. 2d 248 (5th Cir. 1987).

94. 401 U.S. 745 (1971).

95. *Warden v. Hayden,* 387 U.S. 294, 298–299 (1967).

96. *Michigan v. Tucker,* 417 U.S. 433 (1974).

97. *Payne v. Arkansas,* 356 U.S. 560 (1958); *Ashcraft v. Tennessee,* 322 U.S. 143 (1964); *Brown v. Mississippi,* 297, U.S. 278 (1936).

98. 389 F. Supp. 931 (N.D. Ga. 1974), aff'd., 523 F. 2d 1053 (5th Cir. 1975).

99. 222 F. 2d 698 (2d Cir.) cert. denied, 350 U.S. 896 (1955).

100. 394 U.S. 731 (1969).

101. 363 F. Supp. 788 (W. D. N.Y. 1973).

102. 384 U.S. 436, 453 (1966).

103. 389 F. Supp. 931 (N.D. Ga. 1974), aff'd., 523 F. 2d 1053 (5th Cir. 1975).

104. 360 U.S. 315 (1959).

105. 367 U.S. 568, 602 (1961) (plurality opinion).

106. 394, U.S. 731 (1969).

107. 329 F. 2d 68 (2d Cir.), cert. denied, 377 U.S. 967 (1964).

108. *Smith v. State,* 155 Ind. 611, 58 N.E. 1044 (1900).

109. 97 Ill. 320 (1881).

110. 257 Mass. 195, 153 N.E. 454 (1926).

111. *Pierson v. Post,* 3 Cai. R. 175 (N.Y. Sup. Ct. 1805).

112. *Liesner v. Wanie,* 156 Wis. 16, 145 N.W. 374 (1914).

113. *Reg. v. Roe,* 22 L.T.N.S. England, 414 (1870).

114. *State v. Shaw,* 67 Ohio St. 65 N. E. 875 (1902).

115. 11 Ohio S. & C.P. Dec. 753 (1901).

116. *L. Realty Co. v. Johnson,* 92 Minn. 363, 100 N.W. 94 (1904).

117. *Ohio Oil Co. v. Indiana,* 177 U.S. 190 (1900); *Kellogg v. King,* 114 Cal. 378, 46 Pac. 166 (1896); *Sterling v. Jackson,* 69 Mich. 488, 37 N.W. 845 (1888); *L. Realty Co. v. Johnson,* 92 Minn. 363, 100 N.W. 94 (1904).

118. 92 Minn. 663, 100 N.W. 94 (1904).

119. *Diana Shooting Club v. Lamoreux,* 114 Wis. 44, 89 N.W. 880 (1902).

120. Ibid.

121. *Whittaker v. Stangvick,* 100 Minn. 386, 111 N.W. 295 (1907).

Recommended Reading

1982. How well do you know the law? In *Law Enforcement Bible, No. 2,* ed. R. A. Scanlon. South Hackensack, N.J.: Stoeger Publ. Co.

Cannavale, F. J., and W. D. Falcon. 1978. *Witness cooperation.* Heath, D. C. Co.

Livingston, H. 1967. *Office of the witness stand.* Legal Book Corp.

Williams, T. 1989. Game laws weren't writ for fat cats. *Audubon* (July): 104–113.

Multiple-Choice Questions

Mark only one answer.

1. "The right of the people to be secure in their persons, houses, papers, . . . shall not be violated and no warrants shall issue, but upon probable cause . . . and the persons or things to be seized." This is a statement of which amendment to the Constitution?
   a. one
   b. four
   c. five
   d. ten
   e. fourteen

2. Regarding the Declaration of Independence, which statement is not true?
   a. the Declaration of Independence was very much like that of the Constitution
   b. it shared the French enthusiasm for reason, science, and progress
   c. it subscribed to the notion "the voice of the people is the voice of God"
   d. at the time the Declaration of Independence was written, there was no historical evidence for the proposition that all men were created equal
   e. all evidence of history was against the proposition that meant all people were created equal

3. In *Minnick v. Mississippi:*
   a. the court ruled that once a custodial suspect requests counsel in response to Miranda warning, the officers may not attempt to reinterrogate the suspect unless his counsel is present or he initiates contact with the law
   b. the defendant escaped from a jail but was not involved in any known crimes other than the original charge
   c. the FBI agents gave the defendant his Miranda warnings, and he provided them with much of the information they wanted
   d. the trial court suppressed the defendant's statements to the deputy sheriff from Mississippi
   e. the U.S. Supreme Court did not reverse the Mississippi Supreme Court's ruling

4. In *Massiah v. United States* which statement is not true?
   a. Massiah and Colson were two seamen who were accused of having 3½ pounds of cocaine
   b. Colson, as a part of his cooperation with the agents, agreed to have a portable transmitter put in his car where he would talk to Massiah
   c. later Massiah claimed that his testimony should be withheld since it was obtained through illegal eavesdropping

d. Massiah appealed and alleged that the government had violated his Sixth Amendment right to assistance of counsel by having one of the agents approach him

e. later Massiah's case was argued before the U.S. Supreme Court and his conviction was upheld

5. One of the following statements is not true:

a. the question of when a defendant has waived his or her rights to counsel is not easily answered

b. in *Johnson v. Zerbst,* the Supreme Court held that the Sixth Amendment right to counsel can be waived in certain circumstances

c. the Massiah rule has not been expanded substantially in recent years

d. the Supreme Court in 1972 stated that a defendant no longer must show that he was indicted at the time of his statement in order to make a Sixth Amendment argument

e. a defendant need only show at the time his statement was obtained by the officers, in the absence of counsel, there had been an arraignment or a preliminary hearing and information had been filed or he had otherwise been charged

6. Which of the below statements is not correct?

a. the founding fathers insisted that the individual did not belong to the state, but had a personal and immortal soul that should be beyond the dictates of any totalitarian regime

b. on the above principle they established a constitutional government that was under and not above the law

c. this was and still is the most powerful and political idea in the world

d. the federal Constitution was composed by a group of elite world-renowned scholars

e. the federal Constitution is one of the greatest documents in history

7. Since *Chimel v. California,* the principal area that, incident to arrest, can be searched is:

a. only the arrestee

b. all of the area within ten feet of the arrestee

c. the area within the arrestee's immediate control

d. the arrestee and all unlocked containers within the room

e. the entire house if the arrestee is a known felon

8. Which one of the framers of the Constitution said ''Men are born & grown up in this world with a vast variety of capacityes, strength and abilities both of mind and body—.''

a. Thomas Jefferson

b. John Adams

c. Benjamin Franklin

d. William Manning

e. Theodore Roosevelt

9. One of the following statements is not true:

a. an officer who equates curtilage with the area wherein an individual possesses a reasonable amount of expectation of privacy will generally be correct

b. curtilage is easily identified in many urban communities

c. curtilage is difficult to define in many rural areas

d. the question of where curtilage ends and open fields begins is easily defined

e. generally the doctrine of search and seizure permits the taking of abandoned property

10. One of the following statements is true:

a. the admissibility of a confession is easy to determine

b. the definition of admissibility of confessions has not been modified by the court over the years

c. rules of evidence do not require that confessions, like other evidence, always be reliable

d. the Fifth Amendment requires that a person shall be compelled in any crime to be a witness against himself

e. the Fourteenth Amendment states that no person shall be deprived of life, liberty, or property without due process of law

11. An officer preparing for a consent search should take which of the following steps?

a. determine the premises that are protected by the Fourth Amendment

b. identify the person lawfully entitled to the possession

c. obtain that person's voluntary relinquishment of constitutional rights

d. determine that the search is within the limitations expressed or implied in the consent of the person being searched

e. all of the above

12. One of the following statements is not true:

a. under the scope of privilege, a witness in any proceedings where testimony is legally required may refuse to answer questions that might incriminate him

b. the witness does not have to explicitly claim his constitutional immunity in order to protect himself from self-condemnation

c. an accused who takes the witness stand in his own behalf must submit to cross-examination

d. in *McNabb v. United States,* the Court reversed a conviction of a federal court based on a confession that was obtained by questioning the defendant for prolonged periods in the absence of counsel or friends

e. Justice Frankfurter stated that ''civilized standards of procedure and evidence'' be maintained

Chapter 4

State Jurisdiction over Wildlife

THE TENTH AMENDMENT

"The powers not delegated to the United States by the Constitution, nor prohibited by it to the States, are reserved to the States respectively, or to the people." Some people consider the Tenth Amendment to be a companion to the Ninth. ("The enumeration in the Constitution, of certain rights, shall not be construed to deny or disparage others retained by the people.") That is, they believe that it is stating a principle rather than providing a right. Others believe the framers of the Constitution would not have included the Tenth Amendment if it merely expressed the unremarkable notion that the states should not be wholly devoured by the federal government. Instead these proponents of a strong Tenth Amendment claim it is a source of one of the most powerful forces in our nation's history: state's rights. The principle of state's rights has been used for destructive purposes, but it also can have benevolent aspects. State governments represent the people more directly than does the federal government. At times states have gone further than the federal government to protect the individual; at other times it has been the reverse. In the 1960s, the federal government, supported by the Supreme Court, led in civil rights and social welfare legislation against state opposition (Alderman and Kennedy 1991). In some respects the pendulum swung again twenty years later during the "Reagan revolution" of the 1980s.

This system of government by both federal and state authority is called federalism, that is, each center of power prevents the other from becoming too strong and individual freedom is preserved. The structure of federalism is detailed, not in the Tenth Amendment, but in Article One of the Constitution, which lists the power delegated to Congress and those prohibited to the states. One of Congress' Article One powers is the power to regulate "commerce . . . among the several states."

Case 1

In *Garcia v. San Antonio Metropolitan Transit Authority,*[1] the Court did not include the Tenth Amendment in its discussion of the issues (paying minimum wage to state employees). Instead it looked to Article One of the Constitution to support its broad conception of the federal commerce power. First, the Court analyzed the enormous powers exercised by Congress, such as power to regulate commerce and currency, declare war or raise an army. Next, the Court looked at the powers withdrawn from the states; for example, states are forbidden to enter into treaties, to levy import or export taxes, to keep troops without the consent of Congress. Finally, the states did not need the Tenth Amendment to protect them. According to the Court, state interests were well provided for by their representatives and senators. Given the great benefits the states derived from their participation in the federal systems, the Court found state sovereignty was well protected and the burden of paying a minimum wage to a state employee did not destroy that sovereignty (Alderman and Kennedy 1991).

A pleasant officer
approach generally
earns hunter
cooperation.
*Courtesy Iowa Division of
Fish and Wildlife.*

The struggle over the meaning of the last amendment in the Bill of Rights (Tenth Amendment) will continue far into the future. At the moment, it appears to be a truism once again. However, the Supreme Court may decide it represents instead a reservoir of power waiting to be exercised by states. In any event, the Tenth Amendment, along with the Ninth, will continue to preserve freedoms of the Bill of Rights for the ultimate sovereignty of our constitutional framework: ''We the People'' (Alderman and Kennedy 1991).

Inherent Jurisdiction

We have seen that before the signing of the *Magna Charta* at Runnymede, ownership of wildlife was vested exclusively in the person of the king, and that after the signing, ownership was vested in the office of the king, to be held in sacred trust for the people.

Colonists who settled America carried the common law of England with them. After the Revolution, the question arose as to whether or not the newly independent colonies had a common law. It was decided that the common law of England, plus English statutes in force prior to the Revolution, were applicable where conditions permitted. It was believed that the state had, in effect, acquired the authority of the king, and the state in its sovereign capacity held the game in trust for the people; therefore, no statute was necessary to invest it with the ownership of wildlife. Until recently it has been deemed unnecessary to examine the statutes of the various states with regard to the ownership of wildlife.

Referring to this transfer of power from king to colony, the Court has stated, in part: ''Undoubtedly, this attribute of government to control the taking of animals *ferae*

Cases 2–7

naturae, which was thus recognized and enforced by the common law of England, was vested in the colonial governments, where not denied by their charters, or in conflict with grants of the royal prerogative. . . ."[2]

At the Constitutional Convention of 1787, the state delegates invested the federal government with certain powers, but did not mention the control of wildlife. It was not until the Supreme Court upheld the constitutionality of the Migratory Bird Treaty Act that federal government control over migratory birds was finally decided. The Constitution expressly prohibits the state and federal governments from exercising certain powers. The Tenth Amendment provides that all powers not delegated to the federal government and not prohibited to the states are reserved to the states respectively or to the people.

The right of states to govern the taking of fish and game was established in the famous case of *Geer v. Connecticut.*[3] In this case the right of the state to regulate and control the manner in which wild game may be taken is sustained on two grounds:

(1) the sovereign ownership of wild animals by the state; and

(2) the police power of the state, which flows from its duty to preserve for its people a valuable food supply.

The Court recognized the authority of the state to control the taking of wildlife in the following rulings:

. . . It is . . . certain that the power which the colonies thus possessed passed to the States with the separation from the mother country, and remains in them at the present day, insofar as its exercise may be not incompatible with, or restrained by, the right conveyed to the Federal Government by the Constitution.[4]

It is not to be doubted that the power to preserve fish and game within its borders is inherent in the sovereignty of the State. . . .[5]

It is admitted that, in the absence of Federal legislation on the subject, a State has exclusive power to control wild game within its borders and that the South Dakota law was valid when enacted, although it incidentally affected interstate commerce.[6]

In *Geer,* the U.S. Supreme Court opinion also held that the killing of game is a privilege and not a right and that the ownership differs from that of other property, and that even after it becomes possessed by an individual, it is subject to restrictions. Despite these new nonexport laws, fish and game which was not given protection could still be shipped illegally from one state and sold legally in another.

In *State v. Lessard,*[7] a case involving ownership, action was brought against Edward O. Lessard and others by the state for killing, taking into possession, and holding a whale. This whale was killed in an inland Oregon slough in Multnomah County. The state contended that the whale was a royal sovereign fish. The defendant was acquitted and the state appealed on the following grounds:

(1) the whale was owned by the state, and the defendant had wrongfully seized it;

(2) whales within coastal or inland waters of the state are royal fish, which belong to the state, and, hence, the killing of a whale in an inland slough and keeping possession of its body without consent of the state is actionable.

The Supreme Court of Oregon ruled in part:

(1) the complaint is sufficient;

(2) it is clearly stated in several cases that the whale in inland waters is the property of the state, and killing and holding without state permission is wrong.

Cases 8–12

There is a different rule for fish (as opposed to the whale, a mammal), and the defendant improperly applied the other rule. The Supreme Court reversed the decision in favor of the plaintiff (state) and instructed the lower courts to rectify their mistakes.

The state's trusteeship with regard to wildlife is clearly pointed out in the case of *State of Arkansas v. Mallory,*[8] wherein the Supreme Court of Arkansas held:

> Since then (*Magna Charta*), the ownership of wild animals, so far as vested in the sovereign, has been uniformly regarded as a trust for the benefit of the people; and we think that clearly, in effect, the title and ownership of the sovereign has been held to be only for the purposes of protection, control, and regulation . . . but nowhere do we find that in modern times has the absolute and unqualified ownership of such animals by government been asserted and exercised further than for the purposes of controlling and regulating the taking of the same. On the other hand, we find frequent denials of the right of government to do more.

Police Power

Grounds are frequently advanced to support the right of the state legislature to impose regulations governing the taking of wildlife. A common one is police power, which entitles the state to regulate the public health, safety, morals, and welfare of the people. The cases which contend that the state has the right to regulate the taking of game do not base that right upon absolute ownership of that game by the state, but upon the police power. Thus it follows, that the game belongs to all the people and that the state, sovereign by virtue of its police power, regulates the taking of game for the benefit of all. Consequently, the few who are favorably situated may not reap the benefit of the game to the exclusion of those handicapped by distance and other considerations. In *Geer v. Connecticut,*[9] the Court reviews the matter fully and shows that the various nations have always regarded the game as belonging to all the people in common, and that the state has regulated the killing of game, not on the ground of ownership, but as sovereign, acting for the benefit of all.

In *Missouri v. Holland,*[10] the Court said: "To put the claim of the State upon title is to lean upon a slender reed." Also, in *McKee v. Gratz,*[11] the Court maintained: "It (the Circuit Court of Appeals) rightly, as we think, held that the statutes declaring the title to game and fish to be in the State spoke only in aid of the State's power of regulations and left the plaintiff's interest what it was before."

In some states, private ownership of game, such as bison or waterfowl, is recognized. However, unless state laws make definite exceptions, it should be assumed that wild game is in the custody of the state (as opposed to private ownership).

Limitation of State Jurisdiction

This custodial power of the state may be exercised in any way that does not contravene the provisions of the Federal Constitution. Whenever conflict does arise in this regard, the authority of the state must yield to a valid exercise of power by the federal government. A court ruling to this effect reads: ". . . it is clear . . . that the right to regulate the taking and use of game and fish is, generally speaking, in the State as an attribute of its sovereignty, subject only to valid exercise of authority under the provisions of the Federal Constitution."[12]

The Tenth Amendment at the time the Constitution was adopted, was intended to confirm the understanding of the people, that powers not granted to the United States were reserved to the states or to the people. Pointedly, it says, "The powers not delegated to the United States by the Constitution nor prohibited by it to the States, are reserved to the States respectively, or to the people."

Today it is apparent that the Tenth Amendment does not shield the states nor their political subdivisions from the impact of the authority affirmatively granted to the federal government. Thus, when a city is licensed under the Federal Power Act to construct and operate electric power projects on navigable waters, its capacity to attain that objective, by condemnation of a state fish hatchery, cannot be circumscribed by the law of the state from which it derives its charter.[13] . . . At the suit of the Attorney General of the United States, the sanitary district of Chicago was enjoined from diverting water from Lake Michigan in excess of a specified rate. On behalf of a unanimous court, Justice Holmes wrote: ''This is not a controversy among equals. The United States is asserting its sovereign power to regulate commerce and to control the navigable waters within its jurisdiction. . . . There is no question that this power is superior to that of the States to provide for the welfare or necessities of their inhabitants.''[14]

In *Kansas v. Colorado,*[15] the Court said: ''. . . All legislative power must be vested in either the State or the National Government; no legislative powers belong to a State government other than those which affect solely the internal affairs of the State; consequently all powers which are national in their scope must be found vested in the Congress of the United States.'' In *United States v. Appalachian Power Co.,*[16] the petition to intervene was dismissed on the ground that the authority claimed for the federal government was incompatible with the Tenth Amendment; but this could hardly happen today. Under its superior power of eminent domain, the United States may condemn land owned by a state even where the taking will intervene with the state's own project for water development and conservation.[17] ''That the United States lacks the police power, and that this was reserved to the states by the Tenth Amendment, is true. But it is nonetheless true that when the United States exerts any of the powers conferred upon it by the Constitution, no valid objection can be based on the fact that such exercise may be attended by the same incidents which attend the exercise by a State of its police power.''[18] And in a series of cases, which today seem irreconcilable with *Hammer v. Dagenhart,*[19] it sustained federal laws penalizing the interstate transportation of lottery tickets,[20] of women for immoral purposes,[21] of stolen automobiles,[22] and of tick-infested cattle,[23] as well as a statute prohibiting the mailing of obscene matter.[24] Also, over fish found within its waters and over wild game, the state has supreme control.[25] It may regulate or prohibit fishing and hunting within its limits,[26] and for the effective enforcement of such restrictions, it may forbid the possession within its borders of special instruments of violations such as nets, traps, and seines, regardless of the time of acquisition or the protestations of lawful intentions on the part of a particular possessor.[27] To conserve fish found within its waters, for food, a state, constitutionally, may provide that a reduction plant, processing fish for commercial purposes, may not accept more fish than can be used without deterioration, waste, or spoilage. As a shield against the covert depletion of its local supply, a state may render such restriction applicable to fish brought into it from the outside.[28] Likewise, it is within the power of a state to forbid the transportation outside the state of game killed therein;[29] and to make possession illegal during the closed season even of game imported from abroad.[30]

In summary, state fish and game departments, and most people, have always felt that the title to game, other than migratory birds, has been with the respective states. Events, not as recent as is generally believed, have been steadily and firmly changing this concept. Today the federal government manages the game on such federal lands as wildlife refuges, national parks, and monuments and on Indian lands, on request. There are three primary reasons for this:

(1) a more mobile citizenry and, therefore, one that frequently crosses state lines;

(2) a federal government that is far better financed than state agencies; and

(3) a series of court decisions that favor the federal government.

Roadblocks are one of many law enforcement tools.
Courtesy Utah Division of Wildlife Resources.

Speaking to these issues, it is obvious that political boundaries do not coincide with geographic ones. Many great rivers, for example, with their load of pollution flow through several states. The first Federal Water Pollution Control Act, in actuality, shifted most of the pollution control problem from the respective states to the federal government. While many state agencies resented this loss of power, they, by themselves, were virtually helpless, from a legal standpoint, to act against upstream neighbors because they had neither the funds nor personnel to even collect adequate evidence. Today, most states have laws governing water and air pollution, but they operate on a small budget, and they must rely on federal laws where neighboring states are concerned. Further, state governments are subject to greater political pressures in certain fish and game areas than the federal government. Where thorny problems, such as legislation that is unpopular with some industries, or the expense of water pollution control are concerned, it seems inevitable that the federal government must, and will, assume the leadership as well as a large percent of the financial burden. Another area where the federal government has virtually unchallenged authority is in the construction, or its authorization, of high dams. In both of these examples the authority of the federal government is backed by the Supreme Court decisions, and the welfare, and sometimes the very existence, of wildlife is affected.

The welfare of the people of the United States is of paramount importance, and concessions to wildlife are made only when they are compatible with that of the people. Even when concessions to wildlife can be made by way of funds, these expenditures cannot be so excessive that they are harmful to the welfare of the people.

State v. Federal Government in Resource Management

State fish and game departments, particularly those in the West, have shown increasing concern in recent years over what they consider the too active role of the federal government in the management of nonmigratory game, as well as threatened or endangered species. States have been especially concerned with a number of statements made by the Public Land Law Review Commission (PLLRC) (1970), which was established by Public Law 88–606. The PLLRC report, ''One-third of the Nation's Land,'' deals in part with management impacts, direct and indirect, on the nation's land as related to the fish and wildlife resources.

The PLLRC (1970) has said, ''State policies which unduly discriminate against nonresident hunters and fishermen in the use of public lands through license fee differentials in various forms of nonfee regulations should be discouraged.'' The PLLRC also states, ''a federal land-use fee should be charged for hunting and fishing on all public lands open for such purposes.'' This latter proposal has aroused the concern, if not the ire, of most western wildlife administrators who view this as a scheme for license fee splitting in which they are bound to lose. Many residents of the Midwest and eastern United States hold entirely different views.

The International Association of Game, Fish and Conservation Commissioners (1971) formed a committee which discusses, and in many cases partly refutes, the recommendations of the PLLRC.

Court decisions apparently favoring the federal government over state governments, changing philosophies of both the federal government and the people, greater use of and competition for the resource, and possibly such an apparently remote factor as wage differentials between state and federal employees, are among the many factors that complicate public resource management. There are many facets to the problem.

How much of the redwood forest shall be harvested? Should private enterprise be allowed to develop recreation facilities on public lands? Should the harvesting (killing) of game animals be allowed on public lands? Should additional energy producing plants (fossil and nuclear fuel) be developed before we know more of their environmental impact? These and many more questions are being asked of public administrators and Congress by the general public.

Literature Cited

Alderman, E., and C. Kennedy. 1991. *In our defense: The Bill of Rights in action.* New York: William Morrow and Co., Inc.

International Association of Game, Fish and Wildlife. 1971. *Public land policy impact on fish and wildlife.* Washington, D.C.: Intl. Assn. of Game, Fish and Cons. Comms.

Public Land Law Review Commission. 1970. *One-third of the nation's land.* Washington, D.C.: U.S. Govt. Printing Office.

Notes

1. 469 U.S. 528 (1985).
2. *Geer v. Connecticut,* 161 U.S. 519 (1896).
3. 161 U.S. 519 (1896).
4. Ibid.
5. *Geer v. Connecticut; Ward v. Racehorse,* 163, U.S. 504 (1896).
6. *Geer v. Connecticut; Silz v. Hesterberg,* 211 U.S. 31 (1908); *Carey v. South Dakota,* 250 U.S. 118 (1919).
7. Ore. 29 Pac. 2d 509 (1934).
8. 73 Ark. 236, 83, S.W. 955 (1904).
9. 161 U.S. 519 (1896).
10. 252 U.S. 416 (1920).
11. 260 U.S. 127 (1922).
12. *United States v. 2271.29 acres of land, etc.* 31 F. 2d 617 (W.D. Wis. 1928).
13. *City of Tacoma v. Taxpayers,* 357 U.S. 320 (1958).
14. *Sanitary District v. United States,* 266 U.S. 405 (1925).
15. 206, U.S. 46 (1907).
16. 311 U.S. 377 (1940).
17. *Oklahoma v. Atkinson Co.,* 313 U.S. 508 (1941).
18. *Hamilton v. Kentucky Distilleries Co.,* 251 U.S. 146 (1919).
19. 247 U.S. 251 (1918).
20. *Champion v. Ames,* 188 U.S. 321 (1903).
21. *Hoke v. United States,* 227 U.S. 308 (1913).
22. *Brooks v. United States,* 267 U.S. 432 (1925).
23. *Thornton v. United States,* 271 U.S. 414 (1926).
24. *Roth v. United States,* 354 U.S. 476 (1957).
25. *Bayside Fish Co. v. Gentry,* 297 U.S. 422 (1936).
26. *Manchester v. Massachusetts,* 139 U.S. 240 (1891).
27. *Miller v. McLaughlin,* 281 U.S. 261 (1930).
28. *Bayside Fish Co. v. Gentry,* 297 U.S. 422 (1936).
29. *Geer v. Connecticut,* 161 U.S. 519 (1896).
30. *Silz v. Hesterberg,* 211 U.S. 31 (1908).

Recommended Reading

Christensen, J. A. 1983. Law enforcement sources for wild lands. *Journal of Forestry* 81: 785–787.

Iverson, G. 1964. State enforcement of interstate transportation. In *Proceedings Western Assn. of Game and Fish Comms.* 314–317.

Kramer, W. F. 1979. Wildlife law enforcement—state or federal responsibility? In *Proceedings Western Assn. of Fish and Wildlife Agencies,* 184–189.

Mills, L. 1926. Cooperation of the states, one, either by the enactment of a federal statute or state laws. In *Proceedings Intl. Assn. of Game and Fish Cons. Comms.* 26–33.

Sjostrom, R. O. 1959. How federal laws and U.S. Deputy Game Warden Commissioners aid state game law enforcement. In *Proceedings Western Assn. of Game and Fish Comms.* 405–408.

Multiple-Choice Questions

Mark only one answer.

1. One of the following statements is not true:
   a. The police powers of the state allow it to regulate the taking of wildlife
   b. The state does not have absolute ownership of game
   c. The state is sovereign by virtue of its police power
   d. The state is custodian of all wildlife for the people
   e. In a contest of *State v. Federal,* the state is all-powerful

2. One of the following statements is not true:
   a. The Tenth Amendment does not shield the states from the federal government
   b. Justice Jackson said in regard to *State v. Federal,* this is not a controversy among equals, meaning the federal government predominates
   c. All legislative power must be vested in a state or the federal government
   d. The federal government may condemn land owned by the state
   e. The federal government has the power to control navigable waters of the United States

3. The taking of game or fish is:
   a. a privilege
   b. a right
   c. a modified right
   d. a vested right for landowners
   e. an inherent right

4. The case that established the state's right to govern the taking of wildlife is:
   a. *Missouri v. Holland*
   b. *Geer v. Connecticut*
   c. *Carey v. South Dakota*
   d. *Champion v. Ames*
   e. *Silz v. Hesterburg*

5. A state can:
   a. conserve wildlife for the people
   b. forbid the import or export of wildlife
   c. make possession of wildlife illegal at certain times of the year
   d. manage resident waterfowl in defiance of the federal government
   e. order processing plants to accept no more raw produce than they can handle without spoilage

6. The Public Land Law Review Commission (PLLRC) states:
   a. State policies should not unduly discriminate against nonresident sportsmen in regard to fees
   b. A federal land use tax should be charged to users of public land
   c. Inferred the fee could be split with the states
   d. The federal government should pass laws collecting fees regarding the taking of wildlife
   e. The report came out in 1970

7. The International Association of Game, Fish and Conservation Commissioners states:
   a. court decisions were favoring the federal government
   b. philosophies were changing in favor of the federal government
   c. the IAGFCC were largely in agreement with PLLRC report
   d. greater use was being made of the public lands
   e. fee splitting was a way for the federal government to get part of the state license money

8. Which of the following powers is not relegated to the states?
   a. powers not delegated to the United States
   b. custodianship of the wildlife within their boundaries
   c. levying and collecting taxes
   d. managing nonmigratory wildlife
   e. ownership of wildlife within their borders

9. Which of the following powers is not relegated to the federal government?
   a. a regulation of commerce and currency
   b. the power to declare war
   c. the power to enter into treaties
   d. manage nonmigratory wildlife within the boundaries of the federal lands
   e. levy export/import taxes

Chapter 5

Federal Jurisdiction over Wildlife

Cases 1–2

In early days, American law became expansive, rather than defensive in nature, favoring change more than stability. The Marshall Court upheld the expansive reach of federal authority and the federal supremacy against state encroachment. The supremacy clause of Article VI of the Constitution is the very foundation of the federal system and was first given full affect by the decisions of the Marshall Court. Under these rulings, all state acts are subordinate to the valid exercise of federal authority whether expressed in the Constitution, a congressional enactment, or a decision of the Supreme Court (Schwartz 1974).

In the case of *McCulloch v. Maryland,*[1] the U.S. Supreme Court ruled in favor of the government against the state of Maryland. It also defined a doctrine of applied powers for federal government, which is today a part of American constitutional law.

SOURCE OF FEDERAL AUTHORITY

In 1787, sixty-five delegates from the several states met to draft the Constitution of the United States. Out of work of the Convention came the principle of the division of power between the states and the federal government. Many powers were delegated to the states, but those not specifically delegated reside inherently in the states.

Although among the residual powers is the one giving control of wildlife to the states, the federal government exercises jurisdiction over migratory wildlife. This jurisdiction stems from the authority vested in the government to:

(1) create and regulate a federal government;

(2) make treaties;

(3) regulate foreign and interstate commerce; and

(4) levy and collect taxes.

Jurisdiction over Wildlife Habitats

Item one above refers to the power of Congress "to dispose of and make all needful rules and regulations respecting the territory or other property belonging to the United States. . . ."[2] Under this authority Congress has the power to set aside lands or territory for the creation of national monuments, national parks, and national refuges. This gave the government control over the habitats of all wildlife in these areas as well as in such territories as Puerto Rico, the Philippine Islands before their independence on July 4, 1946, extensive military holdings, and, since 1932, the District of Columbia.

Jurisdiction over Migratory Wildlife

Cases 3–6

Control over migratory wildlife may be exercised by reason of item two, the treaty-making power vested in the president, ''He shall have Power, by and with the Advice and Consent of the Senate, to make Treaties, provided two-thirds of the Senators present concur. . . .''[3]

Ducks, geese, and swans are examples of wildlife that are migratory in habit crossing state and international lines. Control, supervision, and management, therefore, become an international problem and are subject to treaty power.

Authority to Enforce Wildlife Laws

''The Congress shall have Power to lay and collect Taxes, Duties, Imports, and Excises. . . .''[4] This power gives the agents of the division of law enforcement of the Department of Interior U.S. Fish and Wildlife Service the right to enforce federal game laws.

Authority to Manipulate Wildlife Populations

The Constitution has been interpreted to also allow the federal government to manipulate wildlife populations through authorization of the general welfare clause.[5] Under this, Congress has established, as part of the U.S. Fish and Wildlife Service, divisions of Fish Hatcheries, Fisheries Research, Biological Services, and Federal Aid.

Authority to Implement Controls

Finally, the federal government has the power to implement its control over wildlife by reason of the right of Congress ''to make all laws which shall be necessary and proper for carrying into Execution the foregoing Powers, and all other Powers vested by this Constitution in the Government of the United States, or in any Department or Officer thereof.''[6]

FEDERAL WILDLIFE LAWS

Although each state has its own set of wildlife laws, there are federal laws that are common to all states. Inasmuch as all conservation officers will be called upon to enforce them at one time or another, it seems desirable to present the highlights.

Special Acts— An Overview

A brief overview will help understand the history and extent of present day federal wildlife laws affecting both game and nongame animals. The Lacey Act of 1900 was the first federal law directly affecting wildlife. The Federal Migratory Bird Law, passed in 1913, was the first migratory bird hunting regulation adopted. It was replaced by the Migratory Bird Treaty Act in 1918, which was based on the 1916 treaty with Great Britain for Canada. This was the beginning of federal responsibility for protection and conservation of migratory bird resources. The Black Bass Act (now under the Lacey Act) passed in 1916 regulated the interstate commerce on fish, with certain exceptions. In 1934, the Migratory Bird Hunting Stamp Act became effective; in 1937, the Migratory Bird Treaty Act with Mexico was proclaimed by the president. Bald eagles obtained federal protection in 1940. During the next thirty years the only major federal legislation in wildlife protection covered the golden eagle; then, in June of 1970, the Endangered Species Convention Act of 1969 prohibited importation of endangered species. The Airborne Hunting Act was enacted in 1971, and the Bald Eagle Protection Act, amended in 1972, provided higher penalties. A Migratory Bird Treaty Act with Japan was signed in 1972 and the one with Mexico was amended. The Migratory Bird Treaty Act with the Soviet Union entered into force on October 13, 1978. The Marine Mammal Protection Act became law in 1972. In December 1973, the Endangered Species Act became law and greatly expanded federal responsibility for endangered and

Cases 7–8

threatened species. In 1975, the Convention on International Trade in Endangered Species of Wild Fauna and Flora became effective.

Lacey Act of 1900 and Amendments of 1981.[7]   (P.L. 97–79, 95 Stat. 1073, 16 U.S.C. 3371–3378, approved November 16, 1981, and as amended by P.L. 100–653, 102 Stat. 3825, approved November 14, 1988, and P.L. 98–327, 98 Stat. 271, approved June 25, 1984). These amendments repealed the Black Bass Act and sections 43 and 44 of the Lacey Act of 1900 (18 U.S.C. 43 & 44), replacing them with a single comprehensive law.

Under this law, it is unlawful to import, export, sell, acquire, or purchase fish, wildlife, or plants taken, possessed, transported, or sold:

(1) in violation of U.S. or Indian law; or

(2) in interstate or foreign commerce involving any fish, wildlife, or plants taken, possessed, or sold in violation of state or foreign law.

The law covers all fish and wildlife and their parts or products, and plants protected by the Convention on International Trade in Endangered Species, and those protected by state law. Commercial guiding and outfitting are considered to be a sale under the provisions of the act.

Felony criminal sanctions are provided for violations involving imports or exports, or violations of a commercial nature in which the value of the wildlife is in excess of $350. A misdemeanor violation was established, with a fine of up to $10,000 and imprisonment of up to one year, or both. Civil penalties up to $10,000 were provided. However, the Criminal Fines Improvement Act of 1987 increased the fines under the Lacey Act for misdemeanors to a maximum of $100,000 for individuals and $200,000 for organizations. Maximum fines for felonies were increased to $250,000 for individuals and $500,000 for organizations.

Amendments to the humane shipment provisions of Title 18 required the Secretary of the Interior to issue regulations governing such activity. It also prohibits importation of wild vertebrates and other animals listed in the act or declared by the Secretary of the Interior to be injurious to man or agriculture, wildlife resources, or otherwise, except under certain circumstances and pursuant to regulations.

United States v. Alexander [8]

Defendants, Alaskan Native Americans, harvested over a half ton of herring roe in southeastern Alaska and sold it in Canada. Their permit had limited their catch to 444 pounds. Defendants were charged and convicted in the district court of violating the Lacey Act [16 U.S.C. Sec. 3372(a)(2)(A)].

To sustain a conviction under the Lacey Act, the government needed to prove the herring roe was taken in violation of state law. In this case, the pertinent state law reflects a federal statute, the Alaska National Interest Lands Conservation Act (ANILCA). ANILCA protects "the subsistence uses [of fish, and wildlife] on public lands by Native and nonNative rural residents" [16 U.S.C. Sec. 3111 (4)], gives priority to nonwasteful uses of fish and wildlife, and restricts subsistence uses only if fish and wildlife population viability is threatened. ANILCA also contains an opt-in clause for, and utilized by, the state of Alaska. Since Alaska adopted a law consistent with ANILCA, the state law did, and does, control.

In its findings, the court of appeals first found that the ANILCA term "customary trade" [16 U.S.C. Sec. 3113] includes some sales for cash. In this case, defendants showed that their tribal ancestors traditionally engaged in herring roe trade with other

Cases 9–12

tribes. Also, the state law's contradictions did not require striking down the Alaska law. Such ruling might do more harm than good in a regulation which is generally good with only minor discrepancies.

In similar challenges, a defendant only needs to defend the activity which is consistent with ANILCA (but contrary to state law) by motion. The court pointed out that this defense is only available to Alaskan Native Americans for whom such trade is customary and traditional, and in a manner consistent with a subsistence lifestyle.

United States v. Carpenter [9]

Defendants, owner/operators of a successful goldfish farm, were convicted in the district court of killing and acquiring migratory birds, thereby violating the Migratory Bird Treaty Act (MBTA) and the Lacey Act. The court of appeals reversed the conviction under the Lacey Act.

The court of appeals held that in order to violate the Lacey Act under 16 U.S.C. Sec. 3372(a)(1), a person must do something such as sell the pelt or feathers of wildlife that has already been taken or possessed in violation of law.

The court also found that the legislative intent of the Lacey Act is to outlaw interstate traffic in birds and other animals illegally killed in their state of origin; its purpose is to deal with a massive illegal trade in fish and wildlife and their parts and products. The Lacey Act did not intend to duplicate the MBTA by making the very act of killing a bird a crime because the bird fell on the property of the shooter.

United States v. Powers [10]

Defendant was licensed by the state of Idaho as an outfitter of certain animals. Although bears were not one of the enumerated animals on his license, defendant outfitted numerous bear hunting expeditions. Several of his clients were from outside of Idaho and their kills were, therefore, taken out of state upon departure. Defendant was convicted in the district court of violating the Lacey Act.

The court of appeals held that, within the facts of this case, outfitting constitutes "transport" within the meaning of the Lacey Act. Defendant violated Idaho law by providing outfitting services for which he was not licensed. The court felt there was enough evidence to support the conviction in that the defendant transported and assisted others in transporting parts of the bears in interstate commerce.

United States v. Williams [11]

Defendant, a Native American, was convicted of violating the Lacey Act [16 U.S.C. Sec. 3372(a)(2)(A) and 3373(d)(1)(B)] by selling moose meat to an undercover federal agent. On appeal, defendant argued that his conviction should be overturned because the government failed to establish the validity of the use of wildlife laws against a tribal member.

The court noted that a showing of conservation necessity is required only of state laws incorporated through the Lacey Act. On the other hand, there is no requirement of conservation necessity for establishing the validity of tribal wildlife laws incorporated through the Lacey Act. Tribal wildlife laws are per se valid against tribe members.

In effect, the government must establish the validity of the use of wildlife laws against tribe members, but similar laws enacted by the tribe can establish this validity.

United States v. Thomas [12]

Similarity in holdings with *Powers*.

Cases 13–16

United States v. Big Eagle[13]

Only important point is that the Lacey Act applies to Native Americans.

United States v. 594,464 Pounds of Salmon[14]

Only important point is that a regulation by the Taiwanese Board of Foreign Trade constitutes "foreign law" within the Lacey Act prohibition of fish or wildlife imports transported in violation of any law or regulation of any state or in violation of any foreign law [16 U.S.C. Sec. 3372(a)(2)(A)].

United States v. DeMasters[15]

(Compare with *Powers*)

Defendant, a New Mexico hunting outfitter, took clients on illegal hunts within defendant's leased property. The court of appeals held that the mere furnishing of guiding services on a wild animal hunt does not fall within the parameters of the "sale of wildlife" provisions of the Lacey Act [16 U.S.C. Sec. 3372(a)]. The court noted that the outfitter could not have transferred the wildlife since it was not under his control so as to make him a seller. Moreover, there was no guarantee that the guide would actually provide the animal which the hunter could successfully shoot and kill.

Note to Powers and DeMasters.

Recently added 19 U.S.C. Sec.

Migratory Bird Treaty Act of 1918.[16] (16 U.S.C. 703–712; Ch. 128; July 13, 1918; 40 tat. 755) as amended by:

Chapter 634; June 20, 1936; 49 Stat. 1556

P.L. 86–732; September 8, 1960; 74 Stat. 866

P.L. 91–578; October 17, 1968; 82 Stat. 1118

P.L. 93–135; December 5, 1969; 83 Stat. 282

P.L. 93–300; June 1, 1974; 88 Stat. 190

P.L. 95–616; November 8, 1978; 92 Stat. 3111

P.L. 99–645; November 10, 1986; 100 Stat. 3590

The original 1918 statute implemented the 1916 Convention between the United States and Great Britain (for Canada) for the protection of migratory birds. Specific provisions in the statute included:

Establishment of a federal prohibition, unless permitted by regulations, to "pursue, hunt, take, capture, kill, attempt to take, capture or kill, possess, offer for sale, sell, offer to purchase, purchase, deliver for shipment, ship, cause to be shipped, deliver for transportation, transport, cause to be transported, carry, or cause to be carried by any means whatever, receive for shipment, transportation or carriage, or export, at any time, or in any manner, any migratory bird, included in the terms of this Convention . . . for the protection of migratory birds . . . or any part, nest, or egg of any such bird" (16 U.S.C. 703).

This prohibition applies to birds included in the respective international conventions between the United States and Great Britain, the United States and Mexico, the United States and Japan, and the United States and the Soviet Union.

Authority for the Secretary of the Interior to determine, periodically, when, consistent with the conventions, "hunting, taking, capture, killing, possession, sale, purchase, shipment, transportation, carriage, or export of any . . . bird, or any part, nest or egg'' could be undertaken and to adopt regulations for this purpose. These determinations are to be made based on ''due regard to the zones of temperature and to the distribution, abundance, economic value, breeding habits, and times of migratory flight'' (16 U.S.C. 704).

A decree that domestic interstate and international transportation of migratory birds which are taken in violation of this law is unlawful, as well as importation of any migratory birds which are taken in violation of Canadian laws (16 U.S.C. 705).

Authority for Interior officials to enforce the provisions of this law, including seizure of birds illegally taken which can be forfeited to the United States and disposed of as directed by the courts (16 U.S.C. 706).

Establishment of fines for violation of this law, including misdemeanor charges (16 U.S.C. 707).

Authority for states to enact and implement laws or regulations to allow for greater protection of migratory birds, provided that such laws are consistent with the respective conventions and that open seasons do not extend beyond those established at the national level (16 U.S.C. 708).

Authority to take migratory birds exclusively for scientific or propagation purposes, pending the development of federal regulations, provided that the take does not violate state or local laws (16 U.S.C. 709).

A repeal of all laws inconsistent with the provisions of this Act.

Authority for the continued breeding and sale of migratory game birds on farms and preserves for the purpose of increasing the food supply (16 U.S.C. 711). The family corvidae (crows, ravens, jays) was added to the Migratory Bird Convention with Mexico on March 10, 1972. Crow hunting seasons may be established by federal regulation.

The 1960 statute (P.L. 86–732) amended the MBTA by altering earlier penalty provisions. The new provisions stipulated that violations of this act would constitute a misdemeanor and conviction would result in a fine of not more than $500 or imprisonment of not more than six months. Activities aimed at selling migratory birds in violation of this law would be subject to fine of not more than $2,000 and imprisonment could not exceed two years. Guilty offenses would constitute a felony. Equipment used for sale purchases was authorized to be seized and held, by the Secretary of the Interior, pending prosecution, and, upon conviction, be treated as a penalty.

Case 17

Migratory Bird Hunting and Conservation Stamp Act.[17]   (16 U.S.C. 718–718j, 48 Stat. 452), as amended.

The ''Duck Stamp Act'' as this March 16, 1934, authority is commonly called, requires each waterfowl hunter sixteen years of age or older to possess a valid federal hunting stamp. Receipts from the sale of the stamp are deposited in a special treasury account known as the Migratory Bird Conservation Fund and are not subject to appropriations.

Funds appropriated under the Wetlands Loan Act (16 U.S.C. 715k-3–715k-r; 75 Stat. 813), as amended, are merged with duck stamp receipts and provided to the secretary for the acquisition of migratory bird refuges under provisions of the Migratory Bird Conservation Act (16 U.S.C. 715 et seq.; 45 Stat. 1222), as amended, and since August 1, 1958, (P.L. 85–585; 72 Stat. 486) for acquisition of ''waterfowl production areas.''

The Postal Service prints, issues, and sells the stamp and is reimbursed for its expenses from money in the fund. Public Law 94–215, approved February 17, 1976, (90 Stat. 189) amended the act to allow, among other things, the sale of stamps at places other than post offices and authorized consignments to retail dealers. The 1976 amendment also changed the name of the stamp from ''Migratory Bird Hunting Stamp'' to ''Migratory Bird Hunting and Conservation Stamp.''

Cases 18–19

United States v. Van Fossan[18]

Defendant was asked by the city to eliminate numerous pigeons congregating in a vacant lot next to defendant's home. To do this, defendant attempted to shoot them. Finding this method less than efficient, defendant spread corn and wheat laced with strychnine on the property. In addition to killing pigeons, two common grackles and two mourning doves were poisoned. The court stated that although neither species is endangered, they are both migratory and their poisoning violates the Migratory Bird Treaty Act (MBTA).

Defendant contended that the fine of $450 and three years probation was excessive. The court responded that, for this level violation, fines range from $500 to $5,000, in addition to probation and/or up to six months imprisonment.

The court also reiterated the finding of other circuits that violation of the MBTA is a strict liability offense; that is, the person in violation of the MBTA need not have known that the victim was migratory or that his actions would put the bird(s) in peril.

United States v. Darst[19]

Defendant was charged with violating the MBTA by trapping great horned owls on two occasions in attempts to protect his poultry stock. The record noted that on the first violation, a Kansas conservation officer recommended that the defendant contact a federal game officer to acquire a trapping permit. The conservation officer observed a second trapped bird on his next visit to defendant's farm; at that time defendant stated that he had not contacted the federal game officer for the permit.

Defendant was found guilty of MBTA violations by a federal magistrate and fined $125 and assessed $25 for court costs. On appeal, defendant stated his arguments on three points:

   (a) designation of the owl as a migratory bird;

   (b) unconstitutional overbreadth of the statute;

   (c) unconstitutional vagueness of the statutory migratory bird term and the excessive breadth of the regulation, 50 C.F.R. Sec. 10.13, listing birds designated as migratory.

The district court first held that the statute is not unconstitutionally vague. Section 703 begins with the language, "[u]nless and except as provided *by regulations* made as hereinafter provided, ..." (emphasis added). It is, therefore, reasonable to infer that a person of ordinary intelligence would know what conduct is prohibited.

The court next asserted that the Secretary of Interior's determinations of "migratory birds" cannot be collaterally attacked "as long as the regulation promulgated to that end is facially valid." *Darst* at 287. Defendant offered no evidence as to the designation's impropriety.

Finally, the court found that there is no authority for defendant's argument of an absolute (or otherwise) federal constitutional right to defend one's property from federally protected wildlife. As the court mentions, a recent Ninth Circuit Court of Appeals decision, *Christy v. Hodel,* 857 F.2d 1324 (9th Cir. 1988), has held that the U.S. Constitution does not expressly or implicitly recognize a right to kill federally protected wildlife in defense of property. Moreover, the regulations allow landowner interest to be balanced against the public interest to protect wildlife. Assistance of federal officials is necessary for this caveat. Defendant failed to follow guidelines in not contacting the federal officer for a permit.

Cases 20–21

United States v. Manning[20]

Defendant managed a commercial goose-hunting operation where, on two occasions, corn was found scattered on a field at which hunters were shooting Canada geese. 50 C.F.R. Sec. 20.21 provides for hunting migratory game birds in open seasons. An exception to this provision is that birds may not be lured (baited) for the purpose of hunting them. Defendant was charged with aiding and abetting the taking of migratory birds in violation of the MBTA and related regulations.

Defendant first asserted that the evidence against him was insufficient for a conviction. The court stated that the federal official in charge of the investigation was afforded great deference—''the benefit of all reasonable inference''—in relation to the evidentiary facts of the case. According to the factual record in the decision, it is quite likely that the defendant was attempting to make frivolous challenges to clear evidence showing blatant violations of the baiting prohibition.

Defendant next contended that the federal agent had said in a public meeting that migratory birds could be hunted over a field from which there were ''bona fide agricultural operations or proceedings.'' According to the defendant, he relied on these statements, ignoring the exception that **waterfowl** could not be hunted over such an area; therefore, according to defendant, the government was estopped from prosecution in this case. The court first found that there were no agricultural operations on this field, bona fide or otherwise. Next, it held that the government cannot be estopped absent a showing of ''affirmative misconduct.'' That is, the official must have purposefully misled. This was not shown. Furthermore, the record reflected that the official had made it clear to the defendant that the spreading of corn on the field was considered baiting. As a result, the defendant could not have reasonably relied on the ''misinformation.''

Defendant's final defense was that the regulation was unconstitutionally vague on both its face and as applied. This assertion was in reference to the unpublished ''zone of influence'' determination made by the federal agent. Defendant asserted that determining whether the baited area was a source of attraction for the hunters' advantage was a judgment call, and, therefore, vague. The court found that authority clearly held that bait placed anywhere, where it is an effective attraction or lure for hunting purposes, no matter the distance, is baiting. The extent of such an area will vary depending on many factors such as wind and terrain. The possibility of marginal cases is not impermissibly ambiguous for criminal definitions when the term is as exact as the subject matter permits.

Bald Eagle Protection Act of 1940.[21]   (16 U.S.C. 668–668d, 54 Stat. 250) as amended.

Approved June 8, 1940, and amended by:

P.L. 86–70; June 25, 1959; 73 Stat. 143

P.L. 87–884; October 24, 1962; 76 Stat. 1346

P.L. 92–535; October 23, 1972; 86 Stat. 1064

P.L. 95–616; November 8, 1978; 92 Stat. 3114

This law provides for the protection of the bald eagle (the national emblem) and the golden eagle by prohibiting, except under certain specified conditions, the taking, possession, and commerce of such birds. The 1972 amendments increased penalties for violating provisions of the act or regulations issued pursuant thereto and strengthened other enforcement measures. Rewards are provided for information leading to arrest and conviction for violation of the act.

Cases 22–24

The 1978 amendment authorizes the Secretary of the Interior to permit the taking of golden eagle nests that interfere with resource development or recovery operations. (See also the Migratory Bird Treaty Act and the Endangered Species Act.) Prosecution of eagle violations may occur under the Eagle Act, the Migratory Bird Treaty Act, and, in some instances, under the Lacey Act Amendments of 1981. Consequently, the penalties may vary according to the statute utilized for prosecution.

Andrus v. Allard [22]

Plaintiffs, dealers in Native American artifacts, some of which are composed of eagle parts, claim that the BEPA and MBTA do not apply to parts of eagles killed in a legal manner before the effective date of the act.

The Supreme Court held that both acts were promulgated without regard to when the protected birds were actually taken. Moreover, the BEPA's language is clear, and its "possession or transportation" provisions are the only exceptions. The acts, therefore, do include a ban on the sale of preexisting artifacts.

The Court then stated that no Fifth Amendment property rights violations occurred through the acts in this case. The regulations involved here do not require the surrender of artifacts. Plaintiffs' inventory included other artifacts not regulated by the acts. There was no "taking" therefore under the Fifth Amendment, only a reduction in value.

United States v. Dion [23]

Defendant, Native American, was charged with shooting bald and golden eagles on the reservation and selling eagle parts in violation of the ESA and the BEPA. The Supreme Court held that, even though the ESA is silent as to Native American hunting rights, the BEPA abrogated defendant's right to assert a treaty right defense to the ESA. Pursuant to 16 U.S.C. Sec. 668a, Native Americans may only hunt eagles for religious purposes when issued a permit.

Defenders of Wildlife v. Administrators, EPA [24]

Wildlife and environmental organizations brought suit against the Administrator of the EPA and the Secretary of the Interior, challenging the continued registration of strychnine pesticides and rodenticide for above-ground uses. The court of appeals held that:

(1) The ESA imposes substantial and continuing obligations on federal agencies. Even though an agency may be acting under a different statute, that agency must comply with the ESA.

(2) Plaintiffs could maintain a suit under the citizen suit provision of the ESA giving citizens the right to enjoin violation of the act [16 U.S.C. Sec. 1540(g)(1)(A)].

(3) The EPA's strychnine registrations had a prohibited impact on endangered species. The relationship between the registration decision and the deaths of endangered species is clear. The EPA's registration of strychnine constitutes "takings" of endangered species.

(4) An agency must obtain an incidental taking statement before it takes the protected species [16 U.S.C. Sec. 1536(b)(4), (o)(2)]. A statement does not retroactively excuse the takings that occurred before the secretary issued the statement.

(5) Unlike the ESA, the BEPA nor the MBTA contain provisions for private rights of action. If plaintiffs believe the EPA is not acting in accordance with the law by violating bird acts, defenders could petition the EPA. If the EPA refused, plaintiffs could obtain judicial review in the district court as provided by the Federal Insecticide, Fungicide, and Rodenticide Act.

Case 25

Marine Mammal Protection Act of 1972.[25]   16 U.S.C. 1361–1407, P.L. 92–522, October 21, 1972, 86 Stat. 1027 as amended by:

P.L. 94–265; April 13, 1976; 90 Stat. 360

P.L. 95–316; July 10, 1978; 92 Stat. 380

P.L. 97–58; October 9, 1981; 95 Stat. 979

P.L. 98–364; July 17, 1984; 98 Stat. 440

P.L. 99–659; November 14, 1986; 100 Stat. 3706

P.L. 100–711; November 23, 1988;

P.L. 101–627; November 28, 1990; 100 Stat. 4467.

The 1972 Marine Mammal Protection Act established a federal responsibility to conserve marine mammals with management vested in the Department of Interior for sea otter, walrus, polar bear, dugong, and manatee. The Department of Commerce is responsible for cetaceans and pinnipeds, other than the walrus.

With certain specified exceptions, the act establishes a moratorium on the taking and importation of marine mammals, as well as products taken from them, and establishes procedures for waiving the moratorium and transferring management responsibility to the states.

The law authorized the establishment of a Marine Mammal Commission with specific advisory and research duties. Annual reports to Congress by the departments of Interior and Commerce and the Marine Mammal Commission are mandated.

The 1972 law exempted Indians, Aleut, and Eskimos (who dwell on the coast of the North Pacific Ocean) from the moratorium on taking, provided that taking was conducted for the sake of subsistence or for the purpose of creating and selling authentic native articles of handicraft and clothing. In addition, the law stipulated conditions under which the secretaries of Commerce and Interior could issue permits to take marine mammals for the sake of public display and scientific research.

The 1976 amendments (P.L. 94–265) clarified the offshore jurisdiction of the statute as the two hundred-mile Exclusive Economic Zone.

The 1978 amendments (P.L. 95–316) extended the original five-year authorization through fiscal year 1981.

Amendments enacted in 1981 (P.L. 97–58) established conditions for permits to be granted to take marine mammals ''incidentally'' in the course of commercial fishing. In addition, the amendments provided additional conditions and procedures for transferring management authority to the states, and authorized appropriations through fiscal year 1984.

The 1984 amendments (P.L. 98–364) established conditions to be satisfied as a basis for importing fish and fish products from nations engaged in harvesting yellowfin tuna with purse seines and other commercial fishing technology as well as authorized appropriations for agency activities through fiscal year 1988.

The 1986 amendments (P.L. 99–659) amended section 101 of the original statute to allow the incidental take of depleted marine mammals in activities other than commercial fishing, provided that such take does not result in an unmitigable impact on subsistence harvest.

Several additional amendments were enacted in 1988 (P.L. 100–711). In addition to reauthorized activities through fiscal year 1993, the amendments established a process for commercial fishermen to obtain an exemption from the moratorium on incidental take of marine mammals for a five-year period. The Department of Commerce

Cases 26–29

is authorized to conduct specific related tasks including the granting of exemptions, providing for observer coverage, and collecting data on the extent of incidental take.

The Secretary of Commerce is required to consult with the Secretary of Interior prior to taking actions related to species for which the Interior Department has jurisdiction. The California sea otter is explicitly excluded from the exemption process.

Earth Island Institute v. Mosbacher[26]

Several environmental groups were granted an injunction against the Secretary of Commerce and the National Marine Fisheries Service (NMFS) in connection with bluefin tuna importation from Mexico.

The district court found, and the court of appeals agreed, that tuna embargo reconsideration must be based on a full year's data on compliance of incidental spinner dolphin takings, and not on six months' data. [See 16 U.S.C. 1371(a)(2)(B)(ii)(III) and 50 C.F.R. Sec. 216.24(e)(5)(iv).] The court rejected the agency's argument that the six-month interpretation would provide an incentive for foreign governments to comply with the statute faster. The court felt that, to the contrary, this interpretation allows foreign countries and the NMFS to withhold data until favorable statistics are available. This would result, as in the case with Mexico, in one-day embargoes. The court was also swayed by the long history of the NMFS nonenforcement of congressional directives during the years preceding the 1988 amendments.

United States v. Clark[27]

In *Clark,* defendant, Alaskan Native American and his family, hunted and killed nine walruses. Defendant completely butchered one walrus and removed certain parts from the rest (tusks, flippers, whiskers, oosik). Because of imminently dangerous tidal conditions, defendant left two carcasses on the beach and lost two in tow.

An exemption in the MMPA applicable to Alaskan Native Americans is intended to protect subsistence hunting and use of animal parts for limited cash economy, so long as neither was wasteful [16 U.S.C. Sec. 1371, 1371(b), (b)(2)]. The district court jury found defendant's actions to be wasteful; the court of appeals affirmed.

The court also held that the regulation promulgated under the MMPA prohibiting Alaskan Native Americans from wasting a ''substantial portion'' of walrus carcass (50 C.F.R. Sec. 18.3), was not void for vagueness. Traditionally, Native American Alaskans would also commonly use the blubber, skin, internal organs, and sometimes the meat. None of these were harvested from eight of the nine walruses.

Kokechik Fishermen's Associations v. Secretary of Commerce[28]

Defendant was enjoined from issuing permit allowing foreign commercial fishermen in United States conservation waters. The court of appeals held that the Secretary of Commerce could not issue the permit under MMPA without first determining the extent to which northern fur seals would be affected by fishing operations. The court stated that it is the secretary's duty to take a systematic view of an activity's effect on marine mammals, and may not allow—subject to civil penalty price—illegal takings of other protected marine mammals. The secretary chose to disregard incidental takings as ''negligible''—an undefined and ambiguous standard which is not an exception provided for under the MMPA when the takings are a certainty instead of merely a possibility.

Jones v. Gordon[29]

Plaintiffs brought action against the National Marine Fisheries Service's (NMFS) permit issuance to zoological parks allowing the capture of killer whales for scientific study and public display. The court of appeals found that it was unreasonable for the NMFS to not issue an environmental impact statement.

The court did not agree with defendant's argument that issuance of the permit under the National Environmental Policy Act (NEPA) made preparation of an environmental impact statement impractical because of the time restriction for agency action under the MMPA. The court found that permit issuance and compliance with both NEPA and MMPA was possible. The court stressed that the NMFS did not possess a blanket exemption from statement preparation.

The court of appeals agreed with the district court's decision that even though the issuance was within the categorical exclusion of permits for scientific research and public display, the action also fell within one of the exclusion's exceptions in that it was the subject of public controversy, unknown environmental impacts, and unknown/unique risks. As such, defendant was required to provide a reasoned explanation of its decision of whether to prepare an environmental impact statement.

Wild Horses and Burros Act. (16 U.S.C. 1331–1340). The Wild, Free-roaming Horses and Burros Act of December 15, 1971 (85 Stat. 649) was designed to protect all unbranded and unclaimed horses and burros on public lands of the United States (P.L. 92–195). These animals were to be free from capture, branding, harassment, or death. The act also provides that all horses and burros on the public lands administered by the Secretary of the Interior through the Bureau of Land Management and the National Park Service or by the Secretary of Agriculture, through the Forest Service, are to be committed to the jurisdiction of the respective secretaries who are directed to protect and manage the animals as components of the public lands. The law further states they are managed in a manner that is designed to achieve and maintain a thriving natural ecological balance on the public lands. Horses or burros protected under this act which stray from public lands into privately owned lands are protected. However, the owners can call federal marshals or an agent of the secretary who shall arrange to have the animals removed.

Case 30

Endangered Species Act of 1973.[30] (16 U.S.C. 1531–1544, 87 Stat. 884) as amended.

Public Law 93–205, approved December 28, 1973, repealed the Endangered Species Conservation Act of December 5, 1969 (P.L. 91–135, 83 Stat. 275). The 1969 Act had amended the Endangered Species Preservation Act of October 15, 1966 (P.L. 89–669, 80 Stat. 926).

The 1973 Act implemented the Convention on International Trade in Endangered Species of Wild Fauna and Flora (T.I.A.S. 8249) signed by the United States on March 3, 1973, and the Convention on Nature Protection and Wildlife Preservation in the Western Hemisphere (50 Stat. 1353), signed by the United States on October 12, 1940.

The 1973 Endangered Species Act provided for the conservation of ecosystems upon which threatened and endangered species of fish, wildlife, and plants depend, both through federal action and by encouraging the establishment of state programs. The act:

Authorizes the determination and listing of species as endangered and threatened.

Prohibits unauthorized taking, possession, sale, and transport of endangered species.

Provides authority to acquire land for the conservation of listed species, using land and water conservation funds.

Authorizes establishment of cooperative agreements and grants-in-aid to states that establish and maintain active and adequate programs for endangered and threatened wildlife and plants.

Authorizes the assessment of civil and criminal penalties for violating the act or regulations.

Authorizes the payment of rewards to anyone furnishing information leading to arrest and conviction for any violation of the act of any regulation issued thereunder.

Section 7 of the Endangered Species Act requires federal agencies to ensure that any action authorized, funded, or carried out by them is not likely to jeopardize the continued existence of listed species or modify their critical habitat.

In 1978, a cabinet-level Endangered Species Committee was established as part of a two-tiered process whereby federal agencies may obtain exemptions from the requirements of Section 7. The Tellico Dam project in Tennessee and the Grayrocks project in Wyoming were to receive expedited consideration by the committee.

The Secretary of Defense is authorized to specify exemptions from the act for reasons of national security. The consultation process under Section 7 was formalized and strengthened, and now includes the requirement that federal agencies prepare biological assessments in cases where the Secretary of the Interior has advised that a listed species may be present.

The 1978 amendments also oblige the secretary to consider the economic impact of designating critical habitat, and to review the list of endangered and threatened species every five years. Public notification and hearing requirements, prior to the listing of a species or its habitat, are specified.

Public Law 100–478, enacted October 7, 1988, (102 Stat. 2306) included in part:

Provides equal opportunity to departments of Interior and Agriculture for enforcing restrictions on import/export of listed plants.

Requires the Secretary of the Interior to monitor all petitioned species that are candidates for listing and specifies emergency listing authority.

Directs the Secretary of the Interior to develop and review recovery plans for listed species without showing preference for any taxonomic group.

Requires Administrator of the Environmental Protection Agency, in cooperation with secretaries of Interior and Agriculture, to conduct a study for identifying reasonable and prudent means to implement endangered species pesticide labeling program, and to report to Congress one year after enactment of this act.

Case 31

Lujan v. Defenders of Wildlife [31]

Plaintiff, environmental groups, filed an action against the Secretary of the Interior claiming that a ruling subsequent to the promulgation of ESA Section 7(a)(2) limited the section's geographic scope to the United States and the high seas. Plaintiffs sought an injunction requiring the secretary to promulgate a new rule restoring his initial interpretation that the geographical area of ESA section 7(a)(2) include all foreign nations. The court of appeals reversed the district court's dismissal of the suit for lack of standing. Upon remand, the district court granted plaintiffs' motion for summary judgment. The court of appeals affirmed. The U.S. Supreme Court reversed and remanded the case.

The Court found that the plaintiffs did not suffer an injury in fact, i.e., a concrete and particularized, actual, or imminent invasion of legally protected interests. Justice Scalia stated that even if the funded activities abroad threaten certain species, plaintiffs failed to show that one or more of their members would thereby be directly affected, apart from the members' special interest in the subject. An intent to visit project sites at some indefinite future time at which they might not be able to observe certain endangered species is not an "imminent" injury.

Cases 32–35

Mt. Graham Red Squirrel v. Madigan[32]

Environmental organizations sued various United States government officials and the State of Arizona Board of Regents seeking to preclude further construction of a telescope on Mt. Graham due to alleged impact of construction on endangered species. The court of appeals held the plaintiffs were not entitled to halt construction.

Plaintiffs asserted that pursuant to ESA Section 7, prior to engaging in any construction, the Forest Service was required to reinitiate formal consultation with the Fish and Wildlife Service, which concludes with the issuance of a biological opinion (50 CFR 402.16). The court found that under the Arizona-Idaho Conservation Act (Sec. 601 et seq.) (AICA), the entire first phase of the project was exempt from ESA Section 7. The court also held that the AICA precluded the courts from halting construction due to plaintiffs' claim that the Forest Service violated the National Forest Management Act by failing to maintain a minimum viable red squirrel population.

Finally, the court held that plaintiffs were estopped from relying on the claim that the Fish and Wildlife Service violated the ESA by failing to designate a critical habitat for the red squirrel. The court considered this argument moot because Congress had adopted a biological opinion and enacted it into law.

United States v. Ivey[33]

Defendants were convicted of smuggling and of conspiracy to buy bobcat hides brought into the United States from Mexico. The court of appeals held that for purposes of the ESA, endangered species were those appearing on the endangered species list at the time of the offense, not just those at the time of the list's original creation.

The court also stated that since Mexico was a nonsignatory to the Convention on International Trade in Endangered Species of Wild Fauna and Flora, defendants were required to obtain permits from the Mexican government before bringing into the United States hides of species which are on the endangered species list.

Finally, the court found that a violation under the ESA was not a specific intent crime; a showing of an intent to commit the crime was not required for a conviction under the ESA.

Sierra Club v. Yeutter[34]

Environmental groups brought action challenging the Forest Service timber management activities that allegedly had an adverse impact on an endangered species, the red-cockaded woodpecker (RCW). The district court enjoined the violations of the ESA.

On appeal, the court of appeals found that the even-aged management practices violated the ESA. The service's failure to follow RCW wildlife management provisions of its handbook constituted a "taking" of the RCW under ESA Section 9.

The court also held that, although the district court may enjoin a federal agency from continuing activity in violation of environmental protection laws and dictate actions the agency must take until it has submitted an acceptable plan, there are limitations; the court cannot infringe on the agency's consultation with the Fish and Wildlife Service by dictating the principal feature of all significant steps the agency must take with regard to the species.

United States v. Nguyen[35]

Defendant was convicted in the district court for violating the ESA by illegally possessing and importing a threatened species of sea turtle. Only important point in this case is that ESA is not a specific intent crime, it is a general intent crime. No showing was required to show that defendant knew the turtle was a threatened species or that

Cases 36–38

any of his actions were violations of law. It was sufficient that defendant knew he was in possession of the turtle.

Section 3372(c) makes guides and outfitters, among others, sellers within the meaning of the Lacey Act.

City of Las Vegas v. Lujan[36]

Nevada Development Authority and private real estate developers brought action to challenge Department of Interior's regulation which listed Mojave Desert tortoises as endangered species.

The court of appeals found that relocation of an endangered species, or alteration of its habitat during construction, constitutes "incidental taking" prohibited under the ESA, unless the secretary grants a special permit.

Also, the court stated that the quantum of data the secretary must possess to issue an emergency regulation need not rise to the level required for normal rulemaking. In effect, the court felt that the level did not need to reach "substantial evidence." The secretary may not, however, disregard available scientific evidence that is in some way better than the evidence he relies on. If such evidence arises within the 240-day emergency regulation period contradicting his previous findings, it is the secretary's duty to withdraw the regulation.

The potential spread of respiratory disease to the Nevada turtles from the Mojave population was sufficient to warrant emergency listing of the former. Likewise, for the emergency regulation, the secretary is not required to show that his actions will stave off spread of the disease or even address the specific emergency that led to the listing.

FEDERAL VERSUS STATE RESPONSIBILITIES

Even though, by reason of the constitutional provision, the major areas of federal and state authority over wildlife control seem to be well defined, there are several areas wherein exist differences of opinion as to the proper division of responsibility for wildlife control. The views of Connery (1935) respective to the issues involved may be summarized as follows:

1. It is easy to say let either the states or the federal government handle all game matters, but this cannot be done;

2. States have no international rights by which to protect birds, mammals, or fishes outside their respective boundaries;

3. States cannot dictate rules and regulations on federal properties, and game has no priority over timber, oil, minerals, and recreation in being singled out for special consideration;

4. Federal influence should be primarily promotional, with regulations secondary;

5. The states should be chiefly regulatory, but have strong promotional and educational programs to supplement the federal developments.

These statements are indicative of the trends that are actually taking place in the area of wildlife management. There are, however, notable exceptions with respect to division of responsibility.

Wildlife on National Forests[37]

Considerable difference of opinion exists, for example, as to the proper division of authority over wildlife on national forests. The legislation providing for the National Forest Administration Act of June 4, 1897, is based on the power of Congress "to dispose of and make all needful Rules and Regulations respecting the Territory or other Property belonging to the United States. . . ."[38]

Cases 39–45

Legal rulings based on this constitutional provision have been rendered by the various federal courts. With regard to the power held by Congress, Judge Sibley states:

> . . . but the Congress still has the power under the supreme law of the land, to make such regulations as are needful, Congress being the judge of what is needful. It is probable that it is the exclusive judge of what is needful. Certainly any regulation looking to the use or disposal or the safety of the property is needful if Congress so conceives it. It is well settled that Congress, in making regulations, may not only deal with them itself, but may, after providing a general scheme, delegate the details to some officer or commission.[39]

In *Shannon v. United States*[40] the Circuit Court of Appeals for the Ninth Circuit said:

> The Federal Constitution delegates to Congress the general power absolutely without limitation to dispose of and make all needful rules and regulations concerning the public domain, and this independent of the locality of the public land, whether it be situated in a State or in a territory . . . the exercise of which power cannot be restricted or embarrassed in any degree by said legislation. . . . The rights given by the State statutes to the subjects of the State extend only to the lands of the State. They end at the borders of the Government lands. At that border the laws of the United States intervene, and it is within their province to forbid trespass. Such laws being within the power of Congress, it is not necessary to discuss the question whether it is sovereign power or police power, or what be its nature, for there is no power vested in the State which can embarrass or interfere with its exercise.

The Supreme Court of the United States, in referring to this constitutional provision in *Kansas v. Colorado,*[41] held: ''The full scope of this paragraph has never been definitely settled. Primarily, at least, it is a grant of power to the United States of control over its property.'' That is implied by the words 'Territory or other property.' This control over its property gives the government the rights of a private ownership. In *United States v. Light,*[42] the Court maintained: ''The United States can prohibit absolutely or fix the terms on which its property may be used. As it can withhold or reserve the land, it can do so indefinitely.''

In *Utah Power and Light Co. v. United States,*[43] the Supreme Court held:

> And so we are of the opinion that the inclusion within a State of lands of the United States does not take from Congress the power to control their occupancy and use, to protect them from trespass and injury and to prescribe the conditions upon which others may obtain rights in them, even though this may involve the exercise in some measure of what commonly is known as the police power. ''A different rule,'' as was said in *Camfield v. United States,*[44] ''would place the public domain of the United States completely at the mercy of the State legislation.''

Wildlife on the national forests is a resource of the land and waters as are forage and timber. Federal control may, therefore, extend to all phases of wildlife regulation. This control is provided for by the National Forest Administration Act, in defining the duties and powers of the Secretary of Agriculture.[45] The act provides that he may make such rulings and regulations and establish such services as will insure the objectives of such reservations, namely, to regulate their occupancy and use, and to preserve the forest from destruction.

It is believed that the regulation by the Secretary of Agriculture of hunting upon the national forests is permissible as a means of fire prevention and for the protection of forest growth against damage. While the secretary has the authority in periods of extreme fire danger to entirely close national forest areas to all use, nevertheless, it is recognized by the U.S. Forest Service that such action is rather drastic and should be avoided, if possible. It sometimes happens that as a result of protective measures, the

Cases 46–53

game increases to a point where, for the protection of the forest growth and forage and to prevent soil erosion, it is necessary to drastically reduce the game.

In *Hunt v. United States,*[46] the Supreme Court upheld the right of the Secretary of Agriculture to kill large numbers of deer on the Kaibab National Forest and Grand Canyon National Game Preserve since it was shown that such action was necessary to protect the national forest and the game preserve, and also the animals themselves. The deer had increased to such large numbers that the forage was insufficient for their sustenance, with the result that they had greatly injured the lands by overbrowsing— killing the young trees, shrubs, bushes, and forage plants, while at the same time, thousands of deer had died of starvation.

The Court said that the power of the United States, therefore, to protect its property did not admit of doubt. The lower court states in its decree that it should not be construed to permit the licensing of hunters to kill deer within the reserves in violation of the state game laws (19 F. 2d, 624). The Supreme Court said that while the Solicitor General of the United States did not concede the authority of the Court to impose this limitation upon the secretary's authority, he was content to let the decree stand. The Court, therefore, accepted the opinion and decree of the lower court, with the modification that all carcasses of deer and parts thereof shipped outside the boundaries of the reserve should be plainly marked with tags or otherwise, in such manner as the Secretary of Agriculture might provide, to show that they were killed under his authority within the limits of the reserves.

Following the Kaibab decision, the Circuit of Appeals for the Fourth Circuit[47] held that the Secretary of Agriculture could cause surplus deer on the Pisgah National Forest and Game Preserve in North Carolina to be killed for the protection of the lands of the United States without regard to state law. The lands involved in this case were purchased by the government under the Weeks Law.[48] The state, by legislative act of 1915,[49] gave its consent for the federal government to make all such rules and regulations as it should deem necessary with respect to game animals.

Congress, in 1916,[50] authorized the president to establish preserves for the protection of game animals, birds, or fish; and prohibited, under penalty, the hunting, catching, trapping or killing, or the willful disturbance of any kind of game animal, bird, or fish, except under such rules and regulations as the Secretary of Agriculture may, from time to time, prescribe.

The Court ruled that this constituted an acceptance by Congress of the state cession of jurisdiction. It also held, however, that without regard to this cession and acceptance of jurisdiction, the secretary had authority to take the action which he did.

Judge Luse of the U.S. District Court for Wisconsin said that it was within the power of Congress to regulate fishing and the hunting of wildlife, other than migratory birds, as an incident to the acquisition and control of the land for the purpose of the Migratory Bird Treaty. So here the Secretary of Agriculture, having the broad power to regulate the occupancy and use of the national forests for all purposes, certainly can regulate hunting.[51]

There appears to be an implied license to graze on the public lands, which was the basis of many private fortunes.[52] But the Supreme Court did not allow this to interfere with civil and criminal prosecutions for grazing on the national forests contrary to the regulations. In *United States v. Light,*[53] the Court said:

> There thus grew up a sort of implied license that these lands, thus left open, might be used so long as the government did not cancel its tacit consent (*Buford v. Houtz,* 133 U.S. 320). Its failure to object, however, did not confer any vested right on the complainant, nor did it

Cases 54–56

deprive the United States of the power of recalling any implied license under which the land had been used for private purposes (*Steele v. United States,* 113 U.S. 130; *Wilcox v. Jackson* 13 Pet. 498, 513).

And in the *Grimaud* case,[54] which was a criminal action for grazing sheep on a forest contrary to the regulations, the Court held:

> Thus the implied license under which the United States had suffered its public domain to be used as a pasture for sheep and cattle, mentioned in *Buford v. Houtz,* 133 U.S. 320, was curtailed and qualified by Congress, to the extent that such privilege should not be exercised in contravention of the rules and regulations (*Wilcox v. Jackson,* 13 Pet. 498, 513).
>
> If, after the passage of the act and the promulgation of the rule, the defendants drove and grazed their sheep upon the reserve, in violation of the regulations, they were making an unlawful use of the government's property. In doing so, they thereby made themselves liable to the penalty imposed by Congress.[55]

A special act means nothing more than that a particular situation has been called specifically to the attention of Congress, and that Congress has, therefore, acted on that situation. There are special acts relating to some other phases of national forest regulation, but this does not mean that the Secretary of Agriculture cannot exert, as to other similar situations, the full extent of the regulatory power conferred. For example, section 55 of the Penal Code makes it a criminal offense to trespass on what was formerly known as Bull Run National Forest, but is now a part of the Oregon National Forest and the watershed of the city of Portland. That does not prevent the secretary from taking similar action for the protection of watersheds of other states, as he has done in many instances.

In the light of this discussion it would seem that the federal government is within its rights when, in the paramount interests of the people of the United States, it attempts to control wildlife on national forests. The present policy of the federal government appears to be that of allowing the states to control harvests on public lands where there is an open season.

Wildlife on National Parks

A court case that evoked considerable interest and tension in several states involved the killing of deer for an ecological study without a state permit within Carlsbad National Park.

The issue was essentially this: the New Mexico Game Commission was advised by the park superintendent that a deer range ecology study was to be undertaken and it would require the killing of not more than fifty deer. The only objection by the state to the program was that park officials had refused to apply for state permit which had been offered. The park officials, although admitting they were claiming no present serious depredation of browse throughout the park, claimed a right of research to determine the future effect upon its vegetation. They, therefore, refused to request a permit to kill deer within the national park.[56]

The basic question was whether the supervisory powers granted the Secretary of the Interior over the management of the national parks authorized the killing of deer within Carlsbad Caverns National Park for ecological studies to determine deer range conditions within the park. The federal district court concluded that the deer-killing program was not within the authority granted to the secretary or his subordinates and enjoined further killing without a state permit. An appeal was taken to the U.S. Court of Appeals, Tenth Circuit.

This court noted that the United States has no ownership in wild animals within the various states. (It should be noted here that the several states which had joined the case as friends of the court, and in support of the position taken by New Mexico, were concerned as to whether federal authorities were attempting to take over the control and management of wildlife on federal lands.) The secretary did not dispute the court's ruling that the United States has no ownership in wildlife. Court decisions were cited indicating that insofar as wild animals within a state are capable of ownership, they are owned by the state in its sovereign capacity.

The state recognized that the secretary, as determined by a prior court case, has the authority to direct the destruction of a large number of deer when such action is necessary to protect the lands of the United States from injury. The state contended that, in this case, the situation was different since depredation had not yet occurred.

The court determined that authority had been granted by federal statute to the secretary to promote and regulate the national parks so as to conserve the scenery and wildlife therein ''in such manner and by such means as will leave them unimpaired for the enjoyment of future generations.'' The court said: ''anything detrimental to this purpose is detrimental to the park.'' The court further stated that the secretary was authorized to destroy such animals that may be detrimental to any park and stated that the obvious purpose of the language is to require the secretary, when it is necessary, to destroy animals. The court further indicated that the secretary need not wait until the damage through overbrowsing has taken its toll on park plant life and deer herd before taking preventative action.

Previously, it had been pointed out that the depth of the question and the concern of various states was caused by the deliberate refusal of the secretary to apply for the deer-killing permits. This, the states felt, was in furtherance of a scheme of federal authorities to eventually take over the control and management of the wildlife on all federal lands. The court answered this by stating that, if there were a real threat, the remedy was not in the courts.

Federal Regulation of Migratory Birds

Another area of friction between the federal government and the states concerns the jurisdiction over wildlife that is migratory. A waterfowl management treaty, signed by Canada and the United States (with the approval of Great Britain) in 1916, and by Mexico and the United States in 1936, forms the legal basis for the federal control over these birds.

Case 57

After the Canada treaty was signed by the president, it was contested on the grounds that it invaded states' rights. The Supreme Court decision, written by Justice Holmes,[57] sustained the Enabling Act of Congress. Under this act, waterfowl and certain other migratory birds were declared to be under the jurisdiction of the federal government. The law appointed the U.S. Biological Survey as the government agency which should administer the law. Justice Holmes, who gave the decision, emphasized the following legal features:

1. The national government has been delegated powers to make treaties; so this did not contradict the Tenth Amendment;

2. That acts of Congress are the supreme law of the land when made in pursuance of the Constitution;

3. That treaties are likewise when made under authority of the United States;

4. That the treaty did not contravene any prohibitory clause in the Constitution;

5. That it dealt with the subject recognized by international custom as the proper one for treaties;

6. That, therefore, it was valid and the law carrying it into effect was constitutional;

7. That it was recognized that the state in its sovereign capacity owned animals *ferae naturae;*

Cases 58–62

8. That also the states' title exceeded an individual's;

9. A state could never prevent the national government from exercising a right to make treaties regulating this subject.

In *Cochrane v. United States,*[58] the Circuit Court of Appeals for the Seventh Circuit said:

> It is impossible to deal with migratory birds on the basis of their being the property of the state or of the state's having an exclusive property interest in them. Ordinarily, an attribute of property and of property rights is possession. Moreover, it is equally impossible even to identify the birds after they have passed from one state to another and returned to the state where they were hatched. It is unbelievable that the framers of the Constitution intended to leave this form of valuable property, which did not vest in the individual and which could not be controlled by the state, unprotected and fated to total destruction. It is not a matter of sentiment, but of common sense.

Further exemplification of the power of Congress is found in *Missouri v. Holland,*[59] in which the court sustained the act of Congress for the provisions of the Migratory Bird Treaty against an attach upon its constitutionality on the ground that it was an invasion of the sovereignty of the state. There the court said:

> The state, as we have intimated, founded its claim of exclusive authority upon an assertion of title to migratory birds, an assertion that is embodied in statute. No doubt it is true that as between a state and its inhabitants the state may regulate the killing and sale of such birds, but it does not follow that its authority is exclusively of paramount powers.
>
> As most of the laws of the United States are carried out within the states and as many of them deal with matters which in the silence of such laws the state might regulate, such general grounds are not enough to support Missouri's claims. Valid treaties, of course, "are as binding within the territorial limits of the states as they are elsewhere throughout the dominion of the United States."[60] No doubt the great body of private relations usually falls within the control of the state, but a treaty may override its power.
>
> Here a national interest of very nearly the first magnitude is involved. It can be protected only by national action in concert with that of another power. The subject matter is only transitorily within the state and has no permanent habitat therein. But for the treaty and the statute, there soon might be no birds for any powers to deal with. We see nothing in the Constitution that compels the Government to sit by while a food supply is cut off and the protectors of our forests and our crops are destroyed. It is not sufficient to rely upon the states. The reliance is vain, and were it otherwise, the question is whether the United States is forbidden to act. We are of the opinion that the treaty and statute must be upheld.

FEDERAL LAWS AND REGULATIONS BY FORD G. SCALLEY[61]

Forest Service

The Forest Service administers National Forests and National Grasslands and is responsible for the management of their resources. It cooperates with federal and state officials in the enforcement of game laws on the national forests and in the development and maintenance of wildlife resources; cooperates with the state and private owners in application of sound forest management principles and practices, in protection of forest land against fire and in the distribution of planting stock. The Forest Service conducts research in the entire field of forestry and wildlife management.[62]

The Secretary of Agriculture was authorized by an 1897 act to protect the national forests, to make such rules and regulations which would insure the objects for which the forests were set aside, and to regulate their occupancy and use [16 U.S.C. Section 551 (1964)]. This authority was originally granted to the Secretary of Agriculture in 1905. The "objects" of the national forests include regulation of water flow, the improvement and protection of the forest, and the assurance of a continuous supply of timber [16 U.S.C. Section 200.1 (b) (1968)].

Cases 63–64

In 1964, Congress declared that the construction and maintenance of an adequate system of roads and trails within and near the national forests was essential if increasing demands for timber, recreation, and other uses were to be met [16 U.S.C. Section 532 (1964)]. That act authorized the Secretary of Agriculture to provide for the acquisition, construction, and maintenance of forest development roads within and near the national forests and other lands administered by the Forest Service. Those roads are to permit maximum economy in the harvest of timber and to provide for the utilization of other resources, which presumably include fish and wildlife resources, although they are not mentioned specifically. The secretary's authority under this act has been delegated to the Chief of the Forest Service [36 C.F.R. Section 2121.2 (1968)]. The regulations of the department provide that access necessary to allow for effective protection, management, and utilization of Forest Service lands will be acquired as quickly as possible [C.F.R. Section 212.8a (1968)]. In the case where access across nonfederal land or over a nonfederal road or trail cannot be obtained with reasonable promptness through negotiations, condemnation is to be undertaken [36 C.F.R. Section 212.9e (1968)].

The Secretary of Agriculture has also been given authority for the construction and maintenance of forest development roads and trails [23 U.S.C. Section 201–205 (1964)]. The secretary is authorized to allocate funds appropriated for development roads and trails among the national forests according to needs. It is important to note, however, that no acquisition may be made under the latter authority except by provision in an appropriation act or other law.

The Forest Service Manual (FSM) recognizes the possibilities of proper utilization of game through adequate access with state as well as federal participation in access acquisition, including the location and closing of roads and trails, FSM Sections 2644–2647. According to the FSM the acquisition of lands, or rights-of-way across them, may be necessary where private lands block access to national forest lands. The FSM does indicate, however, that Forest Service's acquisition be within the national forest boundaries and state's acquisition be outside those boundaries as the preferred routes to take, FSM Section 2645.

The authority and responsibility for protecting and administering the national forests, the right to acquire lands or interests therein as are necessary to carry out its work, and the right and obligation to undertake condemnation where access across nonfederal land or over a nonfederal road or trail cannot be obtained, are all legal vehicles which can be utilized to obtain access rights.

Animal Damage Control[63]

This program provides operational and technical assistance to states, individuals, public and private organizations and institutions in the resolution of human and wildlife conflicts deemed injurious to agriculture, horticulture, animal husbandry, human health and safety, forest and range resources, and wildlife.

Fish and Wildlife Service[64]

The U.S. Fish and Wildlife Service was renamed by an act of Congress April 22, 1974; previously it was known as the Bureau of Sport Fisheries and Wildlife. The Service aids in conservation of the nation's migratory birds, certain mammals, and sport fishes. This includes application of research findings in the development of a management system of wildlife refuges for migratory birds and endangered species; operation of fish hatcheries; management of populations of migratory game birds through regulation of time, degree, and manner of taking; acquisition and application of technical knowledge necessary for perpetuation and enhancement of fish and wildlife resources; biological monitoring of development projects; enforcement of several laws, including the Endangered Species Act, the Lacey Act, the Marine Mammal Protection Act, and the Migratory Bird Treaty Act. The Service administers federal aid to state governments; provides technical assistance to state and foreign governments; serves as the lead federal agency in international conventions on wildlife conservation; and operates a program of public affairs and environmental education to inform the public of the condition of America's fish and wildlife resources.

Bureau of Land Management

Case 65–66

The Bureau of Land Management (BLM) administers the public lands which are located primarily in the western states, and which amount to about forty-eight percent (over 272 million acres) of all federally owned lands. These lands and resources are managed under multiple-use principles for various outputs, including outdoor recreation, fish and wildlife production, and mineral production including that on the outer continental shelf. The Bureau of Land Management was organized in 1946.[65] Although the BLM has considerable authority to acquire access to public lands under its jurisdiction, there is no statutory authority expressly indicating that the acquisition of access is authorized for recreational purposes such as hunting and fishing. It is important to note, however, that the Secretary of Interior is required to manage the public lands for fish and wildlife development and utilization. Therefore, the Secretary of Interior has provided regulations for the ". . . provision and maintenance of public access to fish and wildlife resources" [43 U.S.C. Section 1411 (1964); 43 C.F.R. Section 1725.3–3 (b) (1968)].

The secretary has further provided that nongrazing district lands leased for grazing purposes must be open for public hunting and fishing. The lessee may not ". . . interfere in any manner with such rights. Neither shall the lessee maintain locked gates, signs, or other devices which prevent or interfere with public use of the leased lands" [43 C.F.R. Section 4125.1–1 (i) (10) (1968)].

Likewise in grazing districts organized under the Taylor Grazing Act, those holding grazing permits may not interfere with public hunting and fishing within grazing districts [43 U.S.C. Section 315 (1964)].

Where public lands outside grazing districts are leased for grazing purposes, leased areas which are suitable for multiple use must be managed for multiple use, and provision must be made for adequate access for other users. The district manager is required to negotiate for such access with applicants for grazing leases, or with the lessees. Where negotiations fail, federal regulations provide that access will be acquired "by appropriate legal procedures" [43 C.F.R. Section 4122.1–1 (1968)].

In the case where some BLM lands, outside grazing districts organized under the Taylor Grazing Act, are permitted to be leased according to a priority system for the purpose of grazing, the regulations provide that where two or more otherwise qualified applicants apply for the use of the same public land, the district manager must allocate the use of the land between applicants on the basis of several factors. One of the factors to consider in awarding such leases is the degree to which each applicant will allow public access across his private land to the public land in question. This factor must be considered only where public access to the particular public land is not otherwise available [43 C.F.R. Section 4121.2–1 (d) (2) (1968)].

Access rights may also be acquired by land exchanges under the Taylor Grazing Act. Under the act, the secretary is authorized to accept title to land within or without the boundaries of a grazing district and to issue a patent on public lands of equal value in exchange whenever "public interests will be benefitted thereby" [43 U.S.C. Section 315g (b) (1964)].

National Park Service[66]

By the authority of a 1916 act, the National Park Service, under the supervision of the Secretary of Interior, has administrative jurisdiction over the three general types of areas of the National Park System [16 U.S.C. Section 1 (1964)]. Those areas include:

(1) "Natural Areas" such as national parks and national monuments of scientific significance;

Cases 67–68

(2) "Historic Areas" such as those set aside because of their archeological or historical significance; and

(3) "Recreation Areas" such as national seashores, national lakeshores, national riverways and waterways, and national recreation demonstration areas.

The 1916 act requires the Park Service to regulate the use of the areas under its jurisdiction in such a manner that the wildlife, as well as the scenery and the natural and historic objects of the areas, will be left unimpaired for the enjoyment of future generations [16 U.S.C. Section 1 (1964)]. Pursuant to 16 U.S.C. Section 8a03 (1964), the Park Service has been given authority to construct and maintain park approach roads. The acquisition of rights-of-way within the established boundaries of any area of the National Park System is authorized by 16 U.S.C. Section 1b (7) (1964). It is important to note, however, that the Park Service has taken a position that blocked access is not a problem in park management.

The Secretary of Interior issued a policy statement on June 16, 1968, which provides in part:

> In the case of the National Parks, National Monuments, and historic areas of the National Park System, the Secretary shall. . . .

(1) Provide, where public fishing is permitted, that such fishing shall be carried out in accordance with applicable state laws and regulations, unless exclusive legislative jurisdiction has been ceded for such areas, and a State license or permit shall be required for such fishing, unless otherwise provided by law;

(2) Prohibit public hunting; and

(3) Provide for consultation with the appropriate state fish and game departments in carrying out programs of control of overabundant or otherwise harmful populations of fish and resident wildlife, or research programs involving the taking of such fish and resident wildlife, including the disposition of carcasses therefrom.

In any case where there is a disagreement, such disagreement shall be referred to the Secretary of Interior, who shall provide for a thorough discussion of the problems with representatives of the state fish and game departments and the National Park Service for the purpose of resolving the disagreement. Department of Interior, Administration Policy for the National Parks and National Monuments of Scientific Significance 24, 25 (1968).

Bureau of Reclamation[67]

The U.S. Department of the Interior's Bureau of Reclamation was created by the Reclamation Act of 1902 to reclaim arid lands in the seventeen western states. This has been accomplished by the development of a system of works for the storage, diversion, and development of water. Today, reclamation provides water to more than ten million acres of land. In addition, Reclamation's multipurpose projects provide municipal and industrial water, hydroelectric power, recreational opportunities, and fish and wildlife enhancement.

Bureau of Indian Affairs[68]

This agency is charged with carrying out the major portion of the trust responsibility of the United States to Indian tribes. This trust includes the protection and enhancement of Indian lands and the conservation and development of natural resources, including forestry, fish and wildlife, outdoor recreation, water, range, and mineral resources. (Created in War Department 1824; transferred to Department of the Interior 1949.)

Corps of Engineers[69]

Cases 69–74

It is the policy of the Corps of Engineers to carry out its Civil Works missions in full compliance with the National Environmental Policy Act (NEPA) and with other federal statutes and guidelines for environmental protection. The Corps strives to both maintain and create conditions under which the human and natural environments can exist in productive harmony and to preserve important historic and archeological resources. The Corps attempts to create new opportunities for the public to use and enjoy a project area.

National Marine Fisheries Service[70]

This is part of the National Oceanic and Atmospheric Administration and provides management, research, and services for the protection and rational use of living marine resources for their aesthetic, economic, and recreational value. The NMFS determines the consequences of the natural environment and human activities on living marine resources. It provides knowledge and services to achieve efficient and judicious domestic and international management, use, and conservation of the resources.

Environmental Protection Agency[71]

This agency is charged with mounting a coordinated attack on air and water pollution, management of solid and hazardous wastes, cleanup of hazardous wastes under SuperFund, regulation of pesticides, toxic substances, and some aspects of radiation. The organization of EPA in 1970 placed under one roof programs which had been scattered throughout several agencies of the federal government. Functions include: setting and enforcing environmental standards; conducting research on the causes, effects, and control of environmental problems; assisting states and local governments.

Federal Energy Regulatory Commission[72]

Established October 1, 1977, pursuant to the Department of Energy Organization Act of 1977, the Federal Energy Regulatory Commission regulates the interstate aspects of the electric power and natural gas industries and establishes rates for transporting oil by pipeline. The commission issues and enforces licenses for construction and operation of nonfederal hydroelectric power projects. The FERC also advises federal agencies on the merits of proposed federal multiple-purpose water development projects.

Air Force[73]

A comprehensive multiple-use natural resource program involving fish and wildlife, forestry, soil and water conservation, and outdoor recreation has been conducted on Air Force lands since the mid-1950s. Current policy requires all installations with suitable land and water areas to develop integrated plans for the management and conservation of renewable natural resources.

U.S. Army Engineering and Housing Support Center, Natural and Cultural Resources Division[74]

Natural and cultural resources professionals are responsible for the management of approximately twelve million acres of land on Army military installations. Management objectives include: compliance with environmental laws, conservation and protection of resources, support to the military mission uses of the lands, and contributions to programs which support the public needs. Resources managed include: land, forest, wildlife, soils, vegetation, and historical and archeological sites.

Literature Cited

Connery, R. H. 1935. *Governmental problems in wildlife conservation.* New York: Columbia Univ. Press.

Schwartz, B. 1974. *The American Heritage history of law in America.* Ed. A. M. Jospehy. New York: American Heritage Publ. Co., Inc.

Notes

1. 4 Wheat. 316 U.S. (1819).
2. *Constitution of the United States,* Art. IV, sec. 3, cl 2.
3. Ibid. Art. II, sec. 2, cl. 2.
4. Ibid. Art. I., sec. 8, cl. 1.
5. Ibid.
6. Ibid. Art. I., sec. 8. cl. 18.
7. Office of Legislative Services. *Digest of Federal Resource Laws.* U.S. Fish and Wildlife Services. 1992.
8. 938 F. 2d 942 (9th Cir. 1991).
9. 943 F. 2d 748 (9th Cir. 1991).
10. 923 F. 2d 131 (9th Cir. 1990).
11. 898 F. 2d 727 (9th Cir. 1990).
12. 887 F. 2d 1341 (9th Cir. 1989).
13. 881 F. 2d 539 (8th Cir. 1989).
14. 871 F. 2d 824 (9th Cir. 1989).
15. 86 F. 2d 327 (10th Cir. 1989).
16. Office of Legislative Services. *Digest of Federal Resource Laws.* U.S. Fish and Wildlife Services. 1992.
17. Ibid.
18. 889 F. 2d 636 (7th Cir. 1990).
19. 726 F. Supp. (D. Kan. 1989).
20. 787 F. 2d 431 (8th Cir. 1986).
21. Office of Legislative Services. *Digest of Federal Resource Laws.* U.S. Fish and Wildlife Services. 1992.
22. 100 S. Ct. 318 (1979).
23. 106 S. Ct. 2216 (1986).
24. 882 F. 2d 1294 (8th Cir. 1989).
25. Office of Legislative Services. *Digest of Federal Resource Laws.* U.S. Fish and Wildlife Services. 1992.
26. 929 F. 2d 1449 (9th Cir. 1991).
27. 912 F. 2d 1087 (9th Cir. 1990).
28. 839 F. 2d 795 (D.C. Cir. 1988).
29. 792 F. 2d 821 (9th Cir. 1986).
30. Office of Legislative Services. *Digest of Federal Resource Laws.* U.S. Fish and Wildlife Services. 1992.
31. 112 S. Ct. 2130 (1992).
32. 954 F. 2d 1441 (9th Cir. 1992).
33. 949 F. 2d 759 (5th Cir. 1992).
34. 926 F. 2d 429 (5th Cir. 1991).
35. 916 F. 2d 1016 (5th Cir. 1990).
36. 891 F. 2d 927 (D.C. Cir. 1989).
37. This material is, in part, from a file prepared by the Senior Attorney Solicitor's Officer, U.S. Forest Service, Ogden, Utah. Permission to reproduce it is contained in a letter to the author from S. R. Standing, Asst. Regional Forester, dated Dec. 27, 1954.
38. *Constitution of the United States* Art. IV, sec. 3, cl. 2.
39. *United States v. Gurley,* 279 Fed. 874 (N.A. Ga. 1922).
40. 160 F. 870 (9th Cir. 1908).
41. 206 U.S. 46 (1907).
42. 220 U.S. 523 (1911).
43. 243 U.S. 389 (1917).
44. 167 U.S. 518 (1897).
45. Act of Feb. 1, 1905, 33 Stat. 628.
46. 278 U.S. 96 (1928).
47. *Chalk v. United States,* 114 F. 2d 297 (4th Cir. 1940).
48. Act of Mar. 1, 1911 (36 Stat. 961; 16 U.S.C. 513).
49. N.C. Code Ann., 1939, Sec. 2099.
50. Act of Aug. 11, 1916 (39 Stat. 446, 476; 16 U.S.C. 683).
51. *United States v. 2271.29 acres of land, etc.,* 31 F. 2d 617 (W. D. Wis. 1928).
52. *Buford v. Houtz,* 133 U.S. 320 (1890); *United States v. Light,* 220 U.S. 523 (1911); *United States v. Mid-West Oil Co.,* 236 U.S. 459 (1915).
53. 220 U.S. 523 (1911).
54. *United States v. Grimaud,* 220 U.S. 506 (1911).
55. Ibid.
56. *New Mexico State Game Comm. v. Udall,* 410 F. 2d 1197 (10th Cir. 1969).
57. *Missouri v. Hollond,* 252 U.S. 416 (1920).
58. 92 F. 2d 623 (7th Cir. 1937), *cert. denied,* 303 U.S. 636 (1938).
59. 252 U.S. 416 (1920).
60. *Baldwin v. Franks,* 120 U.S. 678 (1886).
61. Former Asst. Attorney General, Salt Lake City, Utah. Permission to use given in letter dated June 5, 1980, by Ford G. Scalley. Modified Mar. 1979, and Dec. 1992, by W. F. Sigler.
62. Conservation Directory, National Wildlife Federation, 1992.
63. Office of Legislative Services. *Digest of Federal Resource Laws.* U.S. Fish and Wildlife Services. 1992.

64. Conservation Directory, National Wildlife Federation, 1992.
65. Ibid.
66. Ibid.
67. Ibid.
68. Ibid.
69. Ibid.
70. Ibid.
71. Ibid.
72. Ibid.
73. Ibid.
74. Ibid.

Recommended Reading

Bavin, C. R. 1979. The federal-state partnership in wildlife law enforcement. In *Proceedings Western Assn. of Fish and Wildlife Agencies,* 173–177.

Coyner, J. B. 1961. Enforcement of wildlife laws along state lines. In *Proceedings Western Assn. of Game and Fish Comms.,* 242–249.

Peckumn, J. W. 1963. Enforcement of wildlife conservation laws on federal military reservations. In *Proceedings Western Assn. of Game and Fish Comms.,* 282–286.

Reynolds, T. A. 1979. Law enforcement in the National Park System—where do we go from here? In *Proceedings Western Assn. of Fish and Wildlife Service Agencies,* 178–184.

Whelan, N. 1976. *Law enforcement and vandalism in our National Parks.* U.S. Dept. of Agriculture, Forest Service Pacific S.W. Forest Range Experiment Station.

Windsor, B. W., and W. L. Floray. 1967. Cooperation between civilian and military game law enforcement. In *Proceedings Southeast Assn. of Game and Fish Comms.,* 558–560.

Multiple-Choice Questions

Mark only one answer.

1. Which country was the first signatory to the Migratory Bird Treaty Act?
   a. England for Canada
   b. Mexico
   c. Japan
   d. USSR
   e. China

2. The defendant tried getting rid of pigeons by poisoning them. He also killed four migratory birds. What case was this?
   a. *Idaho v. Johnson*
   b. *U.S. v. Darst*
   c. *Virginia v. Coyote*
   d. *U.S. v. Van Fossan*
   e. None of the above

3. In the Marine Mammal Protection Act of 1972 the Department of the Interior does not protect:
   a. sea otters
   b. pinnipeds
   c. polar bears
   d. dugong
   e. manatee

4. The Federal Government can "take" deer or other animals on federal lands under what circumstances?
   a. any one that it chooses
   b. at no time
   c. if the Federal Government feels hunters are not killing the population down to the prescribed numbers
   d. if the State and Federal biologists agree
   e. when the resource must be protected from the animals because of excessive numbers doing damage to the habitat

5. Big game animals can be hunted on National Parks (excluding Teton) by civilian hunters:
   a. when the Park biologists deem it necessary
   b. never
   c. when the animals damage the Park excessively
   d. when animals endanger civilians
   e. any time

6. A licensed outfitter with no permit to guide hunts for bear was convicted for so doing in the case of:
   a. *U.S. v. Carpenter*
   b. *U.S. v. Alaska*
   c. *Idaho v. Powers*
   d. *New Mexico v. Smith*
   e. *U.S. v. Powers*

7. The Constitution allows the Federal Government to manipulate wildlife populations on Federal lands:
   a. under the general welfare clause
   b. under the Enabling Act
   c. by the U.S. Fish and Wildlife Service
   d. by power of eminent domain
   e. it has no such power

8. Regarding migratory birds, which U.S. Supreme Court Justice said the federal government's power to make treaties gave it control over the migratory birds?
   a. Marshall
   b. Warren
   c. Holmes
   d. Jackson
   e. Douglas

9. Federal control over migratory wildlife may be exercised:
   a. by right of eminent domain
   b. by police powers vested in the President's Office
   c. by the treaty-making powers vested in the Federal Government
   d. by Congress
   e. by the Senate

10. The U.S. Forest Service's failure to follow correct management provisions as provided in its handbook regarding the red-cockaded woodpecker constituted:
   a. a taking under the Endangered Species Act Section 9
   b. there was no taking
   c. nothing, because the handbook was outdated
   d. no taking according to the Court of Appeals
   e. no taking according to the District Court

11. Crows can be hunted by states:
   a. anytime
   b. in late fall to early spring
   c. under Federal guidelines only
   d. when the rancher thinks they are too numerous
   e. at no time

12. In regard to the Migratory Bird Treaty Act one statement is not true:
   a. A state may enact laws more restrictive than those of the Federal Government
   b. a state may follow precisely Federal guidelines
   c. a state may make laws less restrictive than Federal laws
   d. the government can close hunting in a state
   e. the Federal Government consults with a state before setting enforcement and season guidelines

Chapter 6

The Wildlife Law Enforcement Officer

HISTORICAL SKETCH OF POLICING AGENCIES

Originally, our ancestors took turns standing guard over communal property. Later, some of the more enterprising individuals hired others to take turns for them. Still later, these hired guards became full-time night watchmen. This function was later taken over by the police department.

Today, private security is twice the size of public law enforcement. The private sector is an estimated eighteen billion dollar industry employing close to two million people. According to a 1984 survey by the National Institute of Justice, public law enforcement resources have remained relatively flat, with a significant part of law enforcement agencies showing an effective decline in personnel despite growth rates in population and crime (Mangan and Shanahan 1990).

According to Schwartz (1974), before 1800 negligence was not a separate tort. That is, under common law a man acts at his own peril. English law was characterized by absolute ability. That is, it did not differentiate between varying degrees of types of conduct, but branded all equally and meted out similar consequences to those concerned. A person committing trespass was liable regardless of how innocent his crossing of the plaintiff's boundary might have been. Later the maxim of "no liability" without culpability was established.

Police Systems in England

The first laws in England were enforced by military guards. In earlier days, the knights who roamed the country were law enforcement officers of a sort. However, they made, as well as enforced, their own laws. Later, the lords and barons made laws and hired law enforcement officers or guards to enforce them. These guards were known as *Comes Stabuli* or Masters of the Stable. The words were later corrupted to "constable," which is still in use in America and England today.

In England, the head of the shire (comparable geographically to a county in the United States) was responsible for policing the area and maintaining law and order. The head of the shire, known as a reeve, was the chief police officer. The words shire and reeve were later contracted to sheriff, or the chief police officer of the shire.

In 1828, because he believed that the constable system in the England of his day was inadequate, Sir Robert Peel banded together the first modern policing group in the English-speaking world. Parliament, under Peel's recommendation, authorized a police department to cope with the rising crime which, it was noted, was increasing because of faster, four-horse carriage transportation. This meant, in effect, that a criminal was

Case 1

Police Systems in North America

able to commit a crime in greater London and escape outside the city limits before he was apprehended by the constable, whose jurisdiction ended at the boundary line. English police officers, even down to the present in London, are sometimes referred to as Bobbies or Peelers.[1]

The first police system in North America was patterned after that of the Mother Country. In the 1870s, the Royal Canadian Mounted Police came into being. In the West during the 1880s the county sheriff, the U.S. marshall, and the vigilantes all played roles.

A hundred years after Peel, August Vollmer and O. W. Wilson advanced the goal of professionalism in law enforcement. Their idea of improvement was standardization, specialization, synchronization, concentration, maximization, and centralization. This lasted until about the mid-1970s when certain people began to question the value of the bureaucracy and the military model of policing.

The state police system was developed in the United States just after 1900. In 1908, the Federal Bureau of Investigation was organized. Its jurisdiction is comparable to no other organization in the world today.

GAME WARDENS

By the year 1880, there were game laws for all of the states and territories which were enforced by local police officers. Later, these jobs were given to wardens, the earliest of which were the deer wardens of Massachusetts in 1739 and New Hampshire in 1741, followed by moose wardens in Maine in 1852. In 1887, the first salaried wardens were appointed in midwestern states, a system that was in force in forty-one states by 1912 (Bavin 1978).

In 1899, Governor Theodore (Teddy) Roosevelt became concerned about the laxity of the enforcement of game laws and ''the inefficiency of the game wardens and game protectors.'' He urged that the men who worked as protectors in the Adirondacks be appointed from the locality itself, and should, in all cases, be thorough woodsmen: ''I want as game protectors men of courage, resolution and hardihood who can handle the rifle, axe and paddle; who can camp out in summer or winter; who can go on snowshoes, if necessary; who can go through the woods by day or by night without regard to trails'' (furnished by the New York State Department of Environmental Conservation).

One of the differences between wildlife law enforcement officers and officers in some other phase of enforcement work is the wildlife officer encounters people of almost all ages, colors, creeds, and social positions. There is an even wider diversity in wildlife cases than, for example, traffic violations, since the very young, the very old, and people with disabilities normally do not drive automotive vehicles; yet they frequently hunt or fish. It is essential to the success of the enforcement program that the qualifications of the wildlife officer be on a high plane.

RESPONSIBILITIES

In General

Being a professional wildlife law enforcement officer takes a thorough understanding of the law, which means the substantive law, as well as the rules of conduct for law enforcement officers. Laws are created by state and federal legislative codes and administrative orders and by local, state, and federal court decisions. There are two types of laws: code law and case law, and the latter may change as it passes from a lower to a higher court.

Cases 2–6

To keep or discourage from doing something

Professionalism starts with rigid, personal qualification standards and objective selection procedures. Today's officer is selected and promoted through a merit or civil service system, or a rough equivalent, in all fifty states and several Canadian provinces. It is essential for agencies to have an active recruiting program, as well as competitive salaries; and additional screening devices, such as medical background and other examinations to test the applicant's suitability and his or her interest in being a wildlife law enforcement officer (Bavin 1976). Some law enforcement agencies look to the spouse as well as the applicant. This may or may not involve an interview, but in many cases following employment it does include briefing as to what will be expected of the employee and the hazards and rewards of the job. Intense original training programs, followed by yearly refresher courses, are becoming the norm.

Evaluation of officers must consider their enforcement effectiveness as well as their deterrent role. If the criminal justice system is to be an effective deterrent, people must believe the enforcement agency is willing and able to discover and apprehend violators. They must also believe there will be sure punishment for lack of compliance. An organization with a public reputation for efficient, effective, and fair law enforcement will achieve a fairly high level of criminal prevention.

According to Schneider,[2] conservation law enforcement will continue to represent a primary arm of management. Its functional texture will continue to broaden both technical and sociological parameters. In 1991, only nine percent of the state agencies indicated their sole responsibility was the enforcement of wildlife laws (see Appendix C). By virtue of this trend, educational requirements of the conservation officer will be expanded through the college level and into advanced work. Some specialization in investigative work and technological areas will be represented in the enforcement staff. Greater emphasis will be directed toward preventive measures. Such a trend will require a continuing need to recognize the conservation-officer-type individual as a front-line contact representative. In the future there will also be more clear-cut staff and line functions in the enforcement arm, regardless of where it is administratively located.

Krug[3] believes that the unionization of wildlife law enforcement personnel may be good for the financial well-being of the wildlife officer but not for wildlife law enforcement and wildlife. He believes in some cases that the effects may be extremely detrimental to law enforcement. And it will have profound effects on the already strained relationship between law enforcement and management biologists personnel and possible administrators. This has been debated heatedly over the last ten years. Where wildlife officers are also peace officers unionization is almost inevitable.

Some states separate enforcement from the fish and wildlife department. Schneider[4] does not regard this arrangement as administratively sound. He also favors the conservation officer approach over the wildlife-law-enforcement-only officer. That is, a broadly trained individual rather than one highly trained in investigative work with lesser training in management and related fields.

Krug[5] also favors the conservation officer with varied duties over the officer who does nothing but law enforcement. He believes that wildlife agencies should employ a group of specially trained, professional law enforcement people to work on such problems as poaching rings and other illegal commercial activities. These groups would move from area to area within a state, acting as undercover agents, out of uniform and sometimes with cars with out-of-state licenses. Krug[6] is opposed to the wildlife officer becoming a "natural resources police officer" who enforces all laws coming under the purview of a department of natural resources. He believes that under such systems, wildlife law enforcement suffers, and personnel spend much of their time on laws related to park management and tourism. The budget crunch for many states is forcing this to happen. He also believes that wildlife agencies are better off recruiting personnel who are primarily fish and game oriented, rather than strictly law enforcement oriented.

In a 1972 survey, Morse (1973) found that conservation officers nationwide spent an average of fifty-nine percent of their time in enforcement. Arrests per officer per year were 48.7, and the number of hunters and anglers per officer averaged 7,669. Of the ninety-one percent of state natural resources agencies that enforced more than wildlife law in 1991, the average officer spent seventy percent of his time on law enforcement (see Appendix C).

Problems Facing Enforcement

Bavin (1976) believes that a major problem confronting wildlife law enforcement is public apathy. His answer is to effectively and efficiently manage people. Wildlife law enforcement must become fully professionalized through rigid personnel selection standards, intensified training programs, fully developed policies, and increased use of science and technology. Above all, there should be public assurance that the constitutional rights of people are protected. The use of laws which prescribe human conduct in relation to wildlife is only one way to bring about wildlife conservation. In this approach, people are expected to comply with laws and regulations. The goal, therefore, becomes law observance. Total observation is impossible to attain and probably not desirable. The most obvious way to bring about law observation is to have public understanding and support, so that people comply voluntarily. The problem is how to accomplish this. The first step is ensuring that legislative bodies enact laws that are acceptable, enforceable, and relate to the need of the resource. Concerned groups, lobbyists, and wildlife biologists who assist in legislative proposals or develop administrative regulations, also have a responsibility.

The New York State Department of Environmental Conservation has this to say about enforcement problems.

> Wildlife laws are not crimes that people are naturally aware of, but, to almost everyone, rape, murder, and robbery are wrong per se. We know these acts are wrong without knowing specific laws. However, wildlife laws that manage a resource are not based on common sense, but rather on biological information. A size limit for fish, a male-only deer hunt, a no fishing with live bait, are all good management techniques, but they are not common knowledge. This is where public information and education are extremely important in effective fish and wildlife law enforcement. The hunting and fishing codes and proclamations are the main source of public information. Law enforcement officers provide additional education, talks to groups and individuals, and the issuance of tickets or arrests of violators.
>
> Law enforcement officers are geared for arresting or citing the violator who is knowingly exploiting the resource. Their role is important in dealing with this category of individual. However, today, more so than in the past, we are dealing with a large group of individuals who do not take the time to read the codes or proclamations. Officers must deal professionally with people who are carelessly or unknowingly ignorant of the law. Discretion is important in the life of a conservation officer. The spirit of the law versus the letter of the law should always be taken into consideration.

PROFESSIONAL QUALIFICATIONS

Anyone considering a career in wildlife law enforcement should be aware of certain facts. All fifty-three state agencies (three states have two) have wildlife law enforcement officers. The federal government has several agencies that employ special agents, notably the Fish and Wildlife Service, the Forest Service, the Bureau of Land Management, and the National Parks Service. There are a rather substantial number in private industry. The National Wildlife Federation, 1400 16th Street, N.W.,

Handgun training is a
must for wildlife
officers.

Cases 7–8

Washington, D.C., 20036–2266, publishes a Conservation Directory that lists all organizations, agencies, and officials concerned with natural resources and management.

What do state agencies look for in beginning employees? Some require a bachelor of science degree from a recognized college or university. Others stipulate two years of college and/or experience. Some require only a high school diploma. Among the state agencies, many favor applicants who have more formal education than is legally required (see Appendix C).

Kimball[7] believes that state wildlife conservation officers should be trained in all aspects of fish and wildlife management. He further believes that after some experience and in-service training, officers are fully capable of performing any tasks assigned by their respective fish and game department under its legislative mandate to manage the wildlife resources held in trust for the people. An understanding of resource needs is imperative if law enforcement officers are to provide the best possible service to their constituents. Whether or not this is set forth as a stipulation of employment, it adds to the capabilities of the officer. Krug[8] believes that other things being equal, the more educated individual will be more effective in enforcing wildlife laws and in other enforcement duties. He further believes that successful prosecution of wildlife law violations requires a highly educated and well-trained officer.

How much education does a conservation officer need to do the job? The first relevant question actually is, ''How much education does the hiring agency require where one wants to work?'' That easily can be answered by an inquiry. It is highly likely that the minimum requirement will be a bachelor of science in natural resources or biology. Many agencies may want candidates to specialize in some particular aspect of law enforcement at the master of science level. The conservation officer of the twenty-first century is going to need all of the education he or she can get. There are many fine conservation officers today with only a high school education, but the twenty-first century will be different (see Appendix C).

Education at the bachelor of science level in natural resources or biological sciences should have two years of basic science such as chemistry, physics, biology, math, and geology and double courses in arts and letters, including English, writing, speaking, history, culture, political science, and economics. It is a good idea for the university to have an introductory general course in wildlife resources each of the first two years; otherwise, some students may find themselves in the wrong field. The second two years of college should have one-third to one-half in resource management and such sciences as limnology, ichthyology, mammalogy, ornithology, and herpetology. The balance of the time should build on the basic science courses taken during the first two years and courses in supervision, budgeting, and administration (Sigler 1975).

Conservation officers in most, if not all, states will be enforcing laws ranging from wildlife to environmental violations. The reason for this is quite simple. Future state budgets will be even more restrictive than they are today, yet problems in the field will have increased substantially. Governors will ask themselves what work force is on-the-spot. In most cases the conservation officers will be elected.

Halberstam (1991) has this to say, in general, about education. The world moves faster than ever. Not to stay abreast is to fall behind. Work demands ever higher levels of education and competency. This is true in the United States and around the world. Troganowicz and Carter (1990) quote a police executive research forum report that indicates college-level officers exhibit greatest sensitivity to the ethnic diversity that will increasingly become the hallmark of society. This study verified that the officers with at least some college education are not only increasing in numbers in the rank and file but also in police management as well.

Neil L. Rudenstine, the twenty-sixth president of Harvard University, has this to say about a core curriculum, ''it is more process today . . . we now put as much emphasis on how to go about doing it as on learning what it is.'' A liberal education, says Rudenstine, is a set of habits instead of a frame of mind, it is a set of aptitudes, and a capacity to adapt to an intelligent way. Things are changing fast. From somewhere in the 1980s to somewhere around 2010, will have been a transformation period, in terms of what people think of knowledge says the Harvard leader (Cattani 1991).

Eastmond and Kadlec (1977) conclude from a survey of alumni, employers, seniors, and faculty, that traditional emphasis on fish and game training in college is no longer adequate for the professional sport fishing and hunting manager in the field. In addition, they believe the professional should be able to deal realistically with political pressures and cost considerations which are crucial for tomorrow's graduates. Skills in planning and analysis, independent thinking, creativity, and use of judgment in decision making should become an important part of undergraduate training.

The role of women in wildlife law enforcement has received only moderate attention to date from either administrators or women. This will change since there is no reason why women cannot perform as well as men in a wildlife enforcement career, either state or federal. In 1991, only 3½ percent of the field force in fifty-three state agencies were women; this is expected to increase markedly (see Appendix C).

Point of View

The underlying philosophy of modern law enforcement procedures stresses compliance with the law, not because of fear of punishment but because of enlightened self-interest growing out of a program of public information. The professionally-trained wildlife officer of today, therefore, enters the service with a point of view that reflects this basic philosophy. He is not so eager to apprehend as to explain and caution. He would much rather have a patrol district frequented by sportsmen who observe good conservation practices because they understand the issues involved, than one wherein conservation is practiced because of the proximity of the badge.

According to Bavin (1978), officers have to make judgments which are originally colored by personal value systems, the way they grew up, and how they see themselves fitting into the world around them. Officers react to situations not only in a way they have been trained to react, but also according to their inner philosophies. Bavin believes that, in the long run, the inner light will recast and reform much of what the officer originally learned. He believes that there will nearly always be an inner conflict and that to deny it will be to impair growth and halt maturity.

PROFESSIONAL DUTIES

Wildlife Management

A law enforcement officer's work may include wildlife management. This involves, among other things, the making of game counts; the settlement of damage claims, including an evaluation of the amount of damage done; the taking of control measures in the case of predators or other animals doing damage to crops; the supervision of hunts; the operation of checking stations; the taking of bag and creel census; and banding and marking animals.

Educational Programs

Gordon (1977) believes that the logical place to start an educational program is with the hunters. Until recently, hunters have traditionally taken the lead in promoting and supporting the biological aspects of wildlife management and the hunting sport. The hunter must again take the first positive steps in the new arena of sociological management. Until very recently, hunters applauded their own efforts to restore and manage wildlife resources, and as a group concerned themselves primarily with the biological aspects of management at the expense of the sociological management of the hunter. American hunters in the past have been primarily responsible for the growth of professional wildlife conservation and management and now the time has come for them to take the lead in this new dimension.

Hunter education began with gun safety programs, which have been an outstanding success in preventing accidents and loss of life, and adding to the hunting sport ethic. State wildlife agencies, organized sports clubs, the National Rifle Association, and others had encouraged safe firearm handling, hunter safety, and proper conduct for many years, but it was not until 1949 that the first state law was passed requiring all young hunters to receive instructions in the safe handling of firearms before going afield (Gordon 1977). Since then, hunter safety training has been implemented in practically all states and Canadian provinces.

Gordon (1977) believes there is much we can do to improve educational programs and reach realistic goals. One improvement would have the state and provincial wildlife administrators periodically examine and evaluate funding priorities in terms of long-range goals for this type of education. Wildlife agencies should have top-level education departments staffed with professionally trained and experienced employees. Hunter education at the state level must be continually upgraded to incorporate a full spectrum of topics. This includes, in addition to gun safety, such programs as conservation instruction, game laws and regulations, game identification, sportsmen-landowner relations, hunter responsibility, ethics, and other topics which will improve the conduct of the sportsmen and women afield. Gordon also believes that greater emphasis should be placed on programs in public schools. Such programs, as a part of the regular school curricula, would give young people a clear understanding and perspective of conservation matters, including the legitimacy of hunting and the management of nongame, as well as game animals.

Law Enforcement

This phase of a state officer's duties may be divided into general considerations and those of a more specific nature. General considerations include knowledge of state wildlife laws and court rulings, use of uniform, sidearms, and automotive vehicles. Specific considerations include making the arrests; identifying, collecting, and preserving evidence; and appearing in court.

Collecting management data is all in a day's work for some conservation officers. *Courtesy Wyoming Game and Fish Department.*

One of the mundane, but very important, jobs of the wildlife officer is checking hunters and anglers for the proper license and legal number of animals. These checks encourage hunters and anglers to purchase and carry the proper license, and to respect bag and creel limits.

THE UNIFORM

The conservation officer should wear a distinctive, good quality, dress uniform whenever conditions permit. A second, less dressy uniform should also be distinctive and easily recognizable, and of respectable quality. The American public tends to respect and appreciate a well-uniformed, courteous law enforcement officer. Complete uniforms should be worn whenever possible. This includes standard headgear, a tie, footwear, and generally a sidearm.

There is little doubt that dress is strongly linked in some positive manner to behavior of the wearer. There are two ways to view this. The first assumes that when a conservation officer, in this case, puts on a military-style uniform, he or she modifies his or her self-image to correspond to the stereotypes associated with the uniform. This view holds that the uniform actually changes the wearer's behavior. The second position holds that those officers who do not feel their behavior to be in accord with the stereotypes of the uniform will not wear it, and that those officers will not remain on the job long term. In either case, the assumption is that a new style of uniform, one that does not carry a century of stereotypes, modifies officer behavior and citizen perception. The uniform change is the most obvious public symbol of this change of values. Data from the University of Alabama and Auburn University lead to speculation that

Cases 9–10

police forces might increase community acceptance with a less stereotypic uniform. The data from Alabama and Auburn indicate there is more power to the uniform change than the simple novelty of the look. If new uniforms bring about a changed attitude in college students, they could well do likewise with the noncollege-age population (Gundersen 1978).

SELF-DEFENSE

It is common for a conservation officer to confront heavily armed hunters whose attitudes are neither known nor predictable. In addition, the conservation officer may, upon routine inspection or questioning, unmask a dangerous or ruthless criminal or an escaped prisoner. Perhaps most importantly, conservation officers are becoming more a part of the law enforcement community and, as such, are expected to be able to defend themselves and uphold the law in the normal peace officer manner.

Summary

If an officer carries a sidearm, it must never be used in a misdemeanor charge to force a suspect to stop, even though it may mean that the subject will escape. In a misdemeanor, firearms should not be drawn or discharged to frighten a prisoner or a suspect. Such action may lay the officer open to serious criticism, charges, or even prosecution. Only where it is necessary to use a firearm to protect life or limb may one be used.

DEFENSE AGAINST PERSONAL LIABILITY

The possibility of civil litigation against a law enforcement officer has increased several fold in the last few years. Facing a million-dollar lawsuit can be a traumatic experience. Damage suits may claim injury from arrests, search, or imprisonment. Such suits, alleging a violation of the plaintiff's constitutional rights, are brought against state law enforcement personnel pursuant to Title 42, United States Code (U.S.C.), Section 1983 of the Federal Law Enforcement Officer's Code pursuant to the cause of action created in *Bivens v. Six Unknown Federal Narcotics Agents.*[9]

The first defense an officer considers is technical defense, including proper service and venue, and a lack of jurisdiction. If this is not applicable, two other defenses may be used for expeditious resolution of the problem. The first is that the plaintiff has failed to state a claim against a law enforcement officer upon which relief can be granted. In essence, assuming all the plaintiff's allegations are true, the law does not entitle the plaintiff to recovery. The second defense is the qualified immunity defense. This shields law enforcement officers from liability if they are found to have acted reasonably under the laws existing at the time of the incident (Higgenbotham 1985a). *Harlow v. Fitzgerald*[10] is important because it modifies an already existing defense. Prior to *Harlow* the qualified immunity defense had both objective and subjective components and the shield of qualified immunity was not available if a public official, such as a police officer, knew or should have known that the action he took within his sphere of official responsibility would violate the constitutional rights of the plaintiff. The subjective portion of *Harlow* frequently focused on the defendant's state of mind at the time of his action. If the defendant did not act with "permissible intentions," the qualified immunity defense was unavailable. This resulted in long trials, even when the defendant was eventually declared not guilty. The *Harlow* Court did away with the subjective component, leaving the qualified immunity to be judged solely by objective standards. The aim of the Court was to avoid excessive disruption of government and to permit the resolution of many unsubstantial claims to be disposed of at an early date. The *Harlow* Court used the term ''clearly established.'' The Supreme Court, however, did not define the term ''clearly established.'' The key issue in asserting the qualified immunity defense is whether or not the law which a defendant officer is alleged to have violated in committing a constitutional wrong against a plaintiff was clearly established at the time of the incident which gave rise to the civil action. Liability will generally not be imposed if the law was not clearly established, but it will be imposed if the law was clearly established (Higgenbotham 1985a).

In *Harlow,* the Supreme Court added another factor to consider. The Court stated that liability will not be imposed if the officer pleading the defense claims extraordinary circumstances and can prove that he neither knew nor should have known about the relevant legal standards. This creates a second prong in the qualified immunity defense. Assuming that the plaintiff has pleaded a constitutional tort allegedly committed by a defendant officer and has proved that the law allegedly violated by the officer was clearly established at the time of the incident, the defendant may still avoid liability by justifying his conduct on ''exceptional circumstances'' (Higgenbotham 1985b).

Case 11

The urgent need of gun safety and the limits of liability of a bonding company are emphasized in the case of an Arizona game warden.[11] The case follows:

> While in the field, Deputy Game Warden Donnell shot at some animals, and the bullet wounded Plaintiff Truog in the wrist. Truog sued the deputy warden, the state game warden, and the bonding company, the American Bonding Company of Baltimore, for damages. The superior court sustained the defendant's demurrer to the complaint and gave a judgment of dismissal to the defendant. The appeal is on the demurrer and judgment.
>
> The plaintiff contended that the deputy warden was not acting within his scope of duties by being out protecting the game and enforcing the laws by which the game was being hunted; thus the bonding company was liable for acts committed in his capacity. The court ruled that the warden could not have been protecting the game by shooting at it and this act in itself places the warden beyond the scope of his duties. An officer's gun does not protect him in his personal and private affairs, which shooting of game would come under.

Assault

An attempt or an offer, with unlawful force or violence, to do bodily harm to another is assault. It is not necessary that the attempt or offer be consummated. Pointing an unloaded pistol at a person who knows the weapon to be unloaded may not be an attempt to do bodily harm, if the assailant realizes his inability to shoot the victim, and if the potential victim is not frightened. However, such an act may be an assault if the victim is put in reasonable fear of bodily injury by the act. Contrariwise, aiming a loaded weapon at someone who is not aware of the act is nevertheless committing assault, since the intent is to do bodily harm. Raising a fist or club over a person's head is an assault. On the other hand, picking up a club is not in itself an assault if there is no attempt to use it. Threatening words in themselves do not constitute assault. For example, a hunter who waves a shotgun about but proclaims at the same time that while he is sorely tempted he, nevertheless, has no intention of using it, is not committing assault. However, if the gun is at any time pointed at another person, the hunter may be charged with careless handling of firearms. It is not assault if one person fires a weapon at another who is obviously beyond the range of the weapon.

Aggravated Assault

This is an assault committed with a dangerous weapon or by other means or force likely to produce death or grievous bodily harm. Grievous bodily harm means serious bodily injury. That is, the natural and probable consequences of a particular act might result in death. Dangerous weapons are not confined to firearms. It has been held that bottles, broken glass, rocks, pieces of pipe, boiling water, drugs, rifle butts, and so forth, can inflict death or grievous bodily harm. On the other hand, an unloaded pistol, when used as a firearm rather than a bludgeon, is not a dangerous weapon.

Battery

An assault in which the attempt to do or offer to do bodily harm is consummated by inflicting such harm is called battery. This may be defined as an unlawful and intentional or culpably negligent application of force to the person of another by a material agent used directly or indirectly.

It may be a battery to push, spit on, or even cut the clothes of another person, or to cause him to take poison, or to deliberately run into him with an automotive vehicle. Throwing a stone into a crowd of people may be a battery on anyone whom the stone hits. It may be a battery if a person is injured as a result of indirect motion. For example, a person may push a second one who falls against the third and injures him. The first person has committed a battery, if the act was intentional.

If an injury inflicted is unintentional and without culpable negligence, the offense is not committed. It is not a battery to lay hands on another to attract his attention or to seize someone to prevent a fall.

Proof of battery will support a conviction of assault, since an assault is necessarily included in a battery. In order to constitute an assault the act of violence must be unlawful and it must be done without legal justification or excuse and without the lawful consent of the person affected.

THE OFFICIAL AUTOMOBILE

A large percentage of the enforcement personnel will travel in well-marked cars. In much of the Mountain West it is necessary for the officer to have a pickup-type vehicle. However, where this is not the case, a passenger-type vehicle is more desirable because it ranks higher in status in the eyes of the general public. In this day of emphasis on credibility, no point that adds prestige should be overlooked. There are times when it is necessary an officer not be identified by automobile, uniform, or person. In such instances an unmarked, nondescript car (possibly with a license plate from another state), hunter-type clothing, and even personnel from another area should be used to apprehend chronic violators. The entire behavior, approach, and techniques of an undercover agent are quite different from that of a uniformed officer.

PURSUIT DRIVING

High-speed chases of suspects go back at least to prohibition days. The question in the mind of each officer is, "Do I or do I not initiate the chase?" Most "hot pursuits" are for misdemeanors. However, there is an increasing number related to serious crimes.

Police administrators believe that there are two steps in dealing with pursuit driving. Firstly, each agency should clearly define its policy regarding tactical driving. It is as important as the policy concerning the use of deadly force. Secondly is training officers in tactical driving. This includes the danger and statistics resulting from tactical driving; the legal, civil, and sociological considerations; the vehicle and road dynamics and capabilities; and, finally, the driver capabilities, attitudes, and behavior. Training includes such ordinary things as controlled braking, perception reaction, cornering and turning, backing, and open-road driving. Many police administrators believe that officers should not be allowed to be involved in pursuit driving unless they have had tactical training (Cunningham 1986).

The legal aspects of pursuit driving are many and varied. For example, high-speed pursuit may pose a greater risk to innocent citizens than if the police were using a

Cases 12–14

deadly firearm. One person suggests that a motor vehicle is the most deadly weapon in the police arsenal. There are some differences in state laws. However, the basis for most pursuit-related liability is negligence. Pursuit litigation usually focuses on whether the police acted prudently and reasonably under the circumstances (Schofield 1988).

In 1990, according to the National Highway Traffic Safety Administration, 314 people were killed in the United States as a result of high-speed pursuit driving. Many safety advocates believe that the cost of high-speed driving far outweighs the reward and is best left for the make-believe world of Hollywood (Ross 1992). Geoffrey Albert, a professor at the College of Criminal Justice, University of South Carolina, believes that "just as when a police officer fires a gun, we've got to look at it every time a police officer gets involved in a chase." Some states have already created strict guidelines for officers involved in pursuit. In other places, such as Mesa, Arizona, they have curbed high-speed chases dramatically in favor of alternatives such as aircraft surveillance. State lawmakers in Massachusetts are considering one of the most comprehensive laws in the country for high-speed pursuit. The proposal would create a uniform state guideline for all cities and towns and increase penalties for drivers who flee police. It would also mandate record keeping and training. They believe that right now the penalties are very minor for eluding the police, and that stiffer penalties would have a greater deterrent effect (Schofield 1988).

General Principles of Liability

The legal theory underlying most pursuit-related lawsuits is that the police were negligent in conducting the pursuit. A negligent action is based on proof of the following four elements:

(1) the officer owed the injured party a duty to not engage in certain conduct;

(2) the officer's action violated that duty;

(3) the officer's negligent conduct was the proximate cause of the accident; and

(4) the suing party suffered actual and provable damages.

Duty Owed.   Courts must first determine duty owed in a pursuit situation by examining the officer's conduct in the light of relevant laws and regulations of the department. In *Smith v. City of West Point,*[12] the court stated that the police are under no duty to allow motorized suspects a leisurely escape. However, the court said the police do have a duty of responsibility with respect to the manner in which they conduct the pursuit. Most states have laws exempting authorized emergency vehicles from certain traffic regulations but subject to certain qualifications. These privileged emergencies dictate the existence of an actual emergency, use of adequate warning devices, and the continued operation of the vehicle for the safety of others. Such things as reckless driving on the part of the officer invalidates the protection. The courts have translated the reasonable care standard into a duty to "drive with the care which a reasonable, prudent officer would exercise in the discharge of official duties of a like nature" (*Breck v. Cortez*).[13] Duty owed also requires that the driver follow departmental policy.

Proximate Cause.   Liability must be based on proof that police conduct breached a duty owed or was the proximate cause of the pursuit-related accident. Proximate cause is difficult to establish in some cases involving the intervening negligence of other drivers, such as when a fleeing motorist collides with an innocent person. Police may be liable because the accident was the proximate and foreseeable result of their failure to adequately warn other drivers in the pursuit. In *Nelson v. City of Chester, Illinois,*[14]

Cases 15–21

it was held that the city breached its duty to properly train its officers in high-speed pursuit driving, although the majority view in many courts is that police are not liable for accidents caused by the intervening negligence of fleeing violators. In *Dint v. City of Dallas*,[15] the court held that the police violated no legal duty to arrest or apprehend a fleeing motorist who subsequently collided with an innocent citizen. This was because the proximate cause of the accident was the suspect's negligent conduct in fleeing (Schofield 1988).

Immunity. Statutes in most states have a limited sovereign immunity to discretionary as well as ministerial decisions (Schofield 1988). The extent of immunity in a particular jurisdiction can be determined only by reviewing state laws and relevant court decisions.

Federal Civil Rights Act

Pursuit-related liability under the Federal Civil Rights Act, 42 U.S.C. 1983, requires proof that an officer's conduct violated a constitutionally protected right. However, certain techniques employed by police during the pursuit may raise constitutional issues. For example, in *Jamieson v. Shaw*,[16] the court held that the constitutionally permissible use of force standard set forth by the Supreme Court in *Tennessee v. Garner*[17] was violated when the passenger in a fleeing vehicle was hurt because of a "dead-man" roadblock. In *Brower v. County of Inyo*,[18] a high-speed chase of over twenty miles ended when the suspect was killed when his vehicle hit a tractor-trailer which had been placed around a bend where the driver did not have a clear option to stop after the roadblock was seen.

Suits by Injured Officers. Under certain circumstances police officers can sue a fleeing motorist or citizen for injuries incurred during a pursuit. For example, in *City of Redlands v. Sorensen*,[19] the court held that a police officer could recover for his injuries from the driver of a speeding vehicle who had violated a legal obligation to stop in response to a red light and siren.

Criminal Prosecutions. A fleeing motorist is also subject to criminal prosecution if a pursuing officer or other person is killed or injured during the chase. In *Commonwealth v. Berggron*,[20] the court held that a fleeing motorist could be convicted of negligent homicide for the death of an officer during a high-speed chase. Police officers are also subject to criminal prosecution for their conduct under certain circumstances. In *State v. Simpson*,[21] the pursuing officer was convicted of reckless driving for attempting to pass in a no-passing zone (Schofield 1988).

Factors Determining Liability

Pursuit-related litigation usually involves an inquiry into whether the pursuit was conducted in a reasonable manner under the circumstances of the case. Each situation is different and requires a particular assessment.

Purpose of Pursuit. The question to be asked is, "Does the purpose of the pursuit warrant the risks involved?" What is the nature and seriousness of the suspect's offense? Is the fleeing motorist suspected of a serious crime or only a misdemeanor?

Driving Condition. This factor involves a general assessment of equipment, roads, traffic, weather conditions, experience, and personal ability of the driver and other closely related conditions.

Use of Warning Devices. It is imperative that adequate visual and audible warning devices, such as flashing lights and sirens, be used in most, if not all, hot pursuit situations. These are not only for the benefit of the pursued suspect, but also for pedestrians and other innocent bystanders. However, over-reliance on warning devices

Cases 22–23

can cause serious problems. It should be remembered that some drivers have the windows up and the air conditioner and radio on. Others are preoccupied and do not see flashing lights. Many departments prohibit unmarked vehicles and ones not equipped with emergency lights and siren from participating in high-speed pursuit. It would seem that this is the best policy except under the most unusual and exigent circumstances.

Excessive Speed.   The speed during a chase should not be so high that the pursuing vehicle cannot stop for intersections, in case there is someone who has not heard the sirens or seen the flashing lights.

Disobeying Traffic Laws.   While it is understood that certain traffic laws can be disobeyed by pursuing police cars, dangerous and high-risk driving maneuvers must be cautiously executed because the police might be held for any resulting accidents.

Roadblocks.   When using roadblocks, special care is required to ensure that innocent persons are not placed in danger, and that the fleeing motorist is afforded a reasonable opportunity to stop safely. This means the roadblocks must be visible far enough ahead so that the suspect can stop easily and safely.

Use of Force.   Occasionally a pursuing police officer may want to resort to firearms or some other use of force to stop the suspect. Generally, such force should be used only when it has been authorized by a supervisor, who should be in control at all times.

Continuation of Pursuit.   There are times when a decision needs to be made whether or not to continue a pursuit. Often it is advisable to have a departmental policy that this decision rests with the supervisor or personnel directing the chase, rather than the driver. Many states have a ''Fresh Pursuit Law'' which authorizes officers from foreign jurisdictions to enter and continue the pursuit. But this is true only if the officer believes that the fleeing motorist committed a felony in the foreign jurisdiction (Schofield 1988).

The "Fighting Words" Doctrine.   Sticks and stones may break my bones but words will never hurt me. This childhood phrase is generally, but not invariably, true in police work. The First Amendment protects a significant amount of speech directed at law enforcement officers, including some distasteful name-calling and profanity. In *Nelbrader v. Blevins*,[22] a civil suit was filed against police officers by an arrestee claiming he did not engage in ''fighting words'' when he allegedly called the police officer a ''son of a bitch'' just prior to his arrest. The court held that there is a clearly established right against retaliation for constitutionally protected speech, and even if the officer thought the plaintiff did call him a ''son of a bitch,'' not every epithet directed at a police officer constitutes disorderly conduct (Schofield 1992).

An analysis of ''fighting words'' doctrine requires both the content of the spoken word and the context in which they were used to determine if words addressed to the law enforcement officer are protected by the First Amendment. Recent federal and state court decisions reveal four accepted principles that can assist officers in deciding whether or not they should proceed. The first, direct threats to officers, generally constitute ''fighting words'' unprotected by the First Amendment. Second, speech that clearly disrupts or hinders officers in the performance of their duty is not constitutionally protected. Third, the ''fighting words'' exception to the First Amendment protection requires a high standard for communications directed to the officer because professional law enforcement personnel are expected to exercise greater restraint in response to such words than the average citizen. Fourth, profanity, name-calling, and obscene gestures directed at officers do not, standing alone, constitute ''fighting words.'' In 1942, the Supreme Court upheld the conviction of *Chaplinski*[23] and defined ''fighting words'' as ''. . . those words which by their very utterance inflict injury or

Cases 24–29

tend to incite an immediate breach of the peace.'' The Court held that such words are not protected by the First Amendment and can be the basis for criminal prosecution. It should be noted that *Chaplinski* is the only Supreme Court decision upholding a ''fighting words'' conviction. In *Houston v. Hill,*[24] Raymond Hill observed his friend Charles Hill intentionally stopping traffic on a busy street, apparently to enable a vehicle to enter traffic. As two Houston police officers approached Charles Hill and began talking to him, Raymond Hill, in an admitted attempt to divert the officers, began shouting at the officers. After one officer responded, ''Are you interrupting me and my official capacity as a Houston police officer?'' Hill shouted something to the effect that why didn't the officer pick on someone his own size. Raymond Hill was then arrested and convicted under a city ordinance for ''willfully or intentionally interrupting a city policeman . . . by verbal challenge during an investigation.'' The Supreme Court ruled Hill's conviction violated his First Amendment rights. The Court suggested that ''fighting words'' exception to the First Amendment protection requires ''. . . a narrower application in cases involving words addressed to a police officer, because a properly trained officer may reasonably be expected to exercise a higher degree of restraint than the average citizen, and thus be less likely to respond belligerently to 'fighting words.' '' The Houston ordinance unconstitutionally criminalized speech directed to an officer because it broadly authorized police to arrest a person who in any manner verbally interrupts an officer (Schofield 1992). In *City of Bismarck v. Nassif,*[25] three police officers were sent to Nassif's residence after he called police to complain they were not doing anything about his earlier complaint. He also threatened to take the law into his own hands. When the officers arrived Nassif exited his house and, appearing upset, shouted loudly and acted aggressively. After attempting to reason with him, one officer told Nassif they were leaving. Nassif then said, ''You — — — I am going back into the house and get my shotgun and blow you bastards away.'' Based on this threat to their safety, the officers arrested Nassif for disorderly conduct. The Supreme Court of North Dakota concluded that Nassif's statement, along with the circumstances of the encounter with the police, constituted ''fighting words'' and was unprotected by the First Amendment. The North Dakota court relied on language from a Supreme Court opinion in which Justice Douglas wrote that the First Amendment protects a significant amount of verbal criticism and challenge directed at police officers unless that language is ''. . . shown likely to produce a clear and present danger of a serious substantive evil that rises far above public inconvenience, annoyance, or unrest.'' The court found that Nassif's threat to get his shotgun and shoot the officers was sufficient to produce a clear and present danger of a substantive evil.[26] In *Brown v. State,*[27] where an arrestee became loud and abusive and threatened to kill one of the arresting officers after being told to keep quiet, the defendant told one of the officers to take off one of the handcuffs so he could fight them and give the officers a ''Sicilian necktie,'' a rather vicious procedure. The Indiana Appellate Court upheld the defendant's disorderly conduct conviction finding that such threats, insults, and provocation directed solely at the arresting officers clearly falls within the ''fighting words'' category of unprotected speech because they ''. . . were stated as a personal insult to the hearer in language inherently likely to provoke a violent reaction.''[28] In *State v. Fratzke,*[29] the Iowa Supreme Court reversed the defendant's conviction of harassment for writing a nasty letter to a highway patrolman to protest a speeding ticket. The court felt that provocative speech that falls short of a direct threat to the officer's safety is protected by the First Amendment. While the letter to the trooper contained vile and vulgar terms, it emphasized that it was not to be interpreted as anything whatsoever in the way of a threat. The court noted that the threat, if there was one, contained a mode of expression far removed from face-to-face exchange, and that the letter was not

Cases 30–32

mailed to the trooper's home but to the clerk of the court, a neutral intermediary. The defendant was merely exercising his uniquely American privilege to speak one's mind even though it was not in good taste (Schofield 1992). In *Buffkins v. City of Oklahoma*,[30] the U.S. Court of Appeals for the Eighth Circuit found as a matter of law that officers could not have reasonably concluded that they had probable cause to arrest Buffkins for disorderly conduct for using "fighting words." Buffkins called the officers "asshole," and as a suspected drug courier, was detained at the airport. She protested that the officers conduct was racist and unconstitutional. She became increasingly loud during the detention and questioning. The officers eventually informed Buffkins that she was free to leave and told her to "have a nice day" to which she replied "asshole system" or "I will have a nice day, asshole." The officer then arrested her for disorderly conduct. In *Duran v. City of Douglas*,[31] the U.S. Court of Appeals for the Ninth District ruled that the First Amendment protects profanities and an obscene gesture directed toward a police officer and that the officer's subsequent detention and arrest of Duran for disorderly conduct was unconstitutional. Duran had been escorted from a bar for being intoxicated and threatening the bartender. Later, while on patrol, the officer observed Duran directing an obscene gesture toward him through an open window and the officer began following Duran's car. Duran stopped when the officer turned on his emergency lights, and Duran responded with profanities in both Spanish and English and was then arrested for disorderly conduct (Schofield 1992). A similar result was reached by the Supreme Court of North Dakota in *City of Bismarck v. Schoppert*.[32] The defendant walked past a police car and gestured at the officers with his little finger "— bitch cop." One of the officers asked the defendant what was the matter and he replied "— you." The officer then emerged from her car and stopped the defendant by grabbing his left arm. She asked him to identify himself and he again answered with obscene language. The defendant, who allegedly smelled of alcohol, told the officers, "You don't know who you're [expletive] with. You just bought yourself a federal lawsuit." The defendant then took one step toward the officer and was arrested for disorderly conduct.

The above cases indicate the "fighting words" doctrine is very narrow indeed, and that an officer must tolerate more abuse than the average citizen and be more restrained than is personally acceptable. Endangering the peace or a real threat of violence toward the officer appear to be the two aspects of "fighting words" doctrine that do not fall under the protection of the First Amendment.

CYNICISM

Cynicism can be defined as displaying an attitude of contemptuous distrust of human nature and mores. It is no respecter of rank, race, creed, or sex. It can grow within an individual or within an organization. If it is detected early there may be an easy cure. If not, the problem becomes more difficult. There is generally an incubation period that has early warning signs. Young officers entering enforcement frequently have feelings of deep commitment and a sense that they are entering a field which is worthwhile and meaningful to society. Many of them notice a gradual change in their relationship with friends and even relatives. The officer may start withdrawing from former associates and friends and increasingly spend time with other law enforcement personnel. The net result is a slow withdrawal from society. Officers get together and tell each other their problems and frustrations and one builds on the other. It is recognized that officers in general face the worst of society. That is their job. An officer is taught to be constantly on guard and to be suspicious of everyone and to trust no one. Older officers may be frustrated by the American system of justice because they have not been promoted or because they see themselves as being unappreciated by society in general and by their superiors (Behrend 1980).

The first line of defense against cynicism is to admit that it can happen to you or one of your colleagues. Training sessions with both peers and superiors present may be a starting point as are training bulletins and general background information. The best way to avoid and/or cure cynicism is sessions with non–law enforcement personnel. This may be a one-on-one dialogue between an officer and a citizen, or it may involve participation in community events and programs, including public issue meetings or membership in Kiwanians, Lions, and PTA. Every community hosts a number of sporting events, community projects and board appointments, commissions, and committees. Officers who assume an active role in community affairs accomplish two purposes. One, they see another side of the community and, secondly, they become acquainted with the community and its opportunities for departmental members. Officers may, by talking with non–law enforcement personnel officers, develop new friends and acquaintances and form entirely different viewpoints. Administrators should review the internal policies and procedures of their organization to ensure that the philosophy of the organization advocates an awareness of community values and does not foster an isolation attitude with respect to the department and the community. Administrators should encourage, and even insist, that officers attend nonlegal community functions. If this is one of the major objectives of the department the problems of cynicism may never arise (Behrend 1980).

STRESS

Law enforcement officers may be giving a citation one minute and friendly advice the next. This apparent dual personality is one of the many causes of stress. Personal problems, as well as work difficulties, can cause stress which, regardless of the cause, is cumulative. It may build up slowly and subtly, or it may come in an instant from some traumatic experience. The latter may be the witnessing of a bad accident, the death or serious injury of a fellow officer, or a failed rescue attempt. Critical incident stress that strongly affects emotions can reduce ability to cope even with ordinary tasks. The debriefing process should commence at once. The myth that a badge and a gun protect an officer against emotional shock is just that (Conroy 1990).

Critical stress can cause such aberrant behavior as excessive alcoholism, drug abuse, and extreme aggressiveness. Debriefing may be by peer, professional, or more likely a combination, depending on departmental policy. The idea behind debriefing is to encourage free exchange of thoughts, fears, and concerns in a supportive group environment and without losing status among one's peers. Debriefing sessions are much more successful and the feedback more positive when there is peer support. Debriefing allows individuals to gain insight and reframe the event in a different perspective. A short-term initial intervention often aids in preventing some of the long-term cumulative effects caused by traumatic incidents (Conroy 1990). It was said that Cornell Green of the Dallas Cowboys once stated, ''No matter how badly I am beaten on the field, by the time I reach the sideline I am convinced that I did the right thing.'' It may appear that Green was either unrealistic or supremely egotistical. He was neither. He knew that the next time he walked out on the field he could not have a failure hanging over his head and uppermost in his mind, instead of believing that he would succeed.

The same philosophy works for law enforcement officers. They make mistakes. Everyone makes mistakes. Officers cannot let mistakes be uppermost in their minds when they take on a new task. Debriefing after a stressful incident includes:

(1) eating three good meals a day;

(2) avoiding excessive sugars, salts, and fats;

(3) exercising regularly by either following work or leisurely pursuits;

(4) maintaining friendships;

(5) getting enough sleep and rest;

(6) limiting smoking or stopping completely;

(7) limiting alcohol and caffeine intake;

(8) identifying and accepting emotional needs;

(9) pacing oneself for even flow of demands;

(10) recognizing the early behavior or physical signs of stress and taking actions against them;

(11) allocating one's time and energy to allow for periods of rest and stimulation; and

(12) taking proper supplements if needed to balance the diet (Schaefer 1985).

All of this does not mean that we should try, or even hope to avoid, all stress. Stress creeps into our everyday activities. Stress should not distress us as long as we have an acceptable way to boot it out.

CONSENT SEARCH

A consent search is a useful tool when there is reason to suspect unlawful conduct but there is not enough evidence to justify a warrant. The consent search is a law enforcement tool that should be used very carefully and by officers who have a clear understanding of the citizen's rights and who are well trained and sensitive to the citizen's sense of intrusion and to the potential for abuse of police power (Morgan 1991).

When a person is in his home he has a much stronger expectation of privacy than when walking down the sidewalk or traveling in an airplane. He also has a similar strong expectation of privacy when using a telephone. Interestingly enough the courts have observed that we can have no such expectation when using a cordless phone.

This nation's theory of law and government assumes that, in the end, each of us is the best judge of our own self-interest. Citizens can choose when, where, and how to exercise their rights. It is the American tradition. These rights are not self-executing, that is, a person's right to freedom of speech does not compel one to speak. Consent searches, accompanied by probable cause, often provide the best possible case. It is important for law enforcement officers to understand the distinction between consent searches and searches requiring a warrant so that they may uphold a high standard of professionalism in conducting all searches (Morgan 1991).

In conducting consent searches it is crucial for the law enforcement officer to understand the inherent intimidation conveyed. Psychologically, letting any unexpected "guest" into one's home can represent a sort of intrusion. As one law professor commented, "For most people, a consent search is preferable, [however,] very few people would be able to resist a law enforcement request for search." In fact it is astonishing

Cases 33–35

the number of times people who have contraband in their possession consent to a search. Consent searches neither technically nor actually erode the Fourth Amendment's protection against unreasonable search and seizure. Even so, although they are not required to do so, officers should inform citizens of their rights to refuse. There is no advantage gained by threatening to get a warrant and "come back in a bad mood." Such a threat would be disrespectful and unlawful and, practically speaking, why use a threat that would invalidate a subsequent consent (Morgan 1991).

RIGHTS OF THE OFFICER

Police officers have the authority to exercise many discretionary powers. The execution of these broad powers affects the public significantly and often in the most sensitive areas of daily life. The exercise of police authority calls for a high degree of judgment and discretion, the abuse or misuse of which can have serious impacts on the individual.

It is sometimes overlooked that police officers have rights as public employees, and as citizens. First and foremost, they have the protection of the U.S. Constitution and the constitutions of most states (Brancato and Polebaum 1981). Police officers have certain rights under the First Amendment. For example, a police officer may be required to take an oath of allegiance. In general, no broader oath may be required of an applicant for public employment. Examples of impermissible oaths are ones which specifically promote respect for the flag, reverence for law, allegiance to the government, as well as ones which disavow membership in "subversive organizations." In most instances, a police officer may not be denied employment or fired on the basis of membership in a political party. The Supreme Court decision, *Elrod v. Burns,*[33] put an end to firing of nonpolicymaking, nonconfidential government employees who have satisfactorily performed their duties, based solely on the grounds of political belief. *Branti v. Finkel*[34] went even further than *Elrod* in protecting public employees from political dismissal.

The political activity of police officers may be restrictive. One source of restriction may be in the rules and regulations of the individual department. For example, some states deny the officer's right to campaign for candidates for public office or to be a member of a political club. Another source of restraint may be the "Little Hatch Acts" as written by the various state or local political entities. Federal law enforcement officers are prohibited under the Hatch Act from participating in partisan electoral activity. The Hatch Act forbids a federal employee from using his official authority or influence for the purpose of interfering with or affecting the results of an election or from taking an active part in the political management of an election or a political campaign (Brancato and Polebaum 1981).

It is not required of an officer to believe in a god as a prerequisite to being employed. The officer may be a member of a sect or a recognized religion. The question of whether or not a police officer may criticize departmental policy is a qualified yes. Disciplinary action may be taken against an officer if there is specific finding that the officer's exercise of free speech impaired his efficiency as a police officer or the department's effectiveness in fulfilling its responsibilities. In *Pickering v. Board of Education,*[35] the court of appeals noted that ordinarily a court must weigh the officer's freedom of speech against the department's interest in efficiently carrying out public responsibilities. But in *Pickering,* because the district court had not specifically determined that the officer's free speech activities impaired department efficiency, an essential showing whose burden rests on the department and not on the officer is necessary, the court held for the plaintiff. In most instances, a police officer may criticize peers or superiors if he

Cases 36–37

can demonstrate that it does not upset the harmonious relationship needed for a smooth-functioning department. In general, most departments have an internal procedure for handling complaints and this should be the first step. If the officer is dissatisfied with the results of internal procedures and airs the criticism publicly, the final decision will depend on whether or not the harmonious working relationship of the department was upset (Brancato and Polebaum 1981).

In general, the private lives of police officers cannot be governed unless their conduct adversely affects on-duty performance. This is based on a precept that the government must act as a rational employer. However, if officers openly flaunt the local customs and mores of the community by their aberrant behavior, they put themself at risk. Privacy rights have been traced to the "Liberty" due process clause protected by the Fifth and Fourteenth Amendments, to the security of the home provided by the Fourth Amendment, to the freedom of conscience and religion recognized in the First Amendment, the rights are attained by the people under the Ninth Amendment, and more recently penumbrae emanating from the First, Third, Fourth, Fifth, and Fourteenth Amendments. These constitutional provisions have protected matters as wide ranging as the right to travel abroad and the right to conceive (or not conceive) children, to bear (or not to bear) them as one sees fit. Privacy is also protected under the Federal Privacy Act of 1974 and under state laws and some state constitutions. For example, an unmarried officer generally may not be dismissed for living with a person of the opposite sex. However, if an unmarried heterosexual relationship becomes open and notorious it probably is not protected. On the other hand, a court might find that an unmarried police officer's relationship that became notorious or offensive (such as by exploitation on television) reflected a lack of concern for privacy and decorum and undermined the officer's reputation with the community. In such instances, firing the officer probably would be warranted because of the adverse impact of the officer's conduct on job performance. The right to privacy may protect the officer's relationship. For example, a New Jersey Supreme Court case, *State v. Saunders,*[36] struck down an anti-fornication statute as a violation of the criminal defendant's right of privacy. An adulterous relationship generally depends on two factors, the status of the adultery laws in the state, and the circumstances surrounding the adultery. There are jurisdictions where adultery is not criminal and, if discreet, might not be a cause for dismissal. However, if it is open and notorious and deviates from the community norms enough to hinder job performance the situation would be different (Brancato and Polebaum 1981).

In the case of homosexual associations, a forceful argument can be made that consenting adults should be protected by the Constitution from all government interference. However, such a doctrine has not been widely adopted. The degree of openness of an officer's homosexuality will affect his or her rights. The public employee may forfeit protection by misrepresenting or lying about his or her homosexual activities. Blatant or outrageous sexual conduct may be cause for dismissal. In *Fabio v. Civil Service Commission of Philadelphia,*[37] the Pennsylvania Supreme Court upheld the dismissal of a police officer who induced his wife to have sexual relations with a fellow officer, and who himself was having an affair with his wife's sister (Brancato and Polebaum 1981).

In general, police departments can regulate the appearance, grooming, and dress codes of officers. The question of whether or not conduct unbecoming an officer or "conduct prejudicial to good order" is valid. On this the courts are sharply divided. There is no clear answer. There have been dismissals for such things as shooting deer out of season and without a license, visiting illegal gambling clubs, accepting tricks

Turkey populations are increasing in many states.
Courtesy Wyoming Game and Fish Department.

from a local lawyer, and making improper advances to a woman acquaintance (Brancato and Polebaum 1981). By statute police officers owe a general duty to the public as a whole to prevent crime. This does not, in most instances, run to the individual members of society.

PUBLIC RELATIONS

An officer may never in the course of his duties be called upon to have a formal press conference. However, even off-the-cuff meetings with the press need preparation in order to avoid pitfalls. Mount (1987) quotes David Brinkley as saying, "When a reporter asks questions, he is not working for the person being questioned, whether businessman, politician, or bureaucrat, but he is working for the readers and listeners." Politicians may have a staff to prepare briefing books and other information before they meet the press. However, that is a luxury not enjoyed by the wildlife officer. The information given should be accurate and recent. Sooner or later officers are going to be asked questions they cannot answer. When this happens one should not be afraid to say so. One of the things an officer should be prepared for is emotional stress. It is one of the factors that will affect performance. An officer should listen carefully to questions and maintain eye contact with the interviewer. This will help to understand the question and avoid distractions. One of the techniques that is used quite often is a restatement of the question and having it confirmed by the interviewer. This helps assure that the officer, and everyone in the room, understands the question. More importantly, it gives the officer time to think. Do not argue. Do not allow emotional words or phrases to put one in an impossible position. Be prepared for loaded questions. Be sure to keep facts and opinions separate unless the opinions are supported by facts. Answer only one question at a time. If someone asks a multiple question, request that person to ask them one at a time. Many police officers have testified in court and been cross-examined by defense attorneys who have tried to confuse and discredit testimony. These experiences will help when the officer meets the press. Talking to the press and briefing people who are working against a deadline can be a problem. Don't try to sidestep hard questions unless necessary. In all cases be honest, straightforward, and sincere. Remember the reporters are doing their job just as you are trying to do yours (Mount 1987).

PUBLIC SPEAKING

When you are going to give a speech of any length it should be a prepared one. Or if you are going to testify under oath, give a deposition, or speak on any policy and procedure, or address some specific problem, be prepared. There is no substitute. Even if you are a calm and collected off-the-cuff speaker you can easily make mistakes that may later haunt you and the department. If it is available use a large print. You can read it much easier and can keep your place better. If it is not available use a copy machine that has an enlarger, and triple space the draft so that your eye can easily keep track of the words and the place. If one is available, by all means use a teleprompter. Short sentences are easier for you to read and are easier for the audience to understand. This does not mean that all sentences have to be brief. But keep most of them short (Gladis 1989).

Use ample margins that you can write in before presenting the speech or as you speak. Keep paragraphs short, and provide yourself with eyebreak spaces so that you can look at the audience while you are talking. Before the speech, conduct an extensive audience analysis so that you know to whom you are speaking and what their probable interests are, how many there will be, and what level of education they likely will have. It helps to rehearse out loud at least once or twice, and to tape your speech. The first time you listen to the tape you may be a bit shocked. Don't staple your draft. Clip it so you won't be tempted to flip each page, which distracts the audience.

Your delivery is as important as content. The introduction of a speech should be a honeymoon period between the speaker and the audience. It gives both of you a chance to know each other. You may want to personalize your speech by using jokes, short stories, and quotes. Whatever you do, keep your eye on the audience. One of the best speakers I have known, R. W. Eschmeyer, had an extremely effective technique. I asked him, ''Bill, how do you keep the audience with you so well?'' He said, ''Did you see the blond girl in the third row? Whenever she looked a bit confused I backed up and repeated something. If she looked like she understood, I kept going. I was talking only to her.'' You may want to give the audience a preview of what you are going to say so they can easily follow your speech. Gladis (1989) points out that research indicates that we all have two brains, a right one and a left one. The left brain—the side your speech will appeal to—understands words, but the right brain understands verbal gestures. If you use gestures during your speech the audience will get a double dose and will better understand your message.

Happy anglers are
often the wildlife
officer's best friends.
*Courtesy Wyoming Game
and Fish Department.*

Pause throughout the speech and look at the audience or simply take your time on one sentence before you go to the next. Try to make at least a fleeting eye contact with everyone unless the crowd is too large or too far away. Whatever you do, do not rush your speech. Speak slowly, clearly, and be sure that the audience understands. When you finish you should give a conclusion that people will remember, restating your basic points and your theme. Your conclusions, when possible, should be delivered extemporaneously. If you need to improve your public speaking skills, consider joining a club or other organization where you have a chance to practice and gain confidence (Gladis 1989).

AIDS

The risk of contracting AIDS from blood and body fluids of infected persons is very low in law enforcement duties. However, the consequences of an AIDS infection is devastating not only to the officer but to his or her family. In 1981, Michael Gottlieb at the University of California Medical Center treated five young male patients with a rare and deadly form of pneumonia and complete deficient immune systems. The term AIDS, or Acquired Immune Deficiency Syndrome, came into existence from Dr. Gottlieb's diagnosis of these five patients, although it had been described earlier under another name. The positive agent of AIDS is a virus called HIV (Human Immunodeficiency Virus). This virus destroys the host's immune system allowing other microorganisms, many of which are not normally dangerous, to invade and grow in the body. It is possible for the virus to remain dormant in the human cells for a short period of time or up to ten years or more. And even though the person who has been infected may not show AIDS symptoms or that of another disease, he or she is still infectious and can transmit the AIDs virus during this dormant period (Bigbee 1991).

The question that arises in the mind of law enforcement personnel is, "How do I avoid contracting AIDS when arresting someone who may be infected?" Blood and blood products, semen, vaginal secretions, saliva, human milk, urine, tears, and body organs have all been found to harbor HIV. As a precaution when dealing with any body fluid it should be assumed infectious, not only from HIV but also Hepatitis B or sexually transmitted diseases. Exposure to body fluids, such as those of an accident victim, is one possibility for AIDS transmission if the officer handling the person has open wounds on his hands. Another hazard is the prick of a possibly infected hypodermic needle when searching a person or vehicle for evidence. The answer to this in part is, do not put your hands where you cannot see. Devise another method of searching clothing, pockets, upholstery, and other invisible areas (Bigbee 1991).

At times insects have been suspect as transmitters of AIDS. To date no case of insect transmission has been shown according to Bigbee (1991). Tests have shown that the AIDS virus can survive at room temperatures in a blood sample for at least fifteen days. If the sample is refrigerated, the life of the virus is prolonged much further. However, in the dry state the virus can survive no more than three to thirteen days at room temperature, with most experts believing it lasts about seventy-two hours. Saliva has a lower concentration of HIV than that of blood or semen. However, that does not mean that precautions shouldn't be taken against saliva. If an officer is spit upon by a suspect, the saliva should be removed by cleaning with soap and water or rubbing alcohol. If saliva comes in contact with the officer's mouth or eyes, appropriate substances, such as mouthwash and eyedrops, should be used. Hands should be washed even though the officer was wearing latex gloves. These gloves need not be sterile, and they should be disposed of properly. Vinyl gloves are not recommended. If an injury occurs from a hypodermic needle or some other sharp instrument which may bear the blood of infected individuals, the injury should be treated as follows:

(1) allow the wound, unless bleeding severely, to bleed until all flow ceases; cleanse the wound with rubbing alcohol or isopropyl sponges, then soap and water;

(2) seek medical attention as soon as possible; and

(3) advise your supervisor.

The HIV virus is highly susceptible and fragile to common disinfectant, to drying and to heat. Ordinary liquid household bleaches, one part to nine parts water or seventy percent alcohol, will inactivate the virus within one minute. Moderately high heat will kill the virus as will an autoclave. Officers should keep in mind that when using bleaches and alcohol the two should never be mixed (Bigbee 1991).

Literature Cited

Bavin, C. R. 1976. Wildlife law enforcement. Paper presented at the National Symposium on Wildlife in America, U.S. Fish and Wildlife Service, Washington, D.C.

———. 1978. Guidance for new wardens. Paper presented at Orono, Maine, U.S. Fish and Wildlife Service, Washington, D.C.

Behrend, K. R. 1980. Police cynicism: A cancer in law enforcement? *FBI Law Enforcement Bull.* (Federal Bureau of Investigation, U.S. Dept. of Justice, Washington, D.C.) 49 (8): 1–4.

Bigbee, D. 1991. The law enforcement officer and AIDS, 4th edition. *FBI Law Enforcement Bull.* (Federal Bureau of Investigation, U.S. Dept. of Justice, Washington, D.C.).

Brancato, G., and E. P. Polebaum. 1981. *Rights of police officers. An American Civil Liberties Union handbook.* Avon Books.

Cattani, R. J. 1991. Harvard ushers in new president. *The Christian Science Monitor.* Oct. 18, 1991.

Conroy, R. J. 1990. Critical incident stress debriefing. *FBI Law Enforcement Bull.* (Federal Bureau of Investigation, U.S. Dept. of Justice, Washington, D.C.) 59 (2): 20–22.

Cunningham, S. A. 1986. Tactical driving: A multi-faceted approach. *FBI Law Enforcement Bull.* (Federal Bureau of Investigation, U.S. Dept. of Justice, Washington, D.C.) 55 (9): 18–21.

Eastmond, J. N., and J. A. Kadlec. 1977. Undergraduate education needs in Wildlife Science. *Wildlife Society Bull.* 5 (2): 61–66.

Gladis, S. D. 1989. Public speaking from a prepared text. *FBI Law Enforcement Bull.* (Federal Bureau of Investigation, U.S. Dept. of Justice, Washington, D.C.) 58 (3): 16–22.

Gordon, L. S. 1977. Hunter education's challenge for America's third century. In *North American Big Game.* Eds. W. H. Nesbitt and J. S. Parker. Washington, D.C.: The Boone and Crockett Club and the National Rifle Assn. of America 47:4.

Gundersen, D. F. 1978. Police uniform: A study of change. *FBI Law Enforcement Bull.* (Federal Bureau of Investigation, U.S. Dept. of Justice, Washington, D.C.) 47 (4).

Halberstam, D. 1991. *The next century.* New York: William Morrow Incorporated.

Higgenbotham, J. 1985a. Defending law enforcement officers against personal liability in constitutional tort litigation (part 1). *FBI Law Enforcement Bull.* (Federal Bureau of Investigation, U.S. Dept. of Justice, Washington, D.C.) 54 (4): 24–31.

———. 1985b. Defending law enforcement officers against personal liability in constitutional tort litigation (conclusion). *FBI Law Enforcement Bull.* (Federal Bureau of Investigation, U.S. Dept. of Justice, Washington, D.C.) 54 (5): 25–31.

Mangan, T. J., and M. G. Shanahan. 1990. Public law enforcement/private security: A new partnership? *FBI Law Enforcement Bull.* (Federal Bureau of Investigation, U.S. Dept. of Justice, Washington, D.C.) 59 (1): 18–22.

Morgan, R. 1991. Knock and talk: Consent searches and civil liberties. *FBI Law Enforcement Bull.* (Federal Bureau of Investigation, U.S. Dept. of Justice, Washington, D.C.) 60 (11): 6–10.

Morse, W. B. 1973. Law enforcement—one-third of the triangle. *Wildlife Society Bull.* 1 (1): 39–44.

Mount, H. A. 1987. In and out of a question-and-answer period successfully. *FBI Law Enforcement Bull.* (Federal Bureau of Investigation, U.S. Dept. of Justice, Washington, D.C.) 56 (3): 17–22.

Ross, E. 1992. Calling all cars! Police may curtail high-speed chases. *The Christian Science Monitor.* April 9, 1992.

Schaefer, R. B. 1985. Maintaining control: A step toward personal growth. *FBI Law Enforcement Bull.* (Federal Bureau of Investigation, U.S. Dept. of Justice, Washington, D.C.) 54 (3): 10–14.

Schofield, D. L. 1988. Legal issues of pursuit driving. *FBI Law Enforcement Bull.* (Federal Bureau of Investigation, U.S. Dept. of Justice, Washington, D.C.) 57 (5): 23–30.

———. 1992. The "fighting words" doctrine. *FBI Law Enforcement Bull.* (Federal Bureau of Investigation, U.S. Dept. of Justice, Washington, D.C.) 61 (4): 27–32.

Schwartz, B. 1974. *The American Heritage history of the law in America.* Ed. A. M. Josephy. New York: American Heritage Publ. Co., Inc.

Sigler, W. F. 1975. Recommended: B.S. degree for state wildlife law enforcement officers. *Wildlife Society Bull.* 3 (4): 173–175.

Troganowicz, R. C., and D. L. Carter. 1990. The changing face of America. *FBI Law Enforcement Bull.* (Federal Bureau of Investigation, U.S. Dept. of Justice, Washington, D.C.) 59 (1): 6–12.

Notes

1. New Scotland Yard, the police force of greater London might be likened to the detective force of New York City proper, or what is one of the world's largest police forces, the Los Angeles (Calif.) County Sheriff's Dept.
2. P. W. Schneider, Northwestern Regional Executive, National Wildlife Federation, to Thomas L. Kimball, Aug. 29, 1978.
3. Alan S. Krug, Midregional Executive, National Wildlife Federation, to Sigler, Aug. 31, 1978.
4. P. W. Schneider, Northwestern Regional Executive, National Wildlife Federation, to Thomas L. Kimball, Aug. 29, 1978.
5. Alan S. Krug, Midregional Executive, National Wildlife Federation, to Sigler, Aug. 31, 1978.
6. Ibid.
7. Thomas L. Kimball, Executive Vice President, National Wildlife Federation, to Sigler, Oct. 16, 1978.
8. Alan S. Krug, Midregional Executive, National Wildlife Federation, to Sigler, Aug. 31, 1978.
9. 403 U.S. 388 (1971).
10. 102 S. Ct. 2727, 2736 (1982).
11. *Truog v. United States,* 56 Ariz. 269, 107 P. 2d 203 (1940).
12. 475 So. 2d 816 (Miss. 1985).
13. 490 N.E. 2d 88 (Ill. App. 1986).
14. 733 S.W. 2d 28 (Mo. App. 1987).
15. 729 S.W. 2d 114 (Tex. App. 1986).
16. 772 F. 2d 1205 (5th Cir. 1985).
17. 105 S. Ct. 1964 (1985). In this the Supreme Court held that the use of deadly force to apprehend an unarmed fleeing felon was an unreasonable seizure which violated the Fourth Amendment.
18. 817 F. 2d 540 (9th Cir. 1987).
19. 176 Calif. App. 3d 202 (1985).
20. 496 N.A. 2d 660 (Mass. 1986).
21. 732 P. 2d 788 (Kan. App. 1987).
22. 757 F. Supp. 1174 (D. Kan. 1991).
23. *Chaplinski v. New Hampshire,* 315 U.S. 568 (1942).
24. 482 U.S. 451 (1987).
25. 449 N.W. 2d 789 (N. Dak. 1980).
26. See "Constitutional law—First Amendment—North Dakota's disorderly conduct statute: Is it limited to fighting words or unconstitutionally overbroad and vague?" 67 N. Dak. L. Review 123 (1991).
27. 576 N.E. 2nd 605 (Ind. App. 3 Dist. 1991).
28. Ibid.
29. 446 N. W. 2nd 781 (Iowa 1989).
30. 922 F. 2d 465 (8th Cir. 1990 *cert. denied,* 112 S. Ct. 273 (1991).
31. 904 F. 2d 1372 (9th Cir. 1990).
32. 469 N.W. 2d 808 (N. Dak. 1991).
33. 427 U.S. 347,375 (1976).
34. 48 U.S.L.W. 43 (U.S. Apr. 1, 1980).
35. 391 U.S. 563,568 (1968).
36. 75 N.J. 200, 381 A. 2d 333 (1977).
37. 48 U.S.L.W. 2752 (Pa. Apr. 30, 1980).

Recommended Reading

Baird, D. 1983. An Idaho tragedy—what can we learn? In *Proceedings of Western Assn. of Fish and Wildlife Agencies,* No. 63, 65–68.

Bristow, A. P. 1982. *Rural law enforcement.* Boston: Allyn and Bacon, Inc.

Morse, W. B. 1982. Sidearms policy and assaults on wildlife law enforcement officers. Paper presented at Annual Meeting of the Midwest Fish and Game Law Enforcement Assn.

Palmer, C. E. 1975. Wildlife law enforcement: A sociological exploration of the occupational roles of the Virginia Game Warden. Ph.D. diss., Virginia Polytechnic Institute and State Univ.

Walsh, W. F. and E. J. Donovan. 1984. Job stress in game conservation officer. *Journal of Police Science Administration* 12 (3): 333–338.

Multiple-Choice Questions

Mark only one answer.

1. The risk of contracting AIDS in the line of duty for a law enforcement officer is:
   a. low
   b. very low ⌐
   c. medium
   d. high
   e. very high

2. One of the following statements is not true.
   a. probably the most dangerous time for an officer when dealing with an AIDS victim is a blind search where the officer may be pricked with an infected needle
   b. AIDS was first described and named in 1981
   c. the HIV virus may remain dormant in human cells for ten years
   d. body fluids of AIDS patients can be dangerous to an officer who has open wounds
   e. an officer that is spit upon by an AIDS carrier is sure to be infected

3. Regarding cynicism, which is true?
   a. it is an attitude of contemptuous distrust of human nature
   b. it is no respecter of rank, race, creed, or sex
   c. it can grow within an organization as well as within an individual
   d. if it is detected early there may be easy cures
   e. all of the above are true

4. Bobbies in England were named after:
   a. a bob
   b. Sir Robert Armstrong
   c. Sir Robert Peel
   d. the King's guards ⌐
   e. the head of the shire

5. The Royal Canadian Mounted Police came into being in:
   a. 1840
   b. 1850
   c. 1860
   d. 1870 ⌐
   e. 1880

6. A hundred years after Peel, which two people most rapidly advanced the goal of professional law enforcement?
   a. Vollmer and Wilson ⌐
   b. Johnson and Wilson
   c. Vollmer and Johnson
   d. Wilson and Johnson
   e. Roosevelt and Tate

7. Regarding a uniformed officer, which of the following is not true?
   a. officers who do not feel their behavior to be in accord with the stereotypes of uniform will wear them anyway
   b. a distinctive formal uniform commands respect
   c. a more casual uniform may at certain times and places be most appropriate
   d. a uniform identifies an officer as such in an arrest situation
   e. officers may modify their behavior to correspond to the stereotypes associated with the uniform

8. In case of civil litigation against an officer, which of the following is not a valid defense?
   a. if the defendant (officer) did not act with ''permissible intentions'' the qualified immunity defense is available
   b. technical defense includes proper service and venue and a lack of jurisdiction
   c. the first argument that can be made is that the plaintiff has failed to state a claim against a law enforcement officer upon which relief can be granted
   d. qualified immunity defense shields law enforcement officers from liability if they are found to have acted reasonably under the existing laws at the time of the incident
   e. qualified immunity defense has both objective and subjective components and the shield of qualified immunity is not available if a public official knew or should have known the action he took was illegal

9. Some of the dangerous factors of pursuit driving are:
   a. lack of a written policy by the department
   b. no supervisor available to advise or order the pursuing officer how to proceed at any given stage in the pursuit
   c. high speed pursuit at times poses a greater risk to innocent citizens than if the police were using a deadly weapon
   d. the basis for most pursuit-related liability is negligence
   e. in spite of all publicity to the contrary, very few people are killed in the United States as a result of high-speed pursuit driving—it is relatively safe for everyone concerned

10. The legal theory underlying most pursuit-related lawsuits is that the police were negligent at one time or another. A legal action is based on proof of one of the following elements.
    a. the officer owed the injured party a duty not to engage in certain conduct
    b. the officer's action violated that duty
    c. the officer's conduct was not negligent but he is still susceptible to a law suit
    d. the officer's negligent conduct was the proximate cause of the accident
    e. the suing party suffered actual and provable damages

11. In the case of the "fighting words" doctrine, which is not true?
    a. the First Amendment protects a significant amount of speech directed at law enforcement officers
    b. the speech directed at officers may include profanity and distasteful name calling
    c. the "fighting words" doctrine exception to First Amendment protection has very limited application for law enforcement officers
    d. law enforcement officers are expected to exercise a higher degree of restraint to "fighting words" than the average citizen
    e. citizens are protected by the First Amendment even though their actions and words disrupt the performance of officers in their constitutionally protected duty

12. Police officers have certain rights under the First Amendment. Which is not one of these?
    a. the officer does not have to take an oath of allegiance
    b. generally the oath of allegiance may be required of applicants for public employment but the requirement can be no broader than this
    c. impermissible oaths are those which specifically promote respect for the flag, reverence for law, and so forth
    d. in most cases a police officer may not be denied employment or fired on the basis of membership in a given political party
    e. nonconfidential government employees, who have satisfactorily performed their duties, may not be fired on the grounds of political beliefs

Chapter 7

The Wildlife Officer in Court

Case 1

An officer may be required to appear in court either as a witness or as the prosecutor for his own case.[1] In either event, adequate preparations and complete familiarity with the details are absolutely necessary. Since a full treatment of case law is outside the scope of this book, only a few salient points will be introduced. Before going ahead, however, it seems desirable to present in a generalized manner a verbal picture of court procedure.

GENERALIZED PICTURE OF A TRIAL

Briefly and considerably oversimplified, a criminal case is tried according to the following pattern. The prosecuting attorney makes a statement briefly summarizing what he or she expects to prove. The defense attorney makes (or is entitled to make) a statement outlining the defense and what he or she expects to prove or disprove. The prosecution then puts its witnesses on the stand for direct examination. Witnesses can be cross-examined by opposing attorneys, but counsel normally cannot cross-examine their own witnesses, nor can they question the validity of their witnesses' own testimony. After the direct evidence by the prosecution is in, the defense may place its witnesses on the stand. After the defense presents its case, rebuttal may be given by the prosecution. The defense may meet this rebuttal evidence by what is known as rejoinder. After all the evidence is in, each side may address the jury. Finally, the prosecution closes the argument by summarizing the material to the jury. The jury is then charged by the judge and its verdict is returned. The sentence is pronounced by the judge in the event the defendant is found guilty. Motions by the defense for a retrial must be made within a specified time.

THE BRIEF

Although it is not expected that every prosecution will result in a conviction, the officer should have such grounds (in the opinion of the district attorney) as will justify instituting the proceedings. The legal instrument that permits the prosecutor to view these grounds is the brief. It is recognized that, at times, a prosecution is justifiable and desirable even when it is a moral certainty that a conviction will not be had.

The brief discloses the strength and weakness of the prosecutor's position, and the possible lines of defense. This helps to determine if a trial is justifiable under evidence submitted. Officers should always prepare a brief of the evidence for their own and for the district attorney's information.

The average district attorney handles dozens of cases a month and can give a misdemeanor charge only limited time. The attorney may not be familiar with fish and game laws, with fish and game conditions, or with practices prevailing in the field. If the attorney can sit down with a brief of the case, illustrated with sketches and photographs, he or she can better grasp quickly the circumstances and possibilities of action. Furthermore, the attorney will have the brief at hand during the trial as a memory refresher and to better oppose the defense during cross-examination.

The brief should list the code sections involved and the alleged offenses. It should give the who, what, when, where, why, and how of the case, all set forth chronologically. It should give the position, observation, and action of the officer or officers. It should list witnesses, and a summary of the testimonies they can or will give. Photographs and sketches should be attached.

PREPARING TO APPEAR IN COURT

Review of the Violation

A refreshed memory will help the trial go smoothly and will appear to present a vivid recollection of events as they occurred. While it is not expected that an officer will remember intricate sets of figures or complicated details, a constant referring to notes may impair the value of the testimony. Officers should also bear in mind that if they refer to their notes in court, the opposing council may ask to see the notes and may cross-examine on them.

For this reason, the officer should rehearse until mentally able to reconstruct the crime and the evidence pointing to it. It is helpful to arrange the material chronologically; then to review it thoroughly from original notes, sketches, photographs, or from any other material the officer plans to use on the witness stand.

Preliminary Discussion

In the courtroom, the officer-witness and the prosecutor should function as a team. A preliminary discussion before the trial begins helps to decide the approach to the case and the line of questioning.

In approaching the case, serious consideration is given to the types of witnesses, the particular judge, and the impaneled jury. In any approach, it is necessary to act with deference toward both the court and the jury.

Good courtroom technique requires that evidence be presented in a logical, concise manner in order to put over a point. During pretrial, the prosecutor learns what the officer is going to testify to and how. The witness learns what questions the prosecutor is going to ask and how they will be asked. The officer also learns how to react to the questions from both prosecutor and defense attorney.

The preliminary discussion should set the officer at ease and help clarify exactly what any question means. Witnesses can be prepared for affirmative examinations, but they obviously cannot be prepared for all possible cross-examination questions. Some of these questions are used to test the credence and knowledge of the witness, but they may be used also to check consistency of testimony.

The witness should not become excited. He or she should be sure the question is understood before answering it, and should not guess at answers. If the answer is not known, the witness should so state. A witness may answer a question in part or ask for a restatement or further explanation. If a witness later realizes that he has misinterpreted and misanswered an earlier question, the witness should so state his position and correct the response to the earlier question.

An officer testifies in court.
Courtesy Idaho Fish and Game Department.

TESTIFYING IN COURT

When witnesses for the prosecution present their testimony in a simple yet firm and sound manner, it creates a more favorable impression on the court with regard to the charge against the defendant, than a case marked by vacillating witnesses who are uncertain in their testimony, poorly prepared by council, and ignorant or unmindful of accepted courtroom procedure. The observation of a few general procedures will improve the officer's ability to testify effectively.

General Procedure

All statements should be made in a conversational tone, but loud enough for the jury and the court to hear and understand. Most answers should be directed toward the jury. All speech should be courteous. The court should always be addressed as "Your Honor," and attorneys should be answered by "Yes, Sir (or Ma'am)," or "No, Sir (or Ma'am)" to questions that can properly be answered that way. At no time should the officer attempt to be witty or sarcastic. If a question is not understood, the officer should so state and ask for a restatement or rephrasing. An undue amount of time need not be taken to answer questions; on the other hand, answers snapped back too quickly may lead to a lack of confidence in testimony and, worse still, may cause a misstatement which can later be used against the officer.

The truth should always be told, even though it is favorable to the defendant. This does not mean volunteering information which will mitigate against the officer's own case, but it does mean answering questions truthfully. The officer should not hesitate to correct mistakes made in earlier testimony. The officer, after realizing that a mistake has been made, should say that the testimony had been "so and so," and that it is now realized this was an error, and the true statement is "so and so." When a time, distance, or speed cannot be defined exactly, it should be stated as "on or about," or "the speed was about."

It is important that the officer appear to be impartial and fair and not overly anxious to convict. The slightest bias should never be shown toward a defendant. The officer does not assume that the defendant is guilty until it is so stated by the court. On the other hand, the officer should assume that the evidence points to the defendant's guilt. The person charged should always be referred to as the defendant.

Witnesses should not volunteer information over and above that asked or implied by a question from the opposing attorney. The exception is questions that naturally lead to a general discussion of events, times, and places. For example, ''Tell the court what happened on July 25 at 7:00 P.M. near Henrysville.''

A defendant is never on trial for past offenses, and reference to past criminal records should be made only if requested by counsel (see entrapment page 212). Counsel may frequently try to bring out the past criminal record of an individual in order to prejudice the jury. Even though the testimony may be stricken from the record, it cannot be so easily stricken from the minds of the jurors. In this event, it is axiomatic that an officer must go into court with ''clean hands.'' An officer with a record of law violations, even before becoming an officer, will frequently find his or her testimony completely discredited by a clever lawyer.

After the testimony is complete the officer should leave the court, unless asked by the court or prosecuting attorney to stay. Staying in court after testifying may create the impression that the officer is overly anxious to convict or overly concerned about the outcome of the case.

Cross-Examination

During cross-examination the officer should be particularly careful about walking into traps set by opposing counsel. Frequently, cross-examination will be used to divert the jury's attention from real issues and put the officer, rather than the defendant, on trial in the minds of the jury. The opposition may attempt to show that the officer is not worthy of belief because of character or reputation, or that the officer is prejudiced, or that the testimony itself is incompetent or immaterial.

Some cross-examination tricks include misquoting the witness, or demanding a yes or no answer. In the former trick, the defense misquotes the officer's testimony, then glibly proceeds with the questioning. It is important that the officer immediately call the misquote to the attention of counsel by saying, ''I am sorry I have not made myself clear, but,'' then restate the facts as in previous testimony.

The defense attorney may demand a yes or no answer when the question is such that neither one, without qualification, can possibly be a correct answer. The officer should state that the question cannot be correctly answered by either yes or no, and that he is willing to give the facts as he understands them.

Another trick used on witnesses is for the defense attorney to shout, ''Whom have you talked to about this case, and who told you to testify the way you have?'' The witness may answer without thinking, ''No one.'' Actually, there are two questions, and the correct answer is generally that the officer has talked to his superior and that he is telling the truth to the best of his ability, and no one has told him to testify otherwise. If the questioner accuses the witness of talking to other people about the case, then the best answer is to list who has been talked to, such as the prosecuting attorney and superiors.

The officer may also be accused of refreshing his memory, which may be true, and is perfectly legitimate. The officer should state that he has used the process known as ''refreshing and recollecting.'' It may be that he has revisited the scene of the crime or discussed the affair with other witnesses, both of which are proper.

TESTIFYING

Some officers, no matter how experienced they are in testifying, have a tendency to lose their composure and, therefore, diminish their credibility. To help officers overcome this problem, the New York City Transit Police Department initiated a seminar.

At the beginning of each session, the seminar stresses that the witness is on trial just as the defendant is. A case may stand or fall based on how expert witnesses handle themselves during testimony. Course instructors continually remind the detectives that, while they have the same rights as any other witness, more is required of them. Simply by virtue of their position they are expected to testify competently, forthrightly, expertly, and without panic. Each seminar participant receives a manual that outlines the four cardinal rules that detectives should follow when testifying:

(1) bring every police report on the case to court;

(2) read all police reports, grand jury minutes, and hearing minutes before taking the stand;

(3) listen carefully to all questions before answering; and

(4) be patient and courteous, do not be a wise guy (Bratton and Esserman 1991).

The Transit Police also use a training film that shows the wrong way and the right way for a witness to behave. Throughout the seminar the instructor emphasizes the vital importance of accurate and complete notes from the moment a suspect is taken into custody until the trial. This document includes complaint, follow-up, information forms, Miranda rights, certification, field investigative work sheets, lineup forms and arrest sheets. Every step in the process is documented by a complete report or work sheet. Officers who want more details should contact the New York City Transit Police Department.

Statements of Time or of Possession

In cases involving wildlife law violations, the elements of time and possession generally play an important role in helping to establish innocence or guilt. The officer-witness should, therefore, know how to answer questions based on these elements.

Frequently the term "on or about" is used in referring to the time element. For example, "On or about 5:00 P.M. I saw John Doe shoot and kill a drake pintail." If shooting time for migratory waterfowl closed at 5:00 P.M., this type of statement would be inconclusive evidence since opposing counsel could contend that it might just as well have been before 5:00 P.M. as after. In case it was not permissible to have mourning doves in possession between 5 and 15 March of a particular year, the following statement is admissible: "On or about 10 March, but between 5 and 15 March, I found John Doe in possession of ten mourning doves."

THE ROLE OF VIDEOTAPE IN COURT

Many cases are lost because the crucial witness is no longer available or is unwilling to testify at the time of the trial. Prosecutors try to combat this problem with the use of videotaped testimony. In some instances videotaped interviews are allowed in lieu of the victim actually appearing before the grand jury. Expert witnesses may be allowed to videotape their testimony rather than appear in person. In order to overcome defense objections, courts usually allow both the prosecutor and the defense attorney to participate in a joint deposition with the expert witness. For many years police have been recording suspects' confessions on audiotape. However, how this was obtained often becomes an important issue during the trial. Besides saving court time and relieving victims and other witnesses from continuously repeating testimony, videotape presents prosecutors with an additional advantage. Perhaps most important is the fact that the video presentation may arouse jurors interest and curiosity and may enhance their appreciation for the information being presented. It may also refresh the officer's memory (Giacoppo 1991).

Perjury

Perjury is committed by willfully and corruptly giving, in a judicial proceeding or court of justice and upon a lawful oath or in any form allowed by law to be substituted for an oath, any false testimony material to the issue or matter or inquiry. Perjury means that false testimony must be given willfully and corruptly, and that the accused did not believe it to be true. A witness may commit perjury by testifying that he knows a thing to be true when, in fact, he knows nothing at all about it, or is not sure about it, or is reasonably sure that it is not true.

A witness may testify falsely as to his belief, remembrance, impressions, opinion, or judgment, and thereby commit perjury. If a witness swears that he does not remember certain matters when, in fact, he actually does, or testifies that his opinion is sure when, in fact, it is otherwise, he commits perjury.

The oath must be required or authorized by law and must be duly administered by one authorized to administer it. When a form of oath has been prescribed, a literal following of such form is not essential, as it is sufficient if the oath administered conforms in substance to the prescribed oath. An oath includes an affirmative when the latter is authorized in lieu of an oath (United States Government 1975).

It is no defense that the witness appeared voluntarily, or that he was incompetent as a witness, or that his testimony was given in response to questions that he could have declined to answer, or even that he was forced to answer over his claim of privilege (United States Government 1975).

CONDUCT IN COURT

The officer-witness should be alert, but relaxed and dignified, on the witness stand. The officer should sit erect, with feet on the floor and with hands folded easily in the lap or placed on the arms of the witness chair. Under no circumstances should the officer allow himself to indulge in nervous mannerisms, or in any way react so that the jury will be disconcerted or annoyed. The officer should maintain an even temper and not be evasive or argumentative with either the opposing council or the court.

The officer's personal appearance should be above reproach. Such items as shined shoes; clean, well-pressed clothes; and freshly shaven face, or a well-trimmed beard are absolutely necessary.

DEFENSE ATTORNEYS

Defense attorneys are sometimes said to fall into two general categories: the friendly type and the brow-beating type. When acting as a witness, the officer should recognize that opposing attorneys are in court to win cases, and should not be taken in by either type. Frequently, the friendly attorney will attempt to lull the witness into a false sense of security and divulge things which he normally would not. The brow-beating attorney may try to intimidate the witness, or to so abuse the witness that he becomes enraged, loses his temper, and fights back, thereby prejudicing the jury against himself and the case. In extreme cases, a witness may ask the court to intervene in his favor.

It should be kept in mind that law cases are no longer largely contests between brilliant lawyers. Trials are conducted in a much greater measure by judges than they were a few decades ago. In dealing with a lay-judge it is not advisable to become too technical, nor to appear in any way to reflect upon the judge's legal knowledge.

Literature Cited

Bratton, W. J., and D. M. Esserman. 1991. Post-arrest training. *FBI Law Enforcement Bull.* (Federal Bureau of Investigation, U.S. Dept. of Justice, Washington, D.C.) 60 (11): 23–25.

Giacoppo, M. 1991. The expanding role of videotape in court. *FBI Law Enforcement Bull.* (Federal Bureau of Investigation, U.S. Dept. of Justice, Washington, D.C.) 60 (11): 1–5.

United States Government. 1975. *Manual for courts-martial.* United States 1969 (Rev. Ed.). Executive Order 11476 (1969) and Executive Order 11835.

Notes

1. In this text only the officer's role as witness is discussed.

Multiple-Choice Questions

Mark only one answer.

1. For an officer and the district attorney to go over testimony in advance of the trial is:
   a. legal
   b. wise
   c. an advantage in presenting the case
   d. practiced widely
   e. all of the above

2. The process known as memory refreshing is:
   a. illegal
   b. legal but unethical
   c. not necessary if an officer is well prepared
   d. legal, ethical, and wise
   e. sometimes used by rookies

3. A district attorney cross-examines his or her own witness:
   a. almost never
   b. anytime
   c. if the witness makes a mistake
   d. if the district attorney decides that the witness is hurting the case
   e. absolutely never under any circumstances

4. As time for a case to be tried approaches, the officer-witness and the district attorney should not:
   a. have a preliminary discussion because it helps the case
   b. consider the witnesses to be called and discuss them
   c. take the attitude of the judge and his or her track record into consideration
   d. consider who the probable jurors will be and how they will react to certain evidence
   e. act with deference to the judge and jury because it will bias them

5. The definition of a brief is:
   a. anything that is brief
   b. papers which are brief
   c. papers prepared by the wildlife officer that are short
   d. a detailed description of a pending case prepared by the officer for himself and the district attorney
   e. a detailed description of the pending case prepared by the officer for himself

6. Which of the following is a near-cardinal sin for an officer-witness?
   a. to tell the truth when it hurts the case
   b. saying one doesn't know, when that is actually the case
   c. to answer the opposing attorney's question and then volunteer considerable information
   d. to wear a well-cared for and neat uniform and shined shoes
   e. to at times look at the jury

7. Do experienced officers ever loose their composure on the witness stand?
   a. yes, occasionally
   b. never
   c. often
   d. as much or more than rookies
   e. so they are human, okay

8. A brief should not include:
   a. the alleged offense and the code
   b. the officer's opinion of the suspect's guilt
   c. the who, what, when, where, why, and how of a case
   d. a list of witnesses
   e. photographs

9. A perjury is:
   a. willfully and corruptly giving false testimony under oath
   b. lying generally, but not necessarily, under oath
   c. swearing to a statement that is not true but believed to be true by the witness
   d. any misstatement under oath
   e. saying one is not sure when that is more or less the case

10. When an officer does not understand the question of the defense attorney he or she should:
   a. so state that to the attorney by saying, ''I do not understand the question''
   b. ask the opposing attorney to please use better English
   c. turn to the judge and insist that the question is not clear
   d. use a head shake and mumble something unintelligible and not understandable
   e. point out to the district attorney that the question is unclear and ambiguous

Chapter 8

Violation of Wildlife Law

HOW AND WHY LAWS ARE BROKEN

Laws are a formal expression of people's belief about right and wrong. If laws are passed for the benefit of society in general, why then do some people continue to break them? This question is under constant discussion in and out of court and will be for a long time into the future. The list of reasons why people break wildlife laws is, fortunately, much shorter than for laws in general. Bruce Johnson, former chief law enforcement officer for the Utah Division of Wildlife Resources, lists five general categories (*The Salt Lake Tribune*, September 1, 1991, page A16). These are:

(1) the thrill killers, i.e., people who kill for an emotional high and generally let the animal lie where it falls;

(2) illegal trophy hunters, i.e., people who want their particular animal or group of animals to display or list as a special trophy. These animals may be taken at any time, including winter, and any place, including national parks. The cost is often no object;

(3) commercial hunters, i.e., often people who operate an illegal guide service. Their only object is a satisfied customer. This group also includes those people who illegally sell animal parts, such as bear gallbladders, deer family antlers in the velvet, and parts or whole eagles;

(4) opportunists, i.e., people who do not go afield with the intention of breaking the law but will do so if the opportunity to take game illegally is too great to pass up; and

(5) subsistence hunters or anglers, i.e., people who want to put food on the table.

There are many more reasons why people break wildlife laws, including ignorance. But the five listed above cover most situations and include those violations most destructive to the resource.

Thrill killers of wildlife do it for the emotional high they have following a kill and of having outsmarted the local game warden. Tom Wharton (1991) interviewed one reformed deer killer in Utah, Kim Jensen, who says in 1972 he and a relative killed, in three nights, sixty-five four points or better bucks. Jensen told Wharton that undercover agents can rarely infiltrate such groups of thrill killers. This is because they are a close-knit family or longtime friends and they do not trust strangers. Jensen says they keep tabs on local wardens and may, at times, even use a dummy poacher to lead the warden astray. Because of their mode of operation and their constituency they are hard to catch and are deadly on wildlife. If a particular family or group is suspect, concentration on its activities may pay off. Informers are another possibility, although finding one in a small town may be quite a problem.

Spotting scopes are
used by hunters and
officers alike.
*Courtesy Wyoming Game
and Fish Department.*

Illegal trophy hunters and sellers of illegal animal parts are caught most often by covert operations. Routine patrols by known law enforcement personnel pay low dividends. Wildlife officers or agents normally infiltrate a known or suspected illegal operation in order to gather evidence that will stand up in court. In the traditional undercover operation, an agent masquerades as a criminal and seeks to involve himself in an already existing and ongoing criminal enterprise. The FBI has begun to rely more on the ''sting'' operations in which the criminal activity itself is bogus. In such operations, the agents establish a criminal enterprise which is supposed to provide criminal opportunities and, thus, attract those ''predisposed'' to engage in such opportunities. Sting operations are costly and also of relatively long duration (Report of the Subcommittee on Civil and Constitutional Rights 1984). The defense in the covert operations is, predictably, entrapment.

Poaching has long been a tradition in much of rural America, but in the last ten to fifteen years it has more often been big business. Now the profits are high and the chances of being caught low. This combination attracts all kinds of people—rich, poor, taxidermists, outfitters, guides, and even organized crime. Opportunistic and subsistence hunters or anglers break laws, but they take relatively little game compared to those people in the first three groups.

Theoretically, all wildlife laws are enforced equally. In practice, there are many factors that modify this ideal philosophy. Violations of laws are not ignored, but some are given a higher priority than others. For example, public sentiment, political climate,

Cases 1–3

available officers, history of the violator, and ecological and economic importance of the violation are all taken into consideration when setting priorities for enforcing wildlife laws.

TYPES OF LAW VIOLATION IN GENERAL

There are two types of violation. Classical writers speak of crimes and of misdemeanors. Crimes comprise acts that are *mala in se,* or wrong in and of themselves. Misdemeanors comprise acts said to be *mala prohibita,* or acts which are wrong because they are prohibited by law. In this category are such acts as traffic or wildlife violations.

MAJOR CLASSES OF WILDLIFE LAW VIOLATIONS AND COURT RULINGS

There are nine major classifications of wildlife law violations which, if interpreted broadly, cover most cases. They are:[1]

(1) taking or attempting to take wildlife out of season;

(2) taking or attempting to take wildlife in an illegal place;

(3) improper or no license;

(4) illegal method;

(5) illegal possession;

(6) illegal procedure;

(7) illegal importation or exportation;

(8) illegal taking or possession or endangering of nongame or endangered species; and

(9) offering for sale wildlife species in violation of state or federal law.

Examples of each are given. However, many violations are not easily categorized and, in practice, citations are more specific. For example, the Michigan Department of Natural Resources lists fourteen commonly occurring types of big game violations; fifteen commonly occurring violations for waterfowl, and sixteen commonly occurring violations for upland game. In each of these three categories they use, in addition, an unnamed number of citations that occur less often.

Out of Season[2]

This violation may result from either fishing or hunting after hours or on days when the season is closed.

In *Utah v. Chindgren,*[3] Steven R. Chindgren appealed his jury conviction of unlawfully taking protected wildlife, a Class B misdemeanor in violation of the Utah Code. When appealed, Chindgren claimed that section 23–13–3 was not applicable to the facts of this case and was void for vagueness, and that the trial court committed several errors. Chindgren, a licensed falconer, on August 12, 1986, entered a field near Layton, Utah, with two dogs and a peregrine falcon, which he unhooded and released. The falcon flew around, captured a duck and landed. Chindgren picked up the falcon, grabbed the duck from its mouth, then ripped off the duck's wing and part of the breast and allowed the falcon to feed on the meat. Two Wildlife Resource officers, who had set up surveillance, observed Chindgren's activities and cited him for taking protected wildlife out of season under Utah Code 23–13–3.

Prior to the trial, Chindgren had filed two motions to dismiss, contending that section 23–13–3 was vague and unconstitutional, and that the facts of the case did not come within the purview of the statute. The trial court denied both motions, finding that the statute applied, and that it was constitutional. The prosecution presented its case to the jury and rested on June 8, 1987. Fourteen days later, trial was reconvened and the defendants presented their case. Chindgren was convicted of violating a Utah Code which said it shall be unlawful for any person to take protected wildlife or for any person to permit his dog to take protected wildlife, except as provided by code or rules and regulations of the Wildlife Board or the Board of Big Game Control. The jury was instructed that to find Chindgren guilty of the crime, it had to find, beyond a reasonable doubt:

(1) that on or about 12 August 1986, the defendant was in the County of Davis, Utah;

(2) that the defendant did, as a party, take a mallard duck;

(3) that the taking was out of season; and

(4) that the defendant was acting knowingly, intentionally, and recklessly.

Chindgren claimed that because section 23–13–3 mentioned only dogs and not falcons or other animals, the legislature did not intend that a person would be liable for the act of an animal other than a dog. Thus, he asserted, the jury was erroneously instructed that Chindgren did, as a party, take a mallard duck. The Utah Code states that to hunt means to pursue, harass, catch, capture, possess, angle, seine, trap, or kill any protected wildlife, or any attempt to commit any of these acts. Further, the law states that hunting means to take or pursue any reptile, amphibian, bird, or mammal by any means. Pursuant to statutory authority, the Wildlife Board promulgates rules and regulations and sets forth the proclamation for raptors, which provides that falconry is the sport of taking quarry by the means of a trained raptor. The court further pointed out that administrative agencies have the power to create rules and regulations which conform to the authorizing statute and do not depart from it. In this case, the state had sought to prove that Chindgren had permitted his peregrine falcon to take a duck out of season. Chindgren, on the other hand, asserted that section 23–13–3 was void for vagueness because it failed to provide adequate notice of the prohibited conduct. One may not, Chindgren said, be held criminally responsible for conduct for which he could not reasonably understand to be proscribed (*United States v. Harriss*).[4]

The statute in this case provides that it is unlawful for any person to take protected game except as permitted by regulations of the Wildlife Board or the Board of Big Game Control. In reviewing sufficiency of evidence, the court said that the evidence supports the conclusion that, at a minimum, Chindgren acted recklessly in releasing a falcon in a field containing an abundance of ducks.

In *Robinson et al. v. State*,[5] the defendant was convicted in county court of violating the state game law by illegally killing a deer. He appealed on the grounds that the evidence was insufficient to sustain the judgment. The evidence is as follows. Two officers working on the forest reserve heard a shot. They investigated and found the defendant and another man armed with rifles. One man had hair and blood on his clothing, apparently from a deer. When the arresting officer followed the tracks of the two defendants, they found a dead deer, still warm, about 175 steps from where they had met the defendants. The deer had been killed with a rifle. The defendant did not take the stand and offered no testimony. The state supreme court ruled that the evidence, though circumstantial, was sufficient to sustain the judgment. The verdict was affirmed.

In *State v. Rathbone*,[6] the killing of elk out of season to protect private property is almost identical to that of the *State v. Burk*.[7]

Illegal Place

Cases 8–11

It is a violation to seek fish and game in closed areas, refuges, in closed streams, and generally on posted private property.

The defendants in *Utah v. Morck,*[8] Morck and Hobbs, were convicted of taking or possessing wildlife without a proper permit by the district court. The defendants appealed and the appeals court affirmed the decision of the district court.

The Utah Division of Wildlife Resources received a phone call from a confidential informant alleging that the defendants were going to the Book Cliffs area near the Ute Indian Reservation in Southern Utah to hunt bear without a valid permit. Two Utah officers went to the location identified by the informant and spotted Morck's truck with fishing poles on the gun rack. On the third day, they observed the defendants, dressed in camouflage clothing, return to the truck carrying rifles. The next night the officers overheard the defendants place a call for a tow truck. The two officers then went to make contact with the defendants, taking with them a search dog. There is no agreement between the officers and the defendants as to how the search dog arrived in the back of the pickup truck. The state claims that the dog leaped into the truck on its own accord, but the defendants claim it was put there by the officers. In any event, bear hides with improper tags for the area were found in the cooler. The defendants were then arrested and charged with taking or possessing protected wildlife without a proper permit. On appeal the defendants argued that warrantless search of their truck did not satisfy the automobile exception rule under the Utah Constitution. The appeals court said that the officers had ample probable cause to believe that evidence of illegal bear hunting would be found in the pickup. The court further stated that it agreed with the trial court in that the dog alerting the officers to the probability of bear skins was not a necessary component of probable cause, but merely added to it. In discussing exigent circumstances, the court said that Utah courts will continue to follow the original exigent circumstances test for warrantless search of automobiles as per *United States v. Carroll.*[9] The appeals court summarized by agreeing with the trial court's finding that both probable cause and exigent circumstances were present. And it found that the warrantless search of the defendant's truck fell within the automobile exception rule as stated by *Carroll.*

In *Aragon v. People,*[10] the plaintiff was a Justice of the Peace of Alamosa County. He convicted several persons of illegally hunting on private land without first obtaining permission from the owner and assessed a $25 fine on each person. The case was appealed to the district court, and writ of prohibition against further action or steps to collect fines was asked for on the grounds that the statute did not provide a penalty for the offense; thus no prosecution was possible. The district court reversed the decision and upheld the writ.

The Justice of the Peace appealed to the state supreme court. This court ruled that even though no penalty was provided in the specific section setting forth the violation, the penalty was provided in a later section of the statute especially put in to cover the violations of the preceding sections. The penalty, as set forth for the preceding violation, was a fine of not less than $25 or more than $250, or by imprisonment in the county jail for not less than ten days or more than three months, or by such fine and imprisonment.

This section adequately covered the section stating the violation; thus the Justice of the Peace did not exceed his jurisdiction, and the judgment of the district court was reversed and cause remanded with direction for further proceedings in accordance with the views expressed by the supreme court.

In *State v. Barnett et al.*[11] the defendant was convicted in district court of hunting game on purportedly posted land. The case was appealed to the Supreme Court of New

Cases 12–14

Mexico. The law states clearly that a person must both post and publish his intentions of closing his land, in English and in Spanish, before the law against trespass will apply. The plaintiff only posted and published his intentions in English, and even though the defendants could read English, the law against trespass was not in effect on the land because the requirements were not met. The law is very clear and strictly construed; thus the decision was reversed, and orders to district court were issued to reverse its decision and release the defendants from its custody.

In *United States v. Brown,*[12] the defendant was convicted before the United States District Court for the District of Minnesota, for violation of national park regulations prohibiting possession of a loaded firearm and hunting ducks in a national park. He appealed. The court of appeals held that the state's active participation in the creation of Voyageurs Park with knowledge that Congress intended that hunting would be prohibited throughout the park was tantamount to a cession of jurisdiction over lands and waters within Park boundaries; that even assuming that State did not cede jurisdiction over such waters, federal regulations prohibiting hunting in such parks were a constitutional exercise of congressional power under the property clause; that congressional power over federal lands includes authority to regulate activities on nonfederal public waters in order to protect wildlife and visitors on the lands; that the regulations were valid proscriptions designed to promote purposes of federal lands within National Park and, under the Supremacy Clause, federal law overrides conflicting state law allowing hunting within the park, and that, thus, such regulations were applicable to defendant who was hunting ducks on waters within the boundaries of Voyageurs National Park. Affirmed.

Improper License

The law is violated by a person who uses a license improperly made out, who uses a resident license when he is entitled only to a nonresident license; who uses a license issued to other than himself; or who fails to display a visible license if required.

The defendant in *Herzig v. Feast et al.*[13] was a game warden who had violated the rights of an alien by arresting him for improper license and illegal possession of deer meat. The alien had lived in Colorado for five years prior to buying a resident hunting license on which no false statements were made. The case hinged on the right of an alien to lawfully purchase a hunting license. The first case had not been prosecuted in justice court; so no disposition was given it. The law made a distinct specification between residents and nonresidents but made no distinction against resident aliens.

The district court had ruled against the alien and for the warden, but the state supreme court reversed the ruling and remanded the case with directions, on the grounds that an alien can lawfully buy a big-game license, kill, and possess game on the license; and the warden's action in this case was wanton and reckless disregard for the alien's rights. The warden should have made sure the law applied to aliens in regard to licenses before subjecting the person to arrest.

In *Washburn v. State,*[14] the defendant was convicted of fishing without a license, and he appealed. The circumstances of the case were odd. Two game rangers saw five men seining a slough. They did not arrest them then but later blocked the road and found six men, two of whom did not have a license, leaving the pond. The defendant's clothing was dry and no evidence was given that he participated in the seining. Furthermore, the pond was landlocked (no inlet or outlet to other waters) and was completely within privately owned land. However, a license is required for seining in waters over which the state has jurisdiction.

The appellate court ruled that the evidence purporting to show the defendant was seining was insufficient and circumstantial at best. They also ruled that the state has no authority to regulate fishing in private, landlocked lakes where fish do not have free

A conservation officer checks a fisherman's license on the Coeur d'Alene River in Idaho. *Courtesy Idaho Fish and Game Department, Jack McNeel.*

Cases 15–16

Illegal Method

access to waters outside the lake and are not free to pass in waters to and from the same. The judgment and sentence of the county court were reversed and the defendant released.

It is against the law in most states to use a 22-caliber rimfire rifle to hunt deer, to use a shotgun that will hold more than three shells when attempting to take migratory waterfowl, or to use too many hooks on fishing gear, or worms on a stream open only to artificial fly fishing.

In *State v. Tyler,*[15] defendant Tyler was tried for illegally taking fish. He was using a pole with a line run through eyes on the pole and with a large hook at the end of the line. The hook could be pulled up and made stationary at the end of the pole and used as a gig. He had snagged one channel catfish. By law, the catfish is a game fish, and the use of a gig is unlawful.

Tyler argued that he was not using the pole as a gig, but witnesses testified differently. He also said everyone else was fishing in this manner and that the fish would have died anyway because of low water. The court ruled that even if others were violating the law, that did not excuse the offense, and that the dying fish were a problem of the state fish and game department, to be handled by them and not by Tyler.

Tyler had been discharged by the county court and in this way escaped further prosecution. However, the case was decided in favor of the state.

In *State v. Baxter,*[16] defendant Baxter was convicted of catching salmon with a net and fined $300. The jury trial was appealed on grounds that the evidence was not sufficient to secure a conviction, that the court erred in giving instructions to the jury on circumstantial evidence, that the nets used in court were not the same ones as those taken from him, and that the defense was not permitted to cross-examine a witness.

The charges were stated by prosecution as follows. Two game wardens saw a man setting nets but did not arrest him at the time. Later, they arrested a man whom they identified as the man seen earlier. The man had three nets in his boat, one containing fish.

The court ruled that the evidence was sufficient for a jury to deliver a verdict of guilty and for a court to sustain it. In cases of this kind, the Washington Supreme Court reviews the case and decides if the evidence is sufficient for a jury to bring a verdict.

Cases 17–21

Very seldom will a supreme court overrule a jury's decision if the evidence is sufficient, even when the supreme court does not agree with the verdict.

The court further ruled that the instructions to the jury about circumstantial evidence were correct because circumstantial evidence if competent evidence, is often as good as actual evidence, and may often be the only type that exists. As for the nets, the three in question were kept separate and were not proven to be the wrong ones. The court said they would have not had great bearing on the case anyhow. The witness that was not cross-examined was later called by the defense and examined completely; so the defendant had no complaint. The verdict of the jury and trial court was affirmed.[17]

In *Kephart v. State*[18] the defendant was convicted in the Custer County court of unlawfully catching channel catfish out of the Washita River by means of a wire trap. He appealed the decision to the Oklahoma Criminal Court of Appeals.

Two statutes could have been used in this case. One prohibits catching game fish by any means other than hook and line, and classifies channel catfish as game fish. The other law prohibits catching certain game fish by means of nets, net seines, gun, wire trap, pot, snare, or gig, but does not name channel catfish as game fish. This last statute was the basis of the defense. However, the court ruled that the information in the complaint, to wit ''unlawfully catch about 25 channel catfish out of the Washita River by means and use of wire trap'' was sufficient to institute proceedings under the statute declaring hook and line as the only legal way to catch game fish.

The defendant claimed he only happened upon the trap and took fish from it but disclaimed ownership of the trap. However, by taking the fish from the stream by the wire trap, with or without ownership, he violated the first statute given above. The judgment and sentence of the county court were affirmed.

In *State v. Schrimer*[19] the defendants were convicted in probate court of attempt to kill deer with the aid of a spotlight, and they appealed to the district court. Here the judgment was reversed and the proceedings dismissed. The state appealed to the supreme court.

The complaint sets forth the acts committed by these men who were using a spotlight on their car after dark to locate deer, which they intended to kill or attempt to kill. They had loaded rifles which they intended to use in an attempt to kill the deer. However, no evidence is presented of any actual attempt to kill deer, and it must be assumed the actual attempt, if not witnessed, did not occur. The statute which applies read, in part: ''Provided, also, that it shall be a misdemeanor to take, kill, or attempt to kill any game with the aid of a spotlight, flashlight, or artificial light of any kind.''

The court ruled the statute, as worded, does not make it unlawful to hunt deer with a spotlight. The complaint cites the statutes and also sets forth the acts committed, but only goes so far as to show hunting the animals and not an attempt to kill them. The facts show hunting the animals and not an attempt to kill them. The facts show only planning or preparation, which is not stated in the statute as a crime. It should be noted that the word ''hunt'' as used by the court actually means ''looking for.'' Also laws have been changed in most or all states to read that ''a person or persons in possession of a firearm and an artificial light capable of being directed, that is (are) in an area at night where deer (game) are known to frequent is guilty of ''spotlighting.''

The defendant cited a number of cases to show precedence.[20] The appellant also cited cases,[21] but these did not apply because they were all from Texas, and the Texas statute involved makes it an offense to hunt game with the aid of a light. The judgment of the district court was affirmed.

A conservation officer patrolling southeast Idaho.
Courtesy Idaho Fish and Game Department, Bob Saban.

Cases 22–24

In *State of New Mexico v. Barber,*[22] Barber and Harris were convicted of hunting by spotlight contrary to New Mexico law. They appealed asserting:

(1) the statute is unconstitutional; and

(2) the state failed to prove criminal intent.

The defendants were observed in a vehicle equipped with additional lights mounted on a roll bar above the cab, driving slowly and in a random fashion. The overhead lights were shining in a wider arc than the headlights and they were in an area where there were big game animals and livestock. The defendants had a loaded rifle and a pistol and were riding in the back of the pickup. The court found that the act was constitutional and neither the title of the act, nor the act itself, requires any criminal intent. The conviction was upheld.

In *State v. Weindel*[23] the defendants were charged with illegally taking fish. The information charged that the defendants ". . . did then and there . . . catch and take fish with their hands . . ." and referred to the statute that reads, in part, "It shall be unlawful for any person to take, catch, or kill any game fish . . ." Nothing is said of nongame fish. The fish caught by the defendants were "mudcats" and not game fish as specified by the statute.

The defendants were tried, and a demurrer to the information was sustained by the county court. The state appealed. The supreme court ruled that the statute does not prohibit taking nongame fish by hand, and the statutes cannot be expanded or extended to cover cases that are not clearly within their scope. The judgment and demurrer were affirmed.

Illegal Possession

In *People v. Miller,*[24] defendant Miller caught five kelp bass in waters off the Mexican Coast. The fish were taken legally in all respects; however, upon landing at San Diego, a game warden asked the defendant to fill out a form declaring his catch, and a permit to bring the fish into California would have been issued. The defendant refused to comply and was arrested, tried, and convicted. The conviction was upheld through all the courts to the Appellate Department of the Superior Court of San Diego County.

The defendant appealed the case on the grounds that the fish and game code did not state that it was a crime, and the state has no right to restrict commerce, since this is the job of the federal government.

Court ruled that the code was very clear on the crime of transporting fish, because an officer was present and the defendant's signature on an application for a permit to import fish was all that was required. The states may pass laws necessary to conserve fish and game or laws pertaining to the bringing in of food, which includes fish and game. The judgment was affirmed.

In *Commonwealth of Massachusetts v. Worth,*[25] defendant Worth was convicted in superior court of illegal possession of the carcass of a deer. He had killed the deer, while driving, by hitting it with his car. He then took the deer home and did not notify the game warden. The state contended, rightly, that the defendant had no right to possess the deer. Worth appealed to the Massachusetts Supreme Judicial Court, contending that the possession of the carcass was lawful, that the possession was not prohibited by statute, and that the wording of the statute ''. . . except as provided in this chapter'' was equivalent to the wording ''. . . in violation of this chapter.''

The court ruled that the possession was not lawful, inasmuch as the property right in game is with the state and no person has a right of possession. Also, the deer was not taken during the lawful season nor while it was destroying property or as a result of property protection. These are the only ways one can legally possess a deer. The fact that the deer was not illegally killed is immaterial to the possession. Also the word ''provided'' is not similar to ''violation'' in this instance. The exceptions asked by the appellant were overruled.

In *State v. Evans,*[26] Officer Rogers was informed by a Forest Service officer that Evans and Acuff, the defendants, had received a fire permit on forest reserve land. When the warden went to check the campers, he found them in the woods wearing bloody clothing. They directed him away from, rather than to, their camp. However, he located the camp and asked the men for permission to search. This permission was denied and enforced by covering the officer with rifles. Later, permission was granted, but the men followed the officer around with their rifles pointed at him. He found a rack, 204 feet from the camp, containing sacks of elk meat and elk meat being smoked. There was a well-worn path between the camp and the rack. A stew pot of elk meat was found still steaming. Parts of elk, including antlers, were nearby. The officer attempted to arrest the men, but they resisted by pointing their rifles at him. Finally Acuff surrendered, but Evans left the camp and was apprehended later in town. The men were convicted by a jury in circuit court and sentenced to a $500 fine and three months in jail, each. They appealed.

The defendants claimed the court should not have allowed the district attorney to say the defendants put the officer ''through an ordeal.'' The court ruled that in view of the circumstances, this was impossible, and on the light side of what could be said. The court ruled that, though circumstantial, the evidence was sufficient to show possession if believed by the jury. The defendants claimed the officer had no search warrant and no reason to believe a crime was committed. The court pointed out that no warrant was necessary to search on forest or public land, and reason was justified by the blood on the men's clothing. The very act of resisting the search and arrest is, by statute, *prima facie* evidence of the violation of the law by the persons resisting. This resistance was also taken into consideration when sentencing the defendants, and the state supreme court ruled the sentence to be within legal bounds. The supreme court could find no error in the proceedings and affirmed the judgment.

Cases 27–31

In *State v. Pulos,*[27] Mr. Pulos was tried and convicted of illegal possession of a duck under the statute which read, in part, "It shall be unlawful . . . at any time between January 15 and September 1 of any year, to take, kill, injure, destroy, or have in possession any wild duck."

The defendant argued that the duck was killed during the lawful season and was stored as food. He was obviously relying on the case of *State v. Fisher.*[28] Also, he contended that it was unfair to allow killing on one day and prohibit as unlawful the possession of the kill on the day after. He claimed the duck was lawfully reduced to his possession, being then his property with the state having no right to take it from him. He appealed to the state supreme court.

The supreme court ruled that the *Fisher* case did not apply because the working of the statute, passed after the *Fisher* case, was very clear on possession at any time other than open season. Court agreed that the law might not be absolutely fair; however, it pointed out that the law does not make a person hunt on the last day of the season nor does it make a person kill more ducks than he or his friends can consume before the seasons ends. Furthermore, the title to wild game is with the state, and no person has an absolute property right to wild game; that hunting and taking game is a privilege subject to the regulations and restrictions of the law-making powers. The court cited a case supporting this position[29] and affirmed the judgment.

In *State v. Fisher,*[30] Fisher was tried and convicted in county court of illegally possessing deer meat after the season closed. The verdict was upheld in district court, and the case was appealed to the state supreme court. The violation was based on a statute that reads, in part, ". . . any person . . . having in possession any deer, or carcass, or part of a deer during the season when it is unlawful to take or kill such deer shall be guilty of a misdemeanor."

Fisher tried to explain the deer was killed legally during the season and was not unlawfully possessed. The district court ruled he was in possession during off-season and instructed the jury in that manner, thus obtaining a judgment of guilty.

The state supreme court ruled that, whether intended or not, the words "such deer" in the statute referred to deer that were taken or killed during the time when it was unlawful to kill deer. Thus, deer that were killed legally did not fall under the category of this statute. The defendant had a right to show legal possession and the judgment was ruled erroneous and reversed. A new trial was ordered.

In *Stewart v. People*[31] the defendant was convicted on two counts, one for killing a deer and one for illegal possession of parts of the deer killed. He appealed on the grounds that he did not kill the deer and he did not have the head of the deer in his possession. Also, an acquittal on an earlier count made him immune to these charges. He contended, in addition, that there was an error in instructions to the jury.

The undisputed evidence in condensed form is as follows: Defendant and two friends planned a deer hunt out of season. One man killed a small deer; another, a large buck; and the defendant, nothing. They carried the deer to a car, the defendant carrying the head of the large deer, which was hidden in a shack and later found by a deputy game warden. Defendant was acquitted on the killing of the small deer.

The state supreme court ruled that the planning and carrying out of the hunt itself made the defendant equally guilty, even though he did not actually kill any deer. Also, he hid a deer head and could be said to be in possession of it, even though it was not on his person. The two acts of killing two deer were absolutely separate acts, unrelated in their commission; thus acquittal on one gave no immunity to the other. The instructions to the jury on accessory to the fact were in error; however, not all errors make judgment reversals necessary. The conviction and sentence of the lower court were affirmed.

Cases 32–35

In *State v. Miles*[32] a complaint was issued, charging defendant Miles with offering a reward for display of a game animal. Miles operated a sporting goods store and offered $10 for the largest deer displayed at his store during the deer season. The Fish and Game Commission had made it illegal to display game by virtue of power granted to it for the taking of game. However, the regulation concerning the Commission's powers to pass rules regulating the taking of game does not mention displaying of game. The court ruled that the displaying of game could not have a logical connection affecting the taking of game, and the Commission had overstepped its authority by ruling against the display of game. The superior court had ruled for the defendant, and the state supreme court affirmed the judgment.

In *State v. Visser*[33] the defendants, the Vissers, were convicted of illegal possession of deer, namely two fawns, during closed season and were fined $250 and costs. They appealed, claiming that the officers had not seen them in possession of the deer or been closer than twenty feet to the deer. The story was that these three men, the Visser brothers, were seen in the woods on state land after a series of shots had been heard. The officer found a car parked and a man, John Visser, approaching the car. Visser carried a rifle that had apparently been recently fired. However, this charge was not pressed. The two other men were seen in the woods hiding, but when told to come forward, they ran into the brush and escaped. The officer found two dead and still warm fawns twenty feet from where the men had been hiding. These two men were later arrested, one wearing clothing that was covered with blood and deer hair. No testimony was offered by the defendant.

The state supreme court ruled that the evidence was sufficient, if believed by a jury, which it was, to reasonably show possession. Even though no one saw the men with the deer, it was illegal to possess these deer, whether observed or not. The circumstantial evidence supported a conviction, and the lower court had given correct instructions to the jury. The verdict and judgment were affirmed.

Illegal Procedure

An example of this would be failure to properly tag a deer.

Illegal Importation or Exportation

Defendant Hughes[34] was convicted in the district court, Jefferson County, Oklahoma, of unlawfully transporting for sale outside the state minnows which were seined or procured within the waters of the state and defendant appealed. The court of appeals held that the statute prohibiting anyone from transporting or shipping for sale, outside of Oklahoma, minnows which were seined or procured within waters of Oklahoma, did not violate the interstate commerce clause of the Constitution.

Illegal Taking or Possession or Endangering of Nongame or Endangered Species

In the well-publicized snail darter case (*TVA v. Hill*)[35] the U.S. Supreme Court held that the Endangered Species Act prohibits the TVA from impounding the river, notwithstanding that the Tellico Dam had been well under construction when the act was passed and when the snail darter had been declared an endangered species, and notwithstanding that Congress, in every year since the starting of the dam (even after the Secretary of the Interior took special action under the act) had appropriated funds for the dam. Such continuance of appropriations not constituting an implied repeal of the act as to the dam and an injunction against completion of the Tellico Dam was the proper remedy.

Offering for Sale Wildlife Species in Violation of State or Federal Law

Cases 36–38

Regulations of the Secretary of the Interior under the Migratory Bird Treaty Act, define migratory birds to include those raised in captivity. This does not contravene the intent of Congress expressed in the act (*United States v. Richards*).[36] The Migratory Bird Treaty Act applies to migratory birds, not wild birds. Proscription in the Migratory Bird Treaty act against the sale of sparrow hawks, captive-raised birds, cannot be said to discourage propagation and to thus support the congressional intent. The defendant, Richards, was convicted before the U.S. District Court for the District of Utah, of the sale of three sparrow hawks, a species protected by the Migratory Bird Treaty Act, and he appealed. The court of appeals agreed with the district court that regulations defining migratory birds to include those raised in captivity, do not contravene the intent of Congress expressed in the act. Regulator prohibition of sale of migratory birds was within the intent of Congress in enacting the Migratory Bird Treaty Act. Defendant's permissive possession of sparrow hawks, now a protected species under the Migratory Bird Treaty Act, did not carry with it the traditional incidences of property rights; accordingly, there was no unconstitutional derivation of property by reason of the prosecution of defendant for selling sparrow hawks in violation of the act.

In *United States v. Gigstead,*[37] the defendant was convicted in U.S. District Court for the District of Minnesota of unlawfully offering for sale and unlawful sales of migratory birds protected by the Migratory Bird Treaty Act of 1918. The defendant appealed. The court of appeals held that the fact the defendant may have been entitled to have protected birds or their stuffed skins in his possession gave him no right to sell them, or to offer to sell them or their skins in ordinary commercial transactions, including transactions with special agents of the Fish and Wildlife Service (who were acting as undercover agents). The facts were essentially as follows. In the summer and fall of 1974, Gilbert Kenneth Gigstead and others were operating a wildlife ranch and charging an admission fee. This ranch included a museum in which were displayed the stuffed skins of wild birds, including migratory birds protected by the Migratory Bird Treaty Act. The defendant held a federal collector's permit which had no expiration date and which had not been recalled or revoked. The defendant claimed that because he had a collector's permit, and that the wildlife ranch had been mentioned as a public scientific and educational institution, and that he offered the birds for sale to government employees who were entitled to acquire them in the line of duty, he was not in violation. He also claimed that the species in question were in abundant supply and were not in any danger of becoming extinct. The court declined to submit the defendant's legal theories to the jury and instructed the jury accordingly. The court said if the jury were convinced by the evidence beyond a reasonable doubt, that the defendant had in fact offered for sale or sold the bird or birds, under consideration, the defendant should be found guilty on that count. Otherwise, he should be found not guilty. The judgment of the district court was affirmed.

Other Cited Cases

In *United States of America, ex rel. Bergen v. Lawrence,*[38] the United States brought action against a cattle rancher alleging wrongful fencing of federal land. The rancher appealed. The court of appeals held that the cattle rancher violated federal law by enclosing public lands, thereby excluding pronghorn antelope from their critical winter range. The district court directed that the cattle rancher's fence be removed or modified to allow passage by antelope. This did not effect impermissible and unconstitutional taking. The decision did not impose a servitude but rather abated a nuisance proscribed by federal law. The rancher's federal grazing leases were not damaged, and the rancher retained right to exclude antelope from his own lands if he could accomplish this without at the same time effecting an enclosure of public lands. The court said that

Cases 39–41

winter forage by antelope was a lawful purpose of public lands and was thus protected from unlawful enclosure of public lands. The cattle rancher, by virtue of the Taylor Grazing Act leases of federal lands enclosed by fence, had only color of leasehold title and not fee title; therefore, the rancher could not assert, as a statutory defense to unlawful enclosure. The court said that the mere presence of gates or openings in the rancher's fences, which enclosed public lands, and which excluded pronghorn antelope from their critical winter range, did not mean that the enclosures were lawful. The defendant, Lawrence, had constructed a twenty-eight-mile fence enclosing over 20,000 acres of private, state and federal lands in an area of south central Wyoming known as the Red Rim. Lawrence grazed his cattle on the Red Rim during the spring and summer months for about sixty days. But during the winter, portions of the Red Rim provide critical ranges for Wyoming pronghorn antelope. The fence Lawrence constructed, however, was antelope proof, denying antelope access to this critical winter range. The district court concluded that this case was controlled by *Camfield v. United States*,[39] which dealt with a "virtually identical" situation. Camfield had acquired from the Union Pacific Railroad the rights to several odd-numbered private sections of land. In building the fence complained of, the defendant had constructed it entirely on odd-numbered sections so that it completely enclosed all the government lands aforesaid, but without locating the fence on any part of the public domain. The U.S. Supreme Court considered Camfield's argument that he could do whatever he wished on his own land, and soundly rejected it. The Court found that the Unlawful Enclosures Act had been promulgated just to avoid such an outcome.

In appealing the district court's ruling, Lawrence claimed that its ruling effects an impermissible and unconstitutional taking. The Court rejected this argument for several reasons. The Court said that the ruling did not impose a servitude for antelope, but rather, abated a nuisance proscribed by federal law. The Court also stated that it could find nothing of Lawrence's that had been "taken." And the Court said finally, that Lawrence retained the right to exclude antelope from his own lands if he can accomplish that without at the same time effecting an enclosure of the public lands.

In *Mescalero Apache Tribe v. State of New Mexico*[40] the United States District Court for the District of New Mexico enjoined the state of New Mexico (New Mexico Department of Game and Fish) from enforcing its game laws against non-Indians for acts done on the Mescalero Apache Reservation. The court of appeals held essentially the same view, ruling that the state could not apply game laws to any persons for acts done on the Mescalero Apache Reservation. And in light of the Tribe's exemplary record in wildlife management preservation and improvement of stocks and fish and game, the reservation did not require dual regulation by the State of New Mexico and the Tribe. Absence of tribal criminal jurisdiction over nonmembers, the court said, did not necessitate a cutback in the Tribe's plenary power over reservation wildlife management. In discussing Indian rights of action, the court said the Mescalero Apache Tribe had standing to seek a determination, and that New Mexico game laws were not applicable to non-Indian activity within its reservation. Further, any attempt by the state to exercise regulatory powers within the confines of a federally recognized "semi-independent" Indian reservation is precluded if the subject matter has been preempted by federal law or if the state regulations infringe on the Tribe's right of self-government. Abrogations of any rights taken by Indian treaty, particularly fundamental hunting and fishing rights, must be explicit to be effective.

Further, the appeals court said that even though the Tribe may be seeking to create jurisdictional disputes, no negative inference should be attached to such posture. In fact, a clear intent to preempt state jurisdiction is an element in the Tribe's favor (*Confederated Tribes of Colville Indian Reservation v. Washington*).[41] Although the

Cases 42–47

state argues that wildlife management efficiency requires its jurisdiction over reservation activities, there has been no claim that there are any endangered species (*Puyallup Tribe, Inc. v. Department of Game*).[42] The state is unable to claim that either it or its lands played any significant role in the creation and preservation of the wildlife resources on the reservation. In the state's claim that the Tribe has no standing and that the suit is otherwise not justifiable, the court states that the state's understanding of standing requirements is overly narrow. Federal courts, for the purpose of standing, may consider the principles of elementary economics. The state's imposition of higher cost on an individual hunter or angler would clearly limit the Tribe's ability to raise the price of its own licenses. The sovereign powers of the Tribe in wildlife management are said to be so pervasive that sovereignty here moves from a mere backdrop into a leading role on the litigational stage. The Tribe's inherent authority stems largely from its traditional reliance on wild animals for basic survival. The court further adds that the Tribe's sovereign powers are not limited to control of wildlife. The Tribe's historical powers extend to the territory itself. A second test for determining the propriety of state regulation on Indian lands analyzes the impact of the regulation on self-government. Even if the treaty and statutory scheme, read against the backdrop of sovereignty, are insufficient to create federal preemption, the Tribe's authority is here protected under a tribal self-government analysis. In the landmark case of *Williams v. Lee,*[43] the Court upheld tribal court jurisdiction over non-Indians. The U.S. Supreme Court restated the controlling test of "essentially, absent governing acts of Congress, the question has always been whether state action infringed on the rights of reservation Indians to make their own laws and be ruled by them." To apply a test, a court must seek an accommodation between the interest of the Tribes and the federal government, on one hand, and those of the state, on the other (*Washington v. Confederated Tribes of the Colville Indian Reservation*).[44] The appeals court pointed out that, unlike Colville, we have here a clear state interference with a traditional tribal regulatory power. In summary, the court said that the instant record shows no need for a joint conservation measure, the right of the state to regulate off-reservation possession of game lawfully reduced to possession in accordance with tribal law is foreclosed (*Hughes v. Oklahoma*).[45]

Case.[46] Eight defendants were charged with aiding and abetting for violations of daily bag limits of waterfowl and for wanton waste of waterfowl. The magistrate judge found each defendant guilty. On appeal the district court held that evidence was insufficient to support convictions of all of the defendants who were merely present or negatively acquiesced in activity, with no evidence demonstrating affirmative action, encouragement or assistance, and sufficient evidence did support the convictions of two defendants. The defendants were engaged in hunting as a group. They occupied the same hunting blind and they each possessed a rifle [shotgun]. There was a series of shots and duck calls, and the defendants were dressed in camouflaged hunting outfits. Each defendant possessed a legal limit of ducks when approached by the U.S. Fish and Wildlife Service agents, and the agents discovered eighteen more ducks near the hunting area, which had not been picked up. The district court said in reviewing the case that the court must view the evidence as well as all reasonable inferences flowing therefrom in a light most favorable to the government. In addition, this court must accept the credible findings as determined by the magistrate judge (*Glasser v. United States*; *United States v. Gianni*).[47] In essence, the central question is whether or not the evidence is sufficient to sustain the conviction of each defendant-appellant of aiding and abetting the commission of the offense. The basic facts are essentially these. Two agents of the Fish and Wildlife Service went to Frank Lyon's farm after hearing a volley of shots and decided to investigate the shooting activities. One of the agents approached the hunting area and heard duck calls and observed two vehicles on a levee adjacent to the area. But later the agent discovered eight hunters. One of the defendant-appellants returned with an agent to the hunting area. They followed a trail that the boat had made earlier through the frozen ice. It was apparent that no other boat had been there before them. A total of eighteen ducks was retrieved, one of which was still alive, and the others were limber and appeared to have been freshly killed. The defendant further said that they had started shooting at the legal time, but it was still dark and they had no idea how many ducks were on the water. In *United States v. Kelton,*[48] the court stated that ". . . Mere association, as opposed to participation, is not sufficient to establish guilt . . . Nor is mere presence at the scene of a crime alone sufficient to sustain the burden of proof the government bears. . . . Presence must be accompanied by culpable purpose before it can be equated with aiding and abetting. . . ." The court stated that aiding and abetting may be established by circumstantial evidence, however, the court said that it must be abundantly clear that the factfinder may not indulge in speculation and conjecture as a substitute for proof in such cases. The sum and substance of the government's evidence was that the defendants were engaged in a group activity or party hunting. It was a fact that eighteen ducks in addition to the legal limit that the defendants had were found adjacent to the hunting area, but this does not lead to the logical conclusion or inference that each and every defendant aided and abetted in illegal conduct resulting in its wanton waste and excess over the limit of migratory fowl. The key point in this case is that neither of the agents personally witnessed any of the defendants shooting even one duck. The two defendants whose guilt was not reversed made statements that were to their detriment.

The Wild, Free-roaming Horses and Burros Act is of interest, among other reasons, because of the conflict it created between the State of New Mexico and the U.S. Government and the resultant decision (*Kleppe v. New Mexico*).[49] In brief, New Mexico asserted that the federal government lacked power to control wild horses and burros on the public lands of the United States unless the animals were moving in interstate commerce or damaging public lands, and that neither of these bases of regulation were available under the then-current situation. The district court ruled in favor of New Mexico, but the circuit court overruled the decision. The circuit court said, in part, that the New Mexico Livestock Board had entered upon public lands of the United States and removed wild burros; an action contrary to the provisions of the Wild, Free-roaming Horses and Burros Act. The court stated that it applied the act as a constitutional exercise of congressional power under the property clause to this case. While Congress can acquire exclusive or partial jurisdiction over lands within a state by the state's consent or cession, the presence or absence of such jurisdiction has nothing to do with Congress' powers under the property clause. Absent consent or cession, a state undoubtedly retains jurisdiction over federal lands within its territory, but Congress equally surely retains the power to enact legislation respecting these lands pursuant to the property clause. And when Congress so acts, the federal legislation necessarily overrides the conflicting state laws under the supremacy clause. A different rule said the court would place the public domain of the United States completely at the mercy of state legislation. The court further added that the federal government does not assert exclusive jurisdiction over the public lands in New Mexico, and the state is free to enforce its criminal and civil laws on those lands. But where these state laws conflict with the Wild, Free-roaming Horses and Burros Act, the law is clear—the state law must recede. The court added that appellee's contention that the act violates traditional power over wild animals is on no different footing. Unquestionably, the states have broad trustee and police power over wild animals within their jurisdiction. But these

Cases 50–53

powers exist only insofar as their exercise may not be incompatible with, or restrain, the rights conveyed to the federal government by the Constitution. The judgment of the district court was reversed and the case remanded for further proceedings consistent with the above opinion.

In *Baldwin v. Montana Fish and Game Commission,*[50] elk hunters who were not residents of Montana and a professional hunting guide who was a resident, brought action in U.S. District Court seeking declaratory and other relief against the state's statutory licensing requirements, which discriminate against nonresidents by imposing substantially higher fees than those assessed against residents. And by requiring non-residents, but not residents, to purchase a combination license for other game in order to be able to hunt a single elk. The appellants alleged a denial of their constitutional rights under the privileges and immunities clauses and under the equal protection clause of the Fourteenth Amendment. The plaintiffs were denied relief by the district court. On direct appeal, the U.S. Supreme Court affirmed the action of the district court. It was held that with respect to those privileges and immunities bearing on the vitality of the nation as a single entity, the state must treat all citizens alike whether residents or nonresidents. The Montana statutes are not violating the privileges and immunity clauses since access to recreational elk hunting did not fall into that category and the statutory distinctions between residents and nonresidents did not violate the equal protection clause of the Fourteenth Amendment, because the legislative choice was an economic means and not unreasonably regulated to the preservation of finite resource and a substantial regulatory interest to the state. And because the state does not have to justify to the penny any cost differential it imposes in a purely recreational, non-commercial, nonlivelihood setting.

In *United States v. New Mexico,*[51] the state of New Mexico brought a proceeding in district court seeking a general adjudication of water rights in the Rio Mimbres River, which originates in the upper reaches of the Gila National Forest and flows generally southwest more than fifty miles past privately owned land and provides substantial water for both irrigation and mining. The Supreme Court said, the United States as a result of its setting aside the Gila National Forest, was entitled to reserve water rights in the Rio Mimbres River to the extent necessary to preserve timber and to provide a favorable flow in the forest, such purposes being those for which the national forests were established, but the United States does not have reserve water rights for purposes of recreation, aesthetics, wildlife preservation, or cattle grazing.

In *State v. Leighty,*[52] the defendant was convicted of not having a license to outfit. Fined in district court, the defendant appealed to the Supreme Court which held that:

(1) evidence of revocation of defendant's outfitting without a license;

(2) wherein introduction of such evidence permitted state to approve in a most straightforward and least confusing way, that the defendant did not have a license to outfit when his dealing with an undercover agent commenced; that law enforcement officials were not required to obtain a search warrant before they could use an undercover agent to gather evidence of possible improper conduct through observation or conversation, and hence evidence obtained by undercover agents in business conversation with the defendant in his home was admissible;

(3) evidence of reinvocation of defendant's outfitting license was admissible; and

(4) testimony of various state officials complained of was not prejudicial.

The action of the district court was affirmed.

Defendant Sieminski[53] was convicted in the district court of illegally taking scallops on the high seas, and defendant appealed. The Superior Court, Third Judicial District, set aside the conviction, and the state appealed. The Supreme Court held that evidence was sufficient to support the trial court's finding that regulation of scallop fishing on high seas immediately beyond Alaska's territorial waters was in furtherance of legitimate state interest and therefore a valid exercise of the state's police power. The Court further ruled that evidence was sufficient to support the trial court's findings that the defendant was a resident of Alaska at the time of the charged offense; that the state was authorized to regulate scallop fishing at sea beyond the three-mile limit. That case was remanded to the superior court.

In *Bergh v. State,*[54] nineteen commercial fishermen brought action against the state of Washington and the state Director of Fisheries, seeking damages for unjust enrichment and tortuous interference with economic advantage caused by director's order reducing length of the salmon fishing season. The court of appeals ruled that the act of the state Fisheries Director and other employees, in reducing the salmon fishing season for commercial fishermen from twelve days to four days and prohibiting any season on chum salmon was exercise of discretionary act by conscientious public servants and such act was not tortuous. Thus commercial fishermen had no cause of action against the state or the Director of Fisheries based upon reduction in salmon fishing season.

In *State v. Westside Fish Company,*[55] the state appealed from an order of the circuit court suppressing evidence obtained in three warrantless searches of business premises of licensed food fish canner and licensed wholesale fish dealer. The court of appeals held that the statute providing the director of the Fish and Game Commission, or his authorized agents, may enter and inspect certain premises for purposes of enforcing commercial fishing laws, was constitutional as it applied to licensed commercial or industrial premises, and thus warrantless search of fish dealer's freezer and seizure of his fish were lawful.

In *Rains v. Washington Department of Fisheries,*[56] it was held that protection of fish life is a legitimate state interest and fish and game regulations are a valid exercise of police power. Rains sued the Fisheries Department because he had been denied a permit from the state to do work which would have returned a creek to its original bed and therefore prevented damage to his land.

In *Department of Fisheries v. Chelan County Public Utility,*[57] the Department of Fisheries commenced an action for declaratory judgment seeking to hold the Chelan County Public Utility District No. 1 liable for cost of constructing fish ladders, and both parties moved for summary judgment. The superior court granted the district's motion and the department appealed. The court of appeals held that:

(1) the state under its police power can promulgate regulations conserving and protecting the fish resources and fish runs within its boundaries;

(2) the statute authorizing the director of the department to require dam owners to undertake modification of fish ladders does not deny equal protection; and

(3) the statute does not deal only with maintenance of existing fishways.

The action of the superior court was reversed and remanded.

In *State v. San Luis Obispo Sportsmen Association,*[58] it was held that as used in Provision of California Constitution according people right to fish upon public lands, words "public lands" means state-owned lands, used by the state and is also compatible with use by the public for purposes of fishing; only property which is being used for specific purposes that are incompatible with its use by the public does not fall within the scope of such constitutional provision. The Supreme Court held that the trial court was correct in concluding that fishing should be permitted at the reservoir.

Cases 59–62

In *State v. Brumley,*[59] defendants were convicted in district court of violation of game laws for possession of big game animals out of season and they appealed. The Idaho Supreme Court held that the search of defendant's pickup on which the arresting officer saw an elk carcass in plain view was not initiated at the time the officer saw the vehicle from across the canyon. The intent to investigate and pursue the vehicle, and investigate and stop was formed later and was not an illegal search. The arresting officer, therefore, had probable cause for arresting the defendant and seizure of the carcass. The appellants argued that the officer lacked probable cause to stop and search their vehicle, contending that the search of the vehicle was initiated at the time the officer formed "the intent" to investigate and pursue appellant's vehicle. The court rejected this view.

The court in *Hamblin v. Arzy,*[60] said that the game laws of Wyoming permitted search, without a warrant, only for the seizure of wildlife unlawfully taken. The laws do not authorize seizure of a snow machine.

Defendants were charged with unlawfully throwing and casting the "rays of a vehicle headlight upon the highway in an area wherein big game may reasonably be expected to be while having in their possession and under their control a firearm, to wit: a Ruger .22-caliber semiautomatic rifle whereby any big game could be killed; which weapon was fully assembled and loaded in violation of (the applicable statute)."[61] The state supreme court affirmed the conviction except as to the passenger in the car. It appears that the game wardens heard several shots fired between ten and eleven o'clock at night and observed lights being shown in such a manner as to aid persons illegally hunting deer known to be in that area. The questions involved concerned the rights of the game wardens to stop the men and ask them to exhibit their licenses and to detain them in order to investigate if they were the persons involved.

The court held:

> The evidence of stopping defendant's auto being without objection the officer could properly testify what they saw from where they stood on the public highway as this did not amount to a search within the constitutional provisions cited. . . . There [is] no 'search' of an automobile where a gun [is] visible to an officer when he looks into the automobile. . . . 'It is not a search to see what is patent and obvious.'
>
> . . . The considerations which permit search of an automobile have been distinguished from those governing the search of a home because of the danger, inherent in the mobility of an automobile, that evidence may be lost and what is reasonable must necessarily depend on the facts of each case. . . . Officers are not required to know facts sufficient to prove guilt, but only knowledge of facts sufficient to show probable cause for an arrest or search.
>
> . . . All game animals are the property of the state except they may be used by a person upon killing or taking the same as by law provided, . . . it is unlawful to hunt or shoot at deer except by permission of the Game Commission . . . and a license issued. . . . It is the duty of a licensee at any time to exhibit his license to any person. . . . Game wardens are authorized to make this request of persons lawfully hunting and also to arrest without a warrant persons detected in the act of violating any game or fish law. . . . Here the wardens heard several shots fired between ten and eleven o'clock at night and they observed lights being shone in such a manner as to aid persons illegally hunting deer known to be in that area; as men of reasonable caution they had probable cause to stop the men coming from the place [where] the shots were fired and lights shone, either to ask them to exhibit their licenses or detain them to investigate if they were the persons involved in those acts. This justified a search even though the defendants were not then arrested.

In *State v. Andrus,*[62] a wolf-killing program by the State of Alaska was halted by a U.S. District Court. The facts underlying the dispute are essentially as follows: The state of Alaska, in order to protect the western Arctic caribou, initiated a wolf-killing program in the area where the caribou live and migrate. The program was an effort by the Alaska Department of Fish and Game to protect the caribou herd which allegedly was being ravaged by wolf predation. In addition to its intrinsic worth, as a valuable part of the ecosystem of the area, the caribou had traditionally provided a substantial portion of the animal harvest for the area's native population. The entire area on which the wolf hunt program was to take place is federally controlled land. The U.S. District Court for the District of Alaska said that pursuant to the Bureau of Land Management Organic Act the Secretary of the Interior has the authority to halt wolf-hunting programs on federal lands. The hunt was terminated.

Noncited Cases

Examples will give an idea of the range, magnitude, and seriousness of some poaching activities. Unscrupulous outfitters purchase trapped mountain lions and endangered jaguars for hunters willing to pay high trophy fees. Poachers shoot protected polar bears for collections or for the skin. A Korean may pay $3,000 for the gallbladder of a polar bear. The illegal trade supplies an Asian market with elk antlers and tails, bear parts, seal penises, and even herring spawn. Some wealthy trophy collectors go into national parks and shoot elk, deer, mountain goats, grizzlies, and bighorns so they can be in the record books or have the wall mount and pictures. Bobcat pelts may sell for $200; a bighorn sheep head for $3,000. Bear meat and bear paw soup are becoming special occasion foods in Asia and are gaining popularity as exotic foods in the United States. The importation of thirty-five frozen black bears by South Korean businessmen made the *Korea Herald* because the gallbladders were selling for as much as $18,300 each. Working undercover agents documented the loss of three hundred sixty-six black bears from the Great Smoky Mountain region over a three-year period. A Texas fish study showed that illegal netters are the largest harvesters of redfish, killing more than forty percent of the species where the population is already near collapse from over-fishing. In Florida, poachers squirt gasoline into the branches of a Banyan tree and let the fumes flush out rare indigo snakes which are prized as pets (Poten 1991).

In the Midwest, one Great Lakes investigation found dealers in four states selling illegal trout and salmon marked as whitefish. Some of these were contaminated with PCBs, which spread toxins to the consumers. The volume of illegal fish was so huge that it brought more than $150,000 in penalties for the poachers. Illegal fishing trade in Chicago alone in 1982 on a raid of five fish wholesalers resulted in all dealers being charged. The paddlefish, a primitive fish which produces a large amount of roe is killed for the eggs. One female can produce as much as ten pounds of caviar worth up to $500 a pound. Undercover agents joined paddlefish poachers in Missouri and found that at least 4,000 paddlefish were killed in one year. One poacher boasted of clearing $86,000 in five nights. In Alaska, walruses are killed not for the meat or waterproof clothing, but for the large tusks that bring up to $1,000. Valuable skins of sea otters are surfacing around the world in a quiet, but lucrative, black market. Some buyers pay large amounts for live sea otters for their aquariums (Poten 1991).

Alaskan guide Ron Hayes used airplanes to herd trophy animals toward the hunters. In 1988, Hayes was arrested and pleaded guilty to federal charges. He served thirteen months in prison, paid a fine of $100,000 and forfeited three airplanes.

Lothar Cistiesielski paid a government agent $11,000 and allegedly smuggled gyr-falcons to Germany. He reportedly resold them immediately for $135,000. He remains at large.

Raoul Chaisson killed twenty-nine ducks, almost ten times the limit in Louisiana. He pled guilty and was fined $425 and sentenced to two years probation and ten days of community service. Chaisson said he stopped shooting only because he ran out of shells.

William Heuer killed a trophy elk in a no hunting area in Montana and attempted to transport it out of state. He was given three years probation, 200 hours of community service, and fined $13,300.

Australian, Peter Stapley, allegedly shot an Alaskan brown bear in 1986, breaking three state wildlife regulations and then disappeared. Today he remains a wanted man.

Bill Day, a Texas banker, wanted a trophy so badly that he paid $20,000 for a record book set of white-tailed deer antlers, which he then had mounted on the skull of a Mexican deer. The true story came out when Canadian officials recognized the antlers. They had been stolen from a Canadian taxidermy shop. Day was sentenced to five years probation and given a $20,000 fine.

Robert O. Halsted, until he retired, was one of the U.S. Fish and Wildlife Service's undercover agents. Halsted learned that loons moved up the coast of North Carolina past Cape Lookout by the thousands. It was a very heavy migration. For the poachers, the important factors were the phase of the moon and the direction of the wind. A full moon meant the migration should be in full swing and an east wind meant the birds were being pushed toward shore and within shotgun range. Halsted and his co-workers moved onto the Cape and waited undercover for the hunters to arrive and start shooting. At daybreak the war against loons and shorebirds began in deadly earnest. After about thirty minutes Halsted and his crew moved in on the shooters and arrested seventy-two men who pled guilty and were fined. Nearly 250 loons were killed, another 150 crippled. Many shorebirds were also killed (Phillips 1981).

Halsted tells of another case where an assistant refuge manager and a Delaware dog warden were raiding the traps set by waterfowl biologists who band ducks during the early spring migration. Halsted, with an assistant, watched the violator approach the duck traps warily. He drove only a short distance along the dike, stopping frequently to check the surrounding countryside for possible intruders. With a dip net he quickly netted all the ducks in the trap and put them in a cage, threw them in his pickup and departed. His co-conspirator was arrested later on his way to deliver them (Phillips 1981).

A three-year undercover investigation in the Gulf Coast of Texas brought charges against 210 people for 1,300 violations. This was the biggest waterfowl undercover operation in history. On 13 December 1988, 100 U.S. Fish and Wildlife Service agents, half the entire force, showered papers on the violators. According to the agents, the hunters and their guides did everything illegal. They shot too early and too late in the day, used lead shot and electronic callers, shot over bait, left crippled birds to die, and herded birds with airplanes. It has been going on like this for years they reported. Sometimes these same bored waterfowl hunters blasted kingfishers, killdeer, ibis, and red-tailed hawks illegally killing 2,800 birds.

So many charges were stacked against the LaBove Shooting Resort that its owner entered a plea bargain offering to pay $270,000 in fines, serve five years probation, and forfeit two trucks and an airboat. The U.S. District Judge's sentence was $1,975 in fines, three years probation, and forfeit of the vehicles. This was not a message the agents wanted sent to the public.

VIOLATIONS AND WILDLIFE MANAGEMENT

Violations of wildlife laws may be considered in another way; from the standpoint of management. Certain violations result in the taking of extra fish or game, and others do not.

Hunting out of season takes game which otherwise would not be taken. Hunting and fishing in improper places, such as refuges, probably takes extra game (although

A conservation officer sets up a deer decoy. *Courtesy Idaho Fish and Game Department, Jack McNeel.*

much of it would never become available for legal take), but hunting on property that is posted frequently does not, though it may be in violation of the fish and game law and the civil law.

The use of an improper license, such as a nonresident buying a resident license, probably does not take additional game, but it costs the state additional revenue. Licenses which are made out improperly, but for which the correct fee has been paid, do not take extra game. Improperly tagged animals does not take additional game (unless the tag is reused). The failure to wear a properly purchased, visible license when the law so requires it does not take additional game.

The use of illegal gear may or may not take extra game, depending on the efficiency and use of the gear. Illegal possession, such as holding game overtime, does not take extra game, although it may permit game to be held so long that it is no longer suitable for table use.

THE USE OF DECOYS

The detection and apprehension of illegal deer spotlighting has been one of the most time-consuming and frustrating problems facing wildlife law enforcement officers in recent years. To counter this, animated, lifelike decoys are being widely used across the United States.

Decoys have been used to catch wildlife violators since the 1940s, when Wisconsin introduced the grouse decoy. The first deer decoy was used in Tennessee in 1983. Texas attempts to catch the illegal act on videotape, and the footage has been instrumental in obtaining a hundred percent conviction rate. New Mexico uses animated, remote controlled deer, elk, and antelope decoys. The improved models have styrofoam bodies of realistic size that can be activated from up to 300 yards. Wyoming uses big game, grouse, and wild turkey decoys. Colorado places decoys in known spotlighting hot spots, often at the urging of local sportsmen and landowners. Some states (Idaho, for example) require only the act of hunting with the use of an artificial light to be in

Training K-9 dogs is all in a day's work for the wildlife officer. *Courtesy New York Department of Environmental Conservation, John Toerg.*

violation. It does not require that an animal be killed or even be present in order for the act to be a violation. In decoy-induced cases, the actual charges may be shooting from a public road, shooting across a road, shooting from a car, having a loaded firearm in a vehicle, or hunting with an artificial light.

The use of decoys is not only an effective way to stop road hunters, it also acts as a deterrent to would-be violators. Another plus is that the enforcement officers' time can be used to much better advantage than when they were seeking violators on a blind routine. Predictably, the common defense against decoys is entrapment, but, to date, no one has won a court case using this defense (Scripts Howard 1991; Cox 1991).

DOG DETECTOR UNITS

The following information was furnished by New York State Department of Environmental Conservation, Division of Law Enforcement, Dog Detector Unit. It describes, in part, New York's K-9 Unit. Only male German shepherd dogs are used. There is an instructor and a supervisor who coordinate all training, maintain active files, and keep records of operations and inventory. The dogs are owned by the Department of Environmental Conservation, Division of Law Enforcement. The Division is responsible for the dogs' food, veterinary bills, equipment, and liability. Dog licenses are furnished by the state. The equipment includes a chain-link fence kennel and doghouse, which is used only for emergencies or short periods. The dogs are required to live in the home most of the time. There are two sets of stainless steel food and water dishes, one for home and one on the road. A padded attack training sleeve and a hidden attack sleeve are included. Tracking harnesses and fifteen-, thirty-, or fifty-foot tracking leads are used in some situations. A wide, leather aggression collar is usually used in the search of buildings. There is a choke training collar with a six-foot lead and a short twenty-four-inch traffic lead.

Many hours go into
K-9 dog training.
*Courtesy Utah Division of
Wildlife Resources.*

The dogs are groomed each day. The handler is responsible for weekly maintenance and training to keep the dog in condition and performing his skills to the highest degree. The division instructor inspects and certifies the dogs every six months. At this time, each dog handler demonstrates his ability in phases of the leash training. Aggression levels are also evaluated. Each dog completes a tracking test and must be successful in the detection work he is trained for. Each year, or every other year, the division instructor sets up a one-week inservice K-9 school for all handlers and dogs. The team must be able to work in canoes, boats, aircraft, trucks, and buses. Each dog is kept socialized and conditioned to all possible encounters such as steel-grated bridges, bridge poles, ladders, catwalks, docks, fire escapes, and so forth. All K-9 handlers participate in public relations with the detector dog, by giving demonstrations and talks.

Environmental conservation officers have at least three years of experience prior to entering the dog detection unit. Dog handlers are always on the lookout for additional recruits. Generally two- to three-year-old dogs in good health and with a good disposition are preferred.

Literature Cited

Cox, J. 1991. Dummy deer dupes poachers. *The Izaak Walton League* 10 (4): 6.

Phillips, J. H. 1981. *Undercover wildlife agent.* Tulsa, Okla.: Winchester Press.

Poten, C. J. 1991. A shameful harvest. *National Geographic* 180 (3): 106–132.

Scripts Howard. 1991. ''Tis'' the Season to Catch Deer Poachers in Colorado. *The Herald Journal,* Logan, Utah, Nov. 8, Page 10.

U.S. Congress. House. 1984. *FBI Undercover Operations.* Subcommittee on Civil and Constitutional Rights. Washington, D.C.: U.S. Govt. Printing Office.

Wharton, T. Poachers use cover of night, neighbors' silence to bag prey. *The Salt Lake Tribune,* Salt Lake, Utah, Sept. 1, 1991.

Notes

1. Many administrators believe that an analysis of any complete set of fish and game regulations will show that the majority of them are of a nature designed to equalize opportunity of taking game by hunters, rather than an effort to reduce the kill.

2. There is a definite relationship between out-of-season and illegal possession. Many persons involved in out-of-season violations are actually charged with illegal possession, since it is difficult to apprehend the violator in the act.

3. *Utah v. Chindgren,* 777 P. 2d 527; (1989).

4. 74 S. Ct. 808, 812, 347 U.S. 612, 617 (1954).

5. *Robinson et al. v. State,* 37 Okla. Crim. 438, 258 Pac. 1073 (Crim. App. 1927).

6. *State v. Rathbone,* 110 Mont. 225, 100 P. 2d 86 (1940).

7. 114 Wash. 370, 195 Pac. 16 (1921).

8. *Utah v. Morck,* 821 P. 2d 1190 (1991).

9. 267 U.S. 132 (1925).

10. *Aragon v. People, ex. rel. Medina et al.,* 121 Colo. 505, 218 P. 2d 744 (1950).

11. *State v. Barnett et al.,* 56 N. Mex. 495, 245 P. 2d 833 (1952).

12. 552 F. 2d 817 (8th Cir. 1977), *cert. denied,* 431 U.S. 949 (1977).

13. *Herzig v. Feast et al.,* 127 Colo. 564, 259 P. 2d 288 (1953).

14. *Washburn v. State,* 90 Okla. Crim. 306, 213 P. 2d 870 (Crim. App. 1950).

15. *State v. Tyler,* 82 Okla. Crim. 112, 166 P. 2d 1015 (Crim. App. 1946).

16. *State v. Baxter,* 16 Wash. 2d 246, 132 P. 2d 1022 (1943).

17. The officers could have saved time and money by placing the man under arrest when they saw him committing the crime. If this were not possible, he could have and should have been kept under constant surveillance until the arrest could be effected. Also, the need for absolute identification of evidence by the officer is shown. The nets should have been, if they were not, very plainly marked and stored in such a manner as to leave no doubt, at a later time, as to their role as competent evidence in this case.

18. *Kephart v. State,* 229 P. 2d 224 (1951).

19. *State v. Schrimer et al.,* 70 Idaho 83, 211 P. 2d 762 (1949).

20. *State v. Addor,* 183 N.C. 687, 110 S.E. 650 (1922); *Dooley v. State,* 27 Ala. 261, 170 So. 96 (1936); *West v. Commonwealth,* 156 Va. 975, 157 S.E. 538 (1931); *State v. Wood,* 19 S. Dak. 260, 103 N.W. 25 (1905); *State v. Hurley,* 79 Vt. 28, 64 A. 78 (1906); *Gustine v. State,* 86 Fla. 24, 97 So. 207 (1923); *State v. Rooney,* 118 Kans. 618, 236 Pac. 826 (1925).

21. *Galloway v. State,* 125 Tex. Crim. 524, 69 S.W. 2d 89 (Crim. App. 1933); *Poteet v. State,* 138 Tex. Crim 9, 133 S.W. 2d 581 (Crim. App. 1939); *West v. State,* 152 Tex. Crim. 7, 210 S.W. 2d 585 (Crim. App. 1948).

22. *State of New Mexico v. Barber,* 91 N. Mex. 764, 581 P. 2d 27 (Ct. App. 1978).

23. *State v. Weindel,* 52 Okla. Crim. 25, 2P. 2d 599 (Crim. App. 1931).

24. *People v. Miller,* 110 Calif. App. 2d Supp. 843, 243 P. 2d 135 (App. Dept. Super. Ct. 1952).

25. *Commonwealth of Massachusetts v. Worth,* 304 Mass. 313 23 N.E., 2d 891 (1939).

26. *State v. Evans,* 143 Oreg. 603, 22 P. 2d 496 (1933).

27. *State v. Pulos,* 64 Oreg. 92, 129 Pac. 128 (1913).

28. 53 Oreg. 38, 98 Pac. 713 (1908).

29. *Sherwood v. Stephans,* 13 Idaho 399, 90 Pac. 345 (1907).

30. *State v. Fisher,* 53 Oreg. 38, 98 Pac. 713 (1908).

31. *Stewart v. People,* 83 Colo. 289, 264 Pac. 720 (1928).

32. *State v. Miles,* 5 Wash. 2d 322, 105 P. 2d 51 (1940).

33. *State v. Visser,* 188 Wash. 179, 61 P. 2d 1284 (1936).

34. *Hughes v. State,* Okla. Cr., 572 P. 2d 573 (1978).

35. 57 L. Ed. 2d 117 (1978).

36. 583 F. 2d 491 (1978).

37. 528 F. 2d 314 (8th Cir. 1976).

38. *United States of America, ex rel. Bergen v. Lawrence,* 848 F. 2d 1502 (1988).

39. 167 U.S. 518, 17 S. Ct. 864, 42 L. Ed. 260 (1897).

40. *Mescalero Apache Tribe v. State of New Mexico,* 630 F. 2d 724 (1980).

41. 591 F. 2d 89, 91 (9th Cir. 1979).

42. 433 U.S. 165, 176–77, 97 S. Ct. 2616, 2623–24, 53 L. Ed. 2d 667 (1977).

43. 358 U.S. 217, 79 S. Ct. 269, 3 L. Ed. 2d 251 (1959).

44. 100 S. Ct. 2069, 2083, 65 L. Ed. 2d 10 (1980).

45. 441 U.S. 322, 327, 335–336, 99 S. Ct. 1717, 1731, 1736–37, 60 L. Ed. 2d 250.

46. *United States v. Doepel,* 755 F. Supp. 249 (E. D. Ark. 1991).

47. 315 U.S. 60, 80, 62 S. Ct. 457, 469, 86 L. Ed. 680 (1942); 678 F. 2d 956, 958 (11th Cir. 1982).

48. 446 F. 2d 669, 671 (8th Cir. 1971).

49. 426 U.S. 529, 543 (1976).

50. 56 L. Ed. 354 (1978).

51. 57 L. Ed. 2d 1052 (1978).

52. 588 P. 2d 526 (1978).

53. *State v. Sieminski,* Alaska 556 P. 929 (1976).

54. Wash. App., 585 P. 2d 805 (1978).

55. Oreg. App., 570 P. 2d 401 (1977).

56. Wash., 575 P. 2d 1057 (1978).

57. Wash. App., 573 P. 2d 378 (1977).

58. 584 P. 2d 1088 (1978).
59. 523 P. 2d 522, 95 Idaho (1974).
60. Wyo., 472 P. 2d 933 (1970).

61. Ibid.
62. 429 F. Supp. 958 (D. Alaska 1977).

Recommended Reading

Andrassy, J. 1970. *International law and the resources of the sea.* New York and London: Columbia Univ. Press.

Anonymous. 1982. How to observe and describe people. In *Law Enforcement Bible No. 2.* Ed. R. A. Scanlan. South Hackensack, New Jersey: Stoeger Publ. Co.

Bramble, L. 1982. Good report writing: A mark of the police professional. In *Law Enforcement Bible No. 2.* Ed. R. A. Scanlan. South Hackensack, New Jersey: Stoeger Publ. Co.

Burkett, B. 1984. Game law enforcement: Effects on wildlife resource. *North Dakota Outdoors* 47 (2).

Calkins, F. 1970. *Rocky Mountain warden.* New York: Alfred A. Knopf, Inc.

Decker, D. J., T. L. Brown, and W. Sarbello. 1981. Attitudes of residents in the peripheral Adirondacks toward illegally killing deer. *New York Fish and Game Journal* 28 (1).

Ellison, C. W. 1982. The stress syndrome of the modern police officer. In *Law Enforcement Bible No. 2.* Ed. R. A. Scanlan. South Hackensack, New Jersey: Stoeger Publ. Co.

Farnsworth, C. L. 1980. A descriptive analysis of the extent of commercial poaching in the United States. Ph.D. diss., Sam Houston State Univ., Huntsville, Texas.

Friedmann, W. 1967. *Legal theory.* 2d ed. Ithaca, N.Y.: Columbia Univ. Press.

Gallman, J. 1963. Preparation of game and fish cases. In *Proceedings Southeastern Assn. of Game and Fish Comms.,* No. 17, 394–397.

Graham, F. Jr. 1991. Is the Endangered Species Act losing its bite? *Audubon (July–Aug.).*

Lerner, M. A., and J. B. Copeland. 1983. Stingtime for poachers. *Newsweek* 101 (59).

Lott, E. 1961. Evidence in wildlife law enforcement. In *Proceedings Southeastern Assn. of Game and Fish Comms.,* No. 15, 475–479.

MacDowell, D. M. 1978. *The law in classical Athens.* Ithaca, N.Y.: Cornell Univ. Press.

Purol, D. A. 1982. Field estimating the legality of harvested deer. Michigan Dept. of Natural Resources, Law Enforcement Division, Report No. 4.

Reisner, M. 1991. *Game Wars–The Undercover Pursuit of Wildlife Poachers.* Viking.

Tennesen, M. 1991. Poaching: Ancient traditions vs. the law. *Audubon (July–Aug.).*

Thomas, M. G. 1983. Poaching impact assessment. In *Human Demographic Impacts on Fish and Wildlife Resources from Energy Development in Rural Western Areas.* Fish and Wildlife Service /obs/ 83/77.

Watkins, T. H. 1980. The thin green line. *Audubon* 82 (5): 86–87.

Multiple-Choice Questions

Mark only one answer.

1. Which type of wildlife law violator is hardest to catch?
   a. commercial hunters
   b. trophy hunters
   c. thrill killers
   d. spur of the moment hunters
   e. subsistence hunters

2. Which type of wildlife violator is probably the most destructive to big game per unit of hunter effort?
   a. commercial hunters
   b. trophy hunters
   c. thrill killers
   d. spur of the moment hunters
   e. subsistence hunters

3. Realistic decoys are used to catch hunters pursuing which game animals?
   a. grouse
   b. deer
   c. pronghorn
   d. elk
   e. all of the above

4. In *Baldwin v. Montana Fish and Game Commission* the U.S. Supreme Court said:
   a. the State must treat all citizens alike, resident and nonresident
   b. the Montana statutes are not violating the privileges and immunity clauses since access to recreational elk hunting does not fall in that category
   c. the statutory distinction between resident and nonresident does not violate the equal protection clause of the Fourteenth Amendment
   d. the legislative choice was an economic means and not unreasonably regulated to the preservation of finite resources and a substantial regulatory interest to the State
   e. all of the above

5. The Wild Free-roaming Horses and Burros Act has created many emotional situations since its passage. In *Kleppe v. New Mexico* which statement is not true?
   a. New Mexico asserted the federal government lacked power to control wild horses and burros on public lands unless they were moving interstate
   b. district court ruled in favor of New Mexico
   c. circuit court overruled the district court
   d. circuit court said that New Mexico livestock board had entered upon public lands and removed burros—an action contrary to the provisions of the Wild, Free-roaming Horses and Burros Act
   e. absent consent or cession, a state undoubtedly retains jurisdiction over federal lands within its territory and Congress does not have the power to enact legislation overcoming this

6. In *Mescalero Apache Tribe v. State of New Mexico* which of the below statements is not true?
   a. the state was able to claim that either it or its lands played a significant role in the creation and preservation of wildlife on the reservation
   b. the state claimed that the tribe had no standing and that the suit was otherwise not justifiable
   c. the court rules that the state's understanding of standing requirements is overly narrow
   d. federal courts, for the purpose of standing, may consider the principles of elementary economics
   e. the sovereign powers of the tribe in wildlife management were said to be so pervasive that sovereignty here moves from a mere backdrop into a leading role on the litigational stage

7. In *United States of America ex rel. Bergen v. Lawrence* which statement is not true?
   a. the United States brought action against cattle rancher Lawrence for allegedly wrongful fencing of federal land
   b. the district court directed that the cattle ranchers fence be removed or modified in such a way as to allow passage by antelope
   c. the court's ruling effected impermissible and unconstitutional taking
   d. the decision did not impose a servitude but rather a day-to-day nuisance prescribed by a federal law
   e. the court said that winter forage by antelope was a lawful purpose of public lands and was thus protected from unlawful enclosures. The court said that the mere presence of gates or openings in the ranchers fence which in effect excluded pronghorn from their critical range did not mean that the enclosure was lawful

8. Steven R. Chindgren was convicted of:
   a. taking pheasants out of season
   b. taking pheasants illegally with a falcon
   c. taking ducks illegally with a dog
   d. taking a mallard duck illegally with a falcon
   e. hunting out of season

9. It is against the law to: (one statement is false)
   a. hunt deer in most states with a .22-caliber rimfire rifle
   b. hunt migratory waterfowl with a shotgun that holds more than three shells
   c. hunt in open and unattended fields
   d. hunt waterfowl with an 8-gauge shotgun
   e. carry a firearm during the bow and arrow deer season

10. In *State v. Evans* which statement is wrong?
    a. when the warden went to check on Evans and Acuff he found them wearing bloody clothes
    b. they directed him away from rather than to their camp and claimed they had not been hunting
    c. Evans and Acuff denied the warden permission to search their camp and enforced it by pointing their rifles at the warden
    d. late permission was granted, but the men followed the officer around with their rifles pointed at him
    e. although there was a substantial amount of circumstantial evidence, the men were acquitted by a circuit court

11. Hunting is generally defined as:
    a. taking or attempting to take game
    b. taking game
    c. attempting to take game out of season
    d. any act involving a lethal weapon where there is game in the vicinity
    e. taking game legally

12. One of the following statements is not true:
    a. a fish and game department is the custodian of game
    b. hunting is a right
    c. the title to game is with the state
    d. in a legal contest between state and federal government the federal government generally wins
    e. game legally reduced to possession is the property of the taker

Chapter 9

Arrests

DEFINITION OF ARREST

Cases 1–4

The definition of arrest, seizure within the meaning of the Fourth Amendment, has been defined in several court cases. In *Tennessee v. Garner,*[1] the U.S. Supreme Court wrote that "whenever an officer restrains the freedom of a person to walk away, he has seized that person." In *Brower v. County of Inyo,*[2] the Court held that a seizure occurs ". . . only when there is a governmental termination of freedom of movement through means intentionally applied." In *Michigan v. Chesternut,*[3] the Supreme Court concluded that seizure within the meaning of the Fourth Amendment occurs only when officers, by means of physical force or show of authority, restrain the liberty of the citizen. In *United States v. Mendenhall,*[4] Justice Stewart wrote the "free to leave" test for determining whether or not a person has been seized. ". . . a person has been 'seized' within the meaning of the Fourth Amendment only if in view of all the circumstances surrounding the incident, a reasonable person would have believed that he was not free to leave." A determination of whether or not police conduct amounts to Fourth Amendment seizure takes into account all of the circumstances surrounding the incident in each individual case (DePietro 1992). Whether or not an encounter is a seizure turns, in part, on what the subject has reason to know or believe rather than in the officer's hidden plans. The subjective intentions of an officer during an encounter are relevant to the Fourth Amendment seizure rule only to the extent that they are conveyed to the suspect.

DePietro (1992) lists eight factors that are relevant in determining whether a particular encounter between police and citizen is a consensual one or a Fourth Amendment seizure.

1. The slightest application of physical force for the purpose of holding or stopping a person may be construed as a seizure. Accidental or unintentional contact is generally not a seizure, but the officers should avoid this whenever possible.

2. The presence of several officers rather than just one may transform what was a consensual encounter into a seizure. Therefore, when the officer is not in jeopardy, an encounter is more likely to be considered consensual if backup officers stay in the background where the citizen may not immediately recognize them.

3. The display or pointing a weapon at a citizen is generally considered by the citizen as compelling compliance rather than one of consensus.

4. If a person's movement is interfered with by officers that position themselves and/or their vehicles where they block the citizen's path or freedom of movement, the citizen generally considers himself not free to leave. Officers wishing to keep an encounter consensual should position themselves so that they leave a clear path for the citizen.

5. Movement from the original site of the confrontation to another location does not, in itself, escalate a consensual encounter into a Fourth Amendment seizure. However, the officers requesting a suspect to accompany them to another location should document that the citizen has a choice.

6. An officer's use of coercive or intimidating language or tone of voice may be interpreted by a reasonable person as one compelling compliance under the meaning of the Fourth Amendment. A uniformed officer repeatedly flashing a badge is an intimidating conduct. A request for consent to search should also be conveyed in a manner that makes it clear that the citizen has a choice and that the compliance is not compelled. Since uncommunicated suspicions generally do not have a bearing on whether or not a particular encounter is consensual, a seizure should be delayed until it is justified.

7. When officers request personal property for examination and ask questions about any discrepancies in these items, the officer should return them promptly since prolonged detention of personal items can transform a consensual encounter into a seizure.

8. Advising citizens they have a right to refuse to consent to a search or to answer questions or to accompany the officer to a different location may prevent such encounters from becoming unlawful seizures.

Admissibility of evidence may depend on whether or not it was seized by officers during a voluntary encounter or a Fourth Amendment seizure. Understanding these things will help an officer ensure that the Fourth Amendment is not implicated until the officer has established sufficient suspicion to justify a seizure.

TYPES OF ARREST

The types of arrest discussed in the following sections are the nonwarrant or sight arrest, the warrant arrest, and the misdemeanor on complaint. The citation plan is treated separately.

Nonwarrant Arrest

A wildlife officer may, without a warrant, arrest a person under the following circumstances:

(1) when a wildlife (or other) law offense has been committed or attempted in the officer's presence;

(2) when a charge is made upon reasonable cause, in the case of a felony, by the arresting officer;

(3) at any time during the day or night when the officer has reasonable cause to believe that the suspect has committed a felony.

In some states misdemeanor arrests can be made at night if committed in the officer's presence. Other misdemeanor arrests at night may also be made only on probable cause. It depends on state law. Full police powers of each state may be different.

Warrant Arrest

A wildlife officer may make an arrest in obedience to a warrant delivered to him or her. A warrant of arrest may be described as a legal instrument, properly executed by competent authority, generally a magistrate, which commands the arrest and apprehension of the person listed therein. Any warrant may be addressed either to a particular police officer or to any police officer, and may be legally served by any police officer to whom it is handed for service.

The telegraphic/telephonic warrant does not require the warrant in hand to make the arrest. There is no requirement to show the actual warrant to the suspect. Officer safety considerations may be jeopardized by so doing. Officers can verbally announce their authority and intent to arrest with a warrant. A suspect can review the actual warrant after being secured and transported to jail.

After making an arrest under a warrant, it is necessary that the subject be taken immediately to court, preferably to the magistrate who issued the order; however, this is not absolutely necessary. After the arrest has been completed, the law requires that the warrant be returned to the issuing official and shall bear on the reverse side of the warrant an endorsement setting forth the date, time, place of apprehension, and signature of the arresting officer.

Misdemeanor on Complaint

In the case of a misdemeanor which has been committed sometime in the past, the wildlife officer should contact the prosecutor's office then go to the justice of the peace or the city judge and request that a complaint be issued. A complaint of this kind may generally be issued anytime within a year of the time of the violation. The county attorney usually prepares the complaint, and the magistrate issues to the officer a warrant or summons which declares that the defendant must come or be brought into court on a certain day (or forthwith, in case of a warrant) and answer the charges against him. In some states this type of arrest cannot be made at night in the case of a misdemeanor, except at the discrimination of the magistrate who so endorses the complaint. In other states there are no day/night restrictions on any type of warrant.

THE SEARCH WARRANT

A search warrant is defined as an order in writing, in the name of the state, and signed by the court, that orders an officer to search certain named buildings and/or premises for a particular piece of personal property and to bring it before the court. This may also include persons. A search warrant for a daylight search may not be used at night. Search warrants cannot be issued on "probable cause" only. They are issued only on the basis of a sworn deposition giving "reasonable cause to believe." A search warrant also has to be returned with endorsement and show or list materials obtained under it.

The Constitution protects a citizen, in general, against search and seizure, and seizure of person and property within his residence. Usually officers use witness statements to obtain search warrants. Officers cannot testify to hearsay or verbal witness statements, but can use sworn statements. Also, the informant may not wish to appear in the case. The officer must examine all evidence and be satisfied that it is adequate before asking for a search warrant.

It is desirable that an officer take at least one officer along for safety and to act as a witness in the event the occupant of the building later accuses the officer of misconduct.

During the search, care should be taken not to damage anything and to replace all material as it was found. When taking property under a warrant, the officer must give a receipt for it, stating in detail what was removed. In the event that no one is present during the search, the receipt should be placed where the material was removed.

THE CITATION

Although a person may be arrested either with or without a warrant, many fish and game violations are handled by issuance of a citation. As commonly employed by fish and game officers, a citation is a notice, issued by an officer directly or by mail, to appear in court at a certain time; it is signed by the officer and acknowledged by the defendant. If a person believed to be a violator is cited rather than arrested, the officer may go about business instead of dropping all work and immediately taking the subject before the magistrate. The legal status of citations may vary from state to state.

Many fish and game violators are happy to sign citations, stating their willingness to appear at a future date, because this is considerably more convenient and avoids the chance of a night in jail or a quick change in plans. Criminals taken on fish and game violations may be anxious to use the citation method rather than be placed under arrest. In addition to obtaining an agreement to appear at a certain date, an officer who properly handles the situation may gain a statement of admission of guilt, which can be used if the subject later decides to plead not guilty. However, recent court decisions on the use of admissions of guilt now apply the same rules of evidence to admissions as that which is applied to confessions. The Miranda requirement on advising the defendant of the right to remain silent and the right to counsel at all times should be kept in mind.

It is generally permissible to seize for the court guns, fishing tackle, and other equipment as evidence in a legal arrest. This adds considerable inducement to the one arrested to appear in court at the proper time. When equipment is seized, it is at times possible to return it immediately and take a receipt for it, if this latter procedure is considered desirable. In many instances this receipt, describing the equipment in detail along with the circumstances of seizure, has received the same weight in federal and state court that the equipment itself would have received.

PRINCIPLES OF ARREST

The basic principles of arrest vary but little from those of a successful military operation. They may be referred to as the four Ss, namely: superiority, simplicity, speed, and surprise. Superiority of manpower has scared even hardened criminals who had previously threatened not to be taken alive. The value of superiority in equipment such as weapons, radios, lights, field glasses, and automotive equipment is obvious. Simplicity makes for an easy and unconfused completion of an arrest plan. Speed and surprise work together to complete a successful arrest. In every arrest situation the officer must be alert. No amount of preparation will be sufficient in an arrest if danger areas are not recognized immediately.

ARREST SITUATIONS

Some of the possibilities in arrest situations are: attempted flight by the subject; attempt by the subject to kill or injure the officer; attempt by the subject to commit suicide; attempt to destroy or hide evidence; danger of injury to innocent bystanders; attempted rescue of the subject by confederates; and public disturbances arising out of the arrest. A well-trained and thoughtful officer will consider these possibilities before attempting arrest.

Here are some of the factors an officer will encounter. Distance: The distance of many armed encounters is less than 10 feet, almost point-blank range. Practice ranges are from 7 to 15 yards. Light: Many shootouts are in such poor light that the officer may not be able to see the sights on his gun. Length of shootout: Shots may be fired for not more than two to three seconds, and the number of rounds fired may be no more than two to three. Location: The assailant may open fire from cover such as a tree, embankment, building, or automobile, or in the open if there is danger recognition, real or imagined. Assailants: An officer may, or may not, know exactly where an assailant is or how many there are. Weapons: An officer must assume that the assailant is armed as well as, or better than, the officer. In many situations a handgun is no match for a shotgun (Adams et al. 1980).

Checking for loaded guns in a vehicle is routine.
Courtesy Utah Division of Wildlife Resources.

ARREST PROCEDURE

When making an arrest, the wildlife officer should observe certain established procedures. These enable the officer to complete the arrest and avoid much of the risk of physical injury, or the chance of a successful defense by the defendant claiming that he was unaware of the officer's identity.

Officer Identification

When approaching a suspect, the officer must first identify himself or herself; state the intention of arresting the subject, if such is the case; and specify the cause of arrest. Although there are admitted exceptions, this is the general procedure. This is important, first for the officer's protection, and second if the officer or the suspect is injured or killed when the arrest is made.

The identification should be made in such a way that the suspect and bystanders, if there are any, will have no doubt as to the officer's official status. Unless it can later be shown that the officer quickly and properly identified himself or herself, a successful defense may be built around the subject's claim of perceived robbery or assault. When the subject is either in the act of committing an offense or making an escape, it is not necessary to notify the accused of the intent to arrest, because it is self-evident. A badge that is plainly visible, or an officer in uniform, would appear to be evidence of official status, but the general rule is the officer should identify himself or herself.

Confident Approach

The question of when an arrest shall be made has always plagued law enforcement officers and probably will continue to do so. After deciding to make an arrest, the officer should proceed with confidence. When the legal right to make an arrest is evident, and it is obviously the officer's duty to do so, then nothing ordinary should prevent him or her from going ahead. But, in certain instances, a delay for more favorable circumstances may be desirable. An immediate arrest for a fish and game violation is certainly not worth the serious injury or death of an officer, hunter, or angler when this same arrest may be accomplished safely at a later time. It is equally obvious that under no circumstances should the officer resort to brutality, profanity, abuse, or threats.

In the case of hardened criminals, officers should never allow themselves to be placed at a physical disadvantage; and unless the officers know otherwise, the only safe procedure is to assume the subject being arrested is dangerous.

Character of Suspect

Unless there are compelling reasons to the contrary, all people should be treated in the same courteous manner at first contact. Adjustments can be made should the situation escalate. Even people with a ''bad character'' expect to be treated in a decent manner in a normal contact. Everyone should be searched, because an officer cannot assume one person is less dangerous than another. Some people never hesitate to impose on an officer who shows the slightest disposition to be lenient, and even long-term acquaintances may be unpredictable when under great stress. It is possible for an officer to be firm, carry out duties to the letter, and at the same time be courteous and considerate.

Body Search

Preliminary and final body search of the suspect should always be made. A preliminary body search at the time of actual arrest is mandatory. When a search is made, it should be thorough and careful. Slipshod methods of searching may not reveal concealed weapons or may give the arresting officer a false sense of security. Some of the usual places where weapons are concealed are the groin area, shoe or boot tops, chest, back and around the belt. The officer should keep in mind that the suspect may have more than one weapon.

As a legal matter, law enforcement agencies arrange for the search of female prisoners to be conducted by females. However, situations may arise in which it is essential, both for the protection of the officer and the prisoner, that if no other female is present, a preliminary search be conducted by a male officer. It is wise to use a witness even if it is not another officer. In this situation a tape recorder is desirable if not absolutely necessary. The occasions on which such circumstances arise are rare. The first search should be complete except for body cavities. A receipt should be issued for all personal property, and the prisoner should sign this receipt. Personal property should be kept separate from evidence which will not be returned.

Use of Handcuffs

There are many techniques used, but most officers employ a speed cuffing technique where the subject spreads his legs, bends over at the waist and thrusts the arms rearward. Of the separate components of police-subject control, handcuffing should have the highest priority. There are several aspects of handcuffing that should be examined. First, use techniques that are effective whether the subject is standing, kneeling, or prone. In brief, handcuffing should have one technique for all occasions. One consideration is the suspect's behavior during handcuffing. The most common behavior is cooperation. The most dangerous subject to handcuff is the uncooperative person who may be intoxicated, on drugs, or an experienced criminal. Individuals may be totally uncooperative both physically and verbally, or wait for an opportune moment to resist. The latter may occur at the moment of touch. In any event, the officer should not attempt to handcuff a person unless he is under control. There is no effective handcuffing method when a subject is throwing a punch (Siddle 1991).

Where handcuffing was deemed by the court to be reasonable, abrasions and minor cuts from the handcuffs have not been listed. Whether the detention is for a misdemeanor or a felony, when the law enforcement officer reasonably believes there is an

Shooting skills help
teach respect for guns.
Courtesy Crossman Guns.

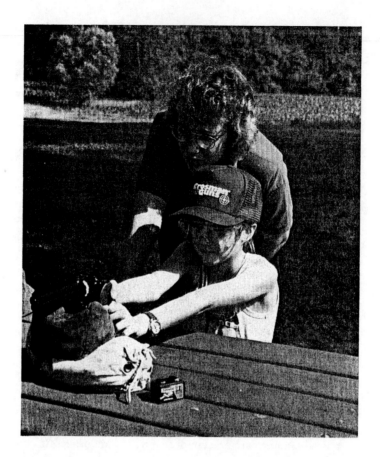

escape risk or a safety hazard, the suspect can be handcuffed without civil repercussions. Officers can handcuff a suspect during an investigation when they believe there is probable cause the detainee has committed a crime and/or may be a threat to the officer. Handcuffs should be double locked. One of the things to keep in mind is handcuffs are at best a temporary restraining device (Siddle 1991). Searches should be made after the suspect is handcuffed. For further information see *PPCT* (Pressure Point Control Tactics) *Defensive Tactics Student Manual,* Series A.

In most cases, the wildlife officer will restrain the dangerous suspect by use of handcuffs. This is particularly true when prisoners must be transported over long distances, when the officer is outnumbered and needs to partially immobilize prisoners to transport them, or when obstinate prisoners threaten to injure themselves or the officer. If the prisoner is to be transported a long distance, it is generally necessary to fasten the handcuffs in front. Front handcuffing can be made more effective by running the prisoner's pants' belt over the chain between the handcuffs, and by fastening the belt in back. To prevent the prisoner from using handcuffs as a weapon, it may be necessary to fasten them behind the prisoner's back, or to run the chain connecting the two links between the prisoner's legs.

A subject must be treated as innocent until proven guilty. While the prisoner is in custody, the officer must provide food and other necessities of life. Care should be taken not to injure the suspect by placing the cuffs on so tightly that the circulation is retarded or cut off. Double locking cuffs ensures that they will not tighten down any further. If the subject complains of tightness, check the cuffs and document your action. If medical attention is required, it must be provided as soon as possible.

When using handcuffs, the officer should always keep in mind that, at best, they are only a temporary restraining device that many experienced criminals are able to slip quite readily. Handcuffs cannot substitute for an alert mind. Officers who relax mentally as soon as the cuffs are on, place themselves immediately at a disadvantage.

FRESH PURSUIT

The act of fresh pursuit, effective in many states, gives the officer the right to cross a state line to make an arrest when in fresh, or initial, pursuit of an alleged criminal. In some states the act of fresh pursuit applies to felonies but not to misdemeanors.

It is recognized that the officer must be the aggressor, and must decide what amount of force is necessary to effect the arrest. If, in the process of pursuing a subject, an officer sees the suspect leave a car and take refuge in a private home, the officer may not break into the home without a warrant. The courts looked at the length/duration of fresh pursuit and decided it is terminated when a suspect is contained in a house—thus necessitating issuance of a warrant. Some states do not allow break-ins without a warrant unless the public at large is endangered. This applies to misdemeanors as well as to felonies.

IMMEDIATE HEARING

Once an arrest has been made the prisoner should not be allowed out of sight or reach, and the officer should be careful that some confederate does not pass weapons or other objects that might be used to injure the officer. The officer must not disregard the prisoner's right to an immediate hearing, and an officer is obliged by law to take the arrested suspect forthwith generally to the nearest police station for booking.

ARRESTING THE DANGEROUS CRIMINAL

Today, more than ever before, the wildlife officer is likely to face a dangerous criminal. It may be as part of a team that has been briefed, or in an unforeseen one-on-one situation. In either case the officer is at a critical disadvantage. The potential assailant recognizes the officer by marked car, uniform or just general activities, and can decide when and whom to assault. There is rarely a concern for innocent bystanders, and no concern for the use of deadly force. A mentally or physically unprepared officer who approaches a dangerous criminal is more likely to be injured or killed. The officer should be trained in tactics and shooting skills to use when under fire (Adams et al. 1980).

ROLE OF INTENSIVE TRAINING

The answer to arrest situation problems is, in part, intensive training. The officer should be trained to the point that he or she will, in a crisis, resort to learned, proper state-of-the-art training techniques without conscious thought. Proper training results in subconscious reactions. Often there is no time to think, only to react. Aside from the legal aspects and the potential danger, this is no different than the situation faced by professional athletes. Some professional football players sit for hours before a game rehearsing what to do in a given situation. They cannot imagine every possibility, but the better trained they are, the higher their performance rating. Law enforcement officers learn from training and experience. But in dealing with armed and dangerous

Conservation officers check hunting licenses at a roadside check station.
Courtesy Idaho Fish and Game Department, Rick Gilchrist.

fugitives, which experiences hopefully will be rare, they must learn mostly from others. They must be mentally and physically ready, they must understand tactics for many situations, and must have their shooting skills honed with sidearm, shotgun, and in some instances, a high-powered, scoped rifle.

FRAME OF MIND

An officer must be in the proper frame of mind before approaching a dangerous assailant. The officer must assume that the person is armed and willing to injure or kill, often without warning. The officer must have internally resolved the legal, moral, and psychological implications of killing another human being. If officers cannot resolve this problem beforehand, they should not be there.

THE APPROACH AND ARREST

Now that you are faced with a dangerous situation, you should, gun in hand, assess the possibilities. Do you know precisely where the assailant is? Is the assailant in an invulnerable place? Is the assailant in the open, such as behind a tree, or in a building, a trailer or a motorhome? If the assailant is in a fixed abode (building, trailer with no attached power unit) you may be able to summon help and surround the place. For those of you who saw the movie *High Noon,* that is not the way to go. Plan your approach, consider possibilities and what your reaction will be to each one. If you have a partner, decide in advance what to do and who will call the moves. Remember, no approach is routine, and the circumstances of no two occasions are identical, even if you have been there before. The circumstances dictate the tactics, not the reverse. Caparatta (1989) believes more officers die from poor tactics than poor marksmanship. Do not act in a routine, predictable manner. You may be approaching someone who is expecting just that, and will be ready. Be confident, but not to the point where it will make you feel like Superman. Use the best available cover, even a little cover is better than none. If cover is too flimsy or nonexistent, consider withdrawing or calling for additional personnel. Keep in mind that the assailant may have moved from the reported

Cases 5–7

site, so consider every potential hiding place as a hazard. Do not duck around a corner without checking. Keep a lookout all around, including behind you. If you take a quick look around a corner, do it at a low level first and a different level every time. Do not stand in front of a door when there may be an armed assailant on the other side. Do not knock with your hand; consider kneeling on one knee before you knock (Adams et al. 1980).

Let's assume the shooting has stopped, and you are approaching the area occupied by the assailant. Be more alert than you have ever been. You do not know, unless you can see, where one or more people are, or what condition they are in. Assume they are lying back and waiting for an easy shot. If you are wrong no harm is done. In the next step assume the assailant you are approaching appears to be disabled. Do not relax, keep your gun in hand, approach the suspect from behind, give verbal commands if hands are hidden, secure the suspect with handcuffs even if you believe the suspect may be unconscious or dead. With your partner covering for you, remove all of the assailant's weapons, seen and unseen. Then try to determine the status of the arrestee; if in need of a doctor, call one. You will need an ambulance in any event. Always move deliberately and carefully, and keep your eyes on the suspect's hands. When the suspect is arrested you should read him or her their rights, and remember, the suspect is your responsibility until in the ambulance (Adams et al. 1980).

THE USE OF FORCE

The Supreme Court has rejected the idea that a "single generic" standard can govern all uses of force by officers. Accordingly, in any case alleging excessive force by enforcement officials, it is first necessary to identity ". . . the specific constitutional right allegedly infringed by the challenged application of force" and then to assess the claim ". . . by reference to the specific constitutional standard which governs that right . . ."[5] The levels of force recognized by many courts range from a simple display of authority to the application of various levels of nondeadly force to the use of deadly force. The appropriate choice in each case is dictated by facts as well as the officer's choice of an option and are, in any case, subject to close scrutiny.

Many courts, as a means of screening an excessive number of force claims, impose the requirement that the plaintiff allege and prove that some "significant injury" resulted from the alleged constitutional violation. Following the premise that "not every push or shove, even if it may later seem unnecessary in the peace of the judge's chambers . . ." violates the constitution.[6] The courts emphasize the need that constitutional claims against force are not trivialized. Accordingly, claims of excessive force are generally dealt with summarily when the plaintiff claims only negligible physical injury or physiological distress. For example, in *Wisniewski v. Kennard,*[7] the plaintiff alleged that the arresting officer placed a gun barrel in his mouth and threatened to blow his head off. He claimed he was frightened and suffered bad dreams as a result. The court did not sustain.

The objective nature of the reasonableness standard was set forth in *Graham,* where the court emphatically rejected the consideration of subjective factors such as the officer's state of mind, when assessing the propriety of the use of force. The court emphasized that the inquiry is "whether the officer's actions are 'objective, reasonable' in light of the fact and circumstances confronting them without regard to their underlying intent or motivation." The Supreme Court has recognized under Fourth Amendment seizure that the right to make an arrest or an investigative stop necessarily carries with it the right to use some degree of physical coercion or threat to effect it. The

Cases 8–10

officer's decision to use force must be viewed by the Court, ''. . . from the perspective of a reasonable officer at the scene, rather than with the 20/20 vision of hindsight. . . .'' (Hall 1992). The Court also recognizes that police officers are often forced to make split second judgments in circumstances that are uncertain, tense, and moving rapidly. This, in relation to the amount of force that is necessary in a particular situation. This point was made in *Sherrod v. Barry,*[8] where an officer shot and killed a robbery suspect who made a quick movement with his hands into his coat, apparently disregarding the officer's repeated command to put his hands up. It was later determined that the suspect was not armed. In reversing the jury verdict against the officer, the appellate court held that the trial court erred in permitting the introduction of evidence concerning the fact that the suspect was unarmed. The court stated that a jury must measure the objective reasonability of an officer's action and stand in the officer's shoes. Only then can the jury judge the reasonableness based on the information the officer had and the judgment he or she exercised in response to that particular situation. In other words, the officer could not know that the man who reached in his pocket was unarmed. In evaluating an officer's use of force under the Fourth Amendment standard, the Supreme Court has instructed that the following factors be considered:

(1) the severity of the crime;

(2) whether the suspect posed an immediate threat to the safety of others and the officer; and

(3) whether or not the suspect was actively resisting arrest or attempting to evade arrest by flight (Hall 1992).

An officer must do three things in the context of the Fourth Amendment:

(1) defend himself or herself and others;

(2) overcome resistance or enforce compliance; and

(3) prevent escape.

The officer should note that the severity of the crime is a factor that can affect judgment in each of these circumstances. Generally speaking, an officer's decision to use physical force is reactive. In other words, the initiative rests with the suspect who decides whether or not and when to commence a threat against the officer. In general, it is necessary for the officer to tailor the use of force to the degree of resistance encountered. As the suspect escalates the level of force, the officer must do likewise. Passive resistance to a seizure presents a different set of problems for law enforcement officers. Passive resistance generally produces ambiguity and frustration on the part of the officer. However, it is important to remember that the Fourth Amendment does not preclude the use of force to effect a seizure, only the use of unreasonable force.

Courts do not view every use of force that results in death as that of deadly force. This point can be important in defending officers whose actions have resulted in the death of a suspect under circumstances that would not constitutionally justify the use of deadly force to prevent escape, as in *Robinette v. Barnes.*[9] On the other hand, law enforcement activities that create a high probability of death are viewed as using deadly force. Firing a loaded firearm is generally considered the use of deadly force even though there is no intent to kill. This because of the relatively high risk of death by the infliction of a wound (Hall 1992).

In *Kellen v. Frink,*[10] a game warden fired a shotgun at an escaping van because he believed that a deer had been illegally killed and placed inside the van. The officer explained later that he fired the shot not to hit anyone but to mark the van for later

Cases 11–16

identification. Unfortunately the slug entered the van and killed one of the passengers. The court held that ". . . firing a loaded shotgun at a vehicle known to be occupied constitutes deadly force as a matter of law . . . [and that because there was no probable cause to believe that the deceased] was a significant threat of serious injury to others, deadly force would never have been appropriate. . . ."

HISTORY OF THE FLEEING FELON RULE

Since the use of deadly force is the most serious situation facing a law enforcement officer, a somewhat detailed discussion is in order. The officer's decision to use deadly force implicates a number of different guidelines that should be used to arrive at the proper decision. In the early days, state law defined police authority to use deadly force. Today police administrators may have departmental policies that are more restrictive than the state law. There is also a federal constitutional standard by which both state law and departmental policy must be measured (Hall 1988).

Most of the early colonists were British; understandably early American customs and laws were British. Historically, the penalty for a felony frequently was death. Therefore, a fleeing felon who had already theoretically forfeited his life by his own decision, could legally be stopped by deadly force. This was known as the "fleeing felon" rule. For most of our 200-year history states exercised their police power with no regard or interference from the federal courts or the Constitution. The Bill of Rights applied only to the federal government. In 1868, the Fourteenth Amendment specifically required "due process of law" before a state can deprive any person of "life, liberty or property." This paved the way for federal legislation designed to enforce provisions of the Fourteenth Amendment in federal courts (i.e., Title 42, U.S. Code, Section 1983). Surprisingly, neither the Fourteenth Amendment nor the enabling legislation affected state police power to any extent until well into the 20th century (Hall 1988).

In the 1930s, the Supreme Court began to accept reviews of state criminal cases in the light of the "due process" requirement of the Fourteenth Amendment. The Court viewed due process as required adherence by the states to the concept of "fundamental fairness." This has been described as a "criminal procedure revolution," wherein virtually all law enforcement activities have been "constitutionalized." This revolution could not have occurred without two Supreme Court decisions in 1961, *Mapp v. Ohio*[11] and *Monroe v. Pape.*[12] Both cases fashioned remedies for alleged violations of the federal constitutional rights by local and state police. *Mapp* requires the suppression of unconstitutionally seized items at state criminal trials; *Monroe* facilitates lawsuits in federal courts against local and state officials for the violation of federal constitutional rights (Hall 1988).

These developments did not immediately impact the "fleeing felon" rule. Further, *Pierson v. Ray*[13] held that a police officer sued under Section 1983 enjoyed a defense of qualified immunity from such suits if the officer was acting in "good faith," with a reasonable belief in the lawfulness of the actions. Two additional cases, *Monell v. Department of Social Services*[14] and *Gwen v. City of Independence,*[15] paved the way for a direct constitutional challenge to the "fleeing felon" rule. That case is *Tennessee v. Garner.*[16] The Supreme Court looked at *Garner* as requiring a determination of "the constitutionality of the use of deadly force to prevent the escape of an apparently unarmed suspected felon." The Court held that such action violates the Fourth Amendment protections against "unreasonable" seizures. The Court further stated that notwithstanding the principle that an officer may arrest a person if he has probable cause to believe the person committed a crime, "he may not always do so by killing him."

Literature Cited

Adams, R. J., T. M. McTernan, and C. Remsberg. 1980. *Street survival, tactics for armed encounter.* Northbrook, Ill.: Calibre Press.

Caparatta, P. 1989. Keep police training on target. *American Rifleman* 137 (1): 30–33.

DePietro, A. L. 1992. Voluntary encounters or Fourth Amendment seizures. *FBI Law Enforcement Bull.* (Federal Bureau of Investigation, U.S. Dept. of Justice, Washington, D.C.) 61 (1): 20–32.

Hall, J. C. 1988. Police use of deadly force to arrest, a Constitutional standard (part 1). *FBI Law Enforcement Bull.* (Federal Bureau of Investigation, U.S. Dept. of Justice, Washington, D.C.) 57 (6): 23–30.

———. 1992. Constitutional constraints on the use of force. *FBI Law Enforcement Bull.* (Federal Bureau of Investigation, U.S. Dept. of Justice, Washington, D.C.) 61 (2): 22–31.

Siddle, Bruce K. 1991. *PPCT Defensive Tactics Student Manual,* Series A. PPCT Management Systems, Inc., Millstadt, Ill.

Notes

1. 471 U.S. 1, at 7 (1985).
2. 486 U.S. 593, at 597 (1989).
3. 486 U.S. 567, 573 (1988).
4. 446 U.S. 544 (1980).
5. *Graham v. Connor,* 490 U.S. 386, at 394 (1989).
6. *Johnson v. Glick,* 481 F. 2d 1028, at 1033 (2d Cir. 1973), *cert. denied,* 414 U.S. 1033 (1973).
7. 901 F. 2d 1276 (5th Cir. 1990).
8. 856 F. 2d 802 (7th Cir. 1988).
9. 854 F. 2d 909 at 912 (6th Cir. 1988).
10. 745 F. Supp. 1428 (S.D. Ill. 1990).
11. 367 U.S. 643 (1961).
12. 365 U.S. 167 (1961).
13. 386 U.S. 547 (1967).
14. 436 U.S. 658 (1978).
15. 445 U.S. 622 (1980).
16. 471 U.S. 1, at 7 (1985).

Recommended Reading

Heinrichs, J. 1982. Cops in the woods. *Journal of Forestry* 80: 722–725, 748.

Peckumn, J. W. 1966. A review of court decisions affecting wildlife law enforcement. In *Proceedings Western Assn. of Game and Fish Comms.* No. 46, 65–73.

Purol, D. A. and T. J. Fournier. 1979. A review of the arrest: Contact ratio as a tool of enforcement using prosecution data from the 1977 Michigan deer hunting season. Michigan Dept. of Natural Resources, Law Enforcement Division. Report No. 1, 22 pages.

Multiple-Choice Questions

Mark only one answer.

1. The definition of arrest has been defined as which of the following?
   a. whenever an officer restrains the freedom of a person to walk away, the officer has seized that person
   b. only when there is governmental termination of freedom of movement through intentionally applied force
   c. arrest occurs when officers by means of physical force or show of authority restrain the liberty of a citizen
   d. when a reasonable person would believe that he or she is not free to leave
   e. all of the above

2. Which of the following does not constitute seizure under the Fourth Amendment?
   a. application of force for the purpose of holding or stopping any person
   b. interference of movement by an officer or the officer's vehicle, as when blocking someone
   c. the presence of several officers around suspect
   d. accidental or unintentional contact with the suspect
   e. the display or pointing of a weapon at the suspect

3. A misdemeanor on complaint may be issued anytime within:
   a. 30 days
   b. 60 days
   c. 90 days
   d. 180 days
   e. 365 days

4. When you as an officer are moving into place in a dangerous situation you should remember:
   a. every situation is routine
   b. many circumstances are identical
   c. if you have been there before you know the answer
   d. most situations are routine
   e. no situation is routine

5. The Supreme Court has recognized, under the Fourth Amendment seizure, that making an arrest carries with it the right to use:
   a. deadly force
   b. close to but not deadly force
   c. some degree of physical force or threat
   d. deadly force to stop a suspect
   e. force is never acceptable

6. Today's "fleeing felon" rule both for states and federal government was made possible by two 1961 cases. One of these was:
   a. *Mapp v. Ohio*
   b. *Pierson v. Ray*
   c. *Tennessee v. Varner*
   d. *Graham v. Ponnor*
   e. *Sherrod v. Barry*

7. In evaluating an officer's use of force under the Fourth Amendment, the Supreme Court considers:
   a. the severity of the crime
   b. the safety of the officer
   c. the safety of bystanders
   d. if the suspect was attempting flight
   e. all of the above

8. Search warrants:
   a. are issued on probable cause
   b. are the collection of evidence that can be used later in court
   c. are issued on sworn deposition giving a "reasonable cause"
   d. are never issued on a holiday
   e. are a last resort

9. The four Ss of sound arrest principles are:
   a. superiority
   b. speed
   c. surprise
   d. simplicity
   e. all of the above

10. The fresh pursuit rule gives an officer:
    a. the right under certain circumstances to cross state lines
    b. probable cause to pursue
    c. the right of unlimited surveillance
    d. no specific right
    e. the right to pursue during daylight hours

Chapter 10

The Stopping and Search of Motor Vehicles

One of the seemingly routine, and potentially most dangerous, actions an officer can take is stopping and searching motor vehicles. The person stopped may be completely harmless, or he may be on the ten most-wanted list. The chances of the latter are remote, but one mistake can be one too many. Even the most law-abiding person can panic or become enraged, and a wanted felon may believe he has been recognized.

Stopping motor vehicles falls into two categories:

(1) one police vehicle stopping another vehicle; and

(2) roadblocks where several officers working on a prearranged plan stop all vehicles.

Not only are the techniques of the two methods different, the laws (codes and court decisions) under which the officer operates are different. Let's look at the one-on-one situation first.

PROBABLE CAUSE

The Fourth Amendment of the U.S. Constitution prohibits "unreasonable searches and seizures." The U.S. Supreme Court has held that warrantless searches "are per se unreasonable under the Fourth Amendment—subject to only a few specifically established and well-delineated exceptions" (Hall 1981a).

The Fourth Amendment reads as follows. "The right of the people to be secure in their persons, houses, papers, and effects, against unreasonable searches and seizures, shall not be violated, and no warrants shall issue, but upon probable cause, supported by oath or affirmation, and particularly describing the place to be searched, and the persons or things to be seized."

Case 1

In 1925, the Supreme Court in *Carroll v. United States*[1] made an historic decision that changed procedures for law enforcement under certain situations. In brief, the *Carroll* decision stated that an officer can stop and search a motor vehicle if there is probable cause to believe a law has been broken. George Carroll and John Kiro had offered to sell two undercover agents whiskey, but did not deliver. Later, the agents were patrolling an area known for its smuggling activities, when they spotted Carroll and Kiro in the vehicle they had been driving when they offered to sell the agents whiskey. They stopped the car and found contraband whiskey under the seats. The whiskey was later introduced as evidence, and a conviction was secured. The defendants appealed the case to the Supreme Court. They contended the search was not incidental to arrest, since they were not arrested until the contraband was discovered.

172

Stopping cars in a
potential arrest
situation is a time to be
alert.
*Courtesy California
Department of Fish and
Game.*

Cases 2–6

In upholding the decision, the Court pointed out the difference between a store, dwelling houses, or other fixed abodes, and a motor boat or automobile that could quickly be moved out of the locality or jurisdiction. The Court said such searches are reasonable where:

(1) ''. . . the search and seizure without a warrant are made upon probable cause . . . ;'' and

(2) ''. . . it is not practicable to secure a warrant because the vehicle can be quickly moved . . .'' (Hall 1981a).

The Supreme Court decision in *Carroll* was the beginning of the Carroll Rule, or what is now more often known as the ''vehicle exception'' rule. Judicial interpretation and exigent circumstances standards are the two factors involved. The exception to the warrant requirements recognized in *Carroll* should not be confused with one of the other major exceptions to the warrant requirement, the search incidental to arrest (Hall 1981a). The latter is dependent upon a lawful custodial arrest, regardless of the probability that weapons or evidence will be found. It encompasses only the arrestee's person, and the area within arrestee's immediate control.

Under the Carroll Rule, the Supreme Court has ruled, until recently, the passenger compartment may be searched, but not the trunk or containers, open or closed. Now the police may search cars and any closed container inside the car without a warrant.[2] This, and another Supreme Court ruling,[3] stated that once a motorist gives police permission to search his car, officers may open bags or containers within the car. This gives the police more leeway than they have had since the days of Chief Justice Taft (Hall 1981a).

The vehicle exception rule, on the other hand, does not depend on the right to arrest. It depends on probable cause, and the vehicle may be searched. But, until the May 1991 ruling, it did not extend to separate containers, unless their contents were in plain view or the contents could be inferred from the nature of the container, or the container was open (*Arkansas v. Sanders, Robbins v. California*).[4]

For many years the states did not invoke the probable cause rule to stop motor vehicles. In 1949, the U.S. Supreme Court applied the Fourth Amendment to the states through the due process clause of the Fourteenth Amendment (*Wolf v. Colorado*)[5] and, in 1961, the Court imposed the exclusionary rule (*Mapp v. Ohio*).[6] Beginning in 1970, there were several cases decided by the Supreme Court which involved the Carroll Rule, and almost all of them involved search of vehicles by state and local officers. It is one of the most effective search and seizure tools available to law enforcement officers today (Hall 1981a). Since the officer(s) on the scene is making the initial judgment, it is necessary to look closely at the requirements of probable cause and exigent circumstances.

Case 7

The Supreme Court in *Carroll* stated the definition of probable cause in part as ". . . facts and circumstances within their knowledge, and of which they had reasonably trustworthy information . . . sufficient in themselves to warrant a man of reasonable caution in the belief. . . ."

Firsthand information is one of the most common, and in many cases, the most important source of knowledge. *Carroll* and *Brinegar*[7] are two good examples. In *Brinegar,* Justice Rutledge delivered the opinion of the U.S. Supreme Court. Briefly this is what happened. Brinegar was convicted of importing intoxicating liquor into Oklahoma from Missouri in violation of the federal statute, which forbids such importations contrary to the laws of a state. His conviction was based, in part, on the use of evidence gained against him when liquor was seized from his automobile in the course of his alleged unlawful importation.

Prior to Brinegar's trial it was moved to suppress evidence gained from the seizure, since the defense contended that the seizure was unlawful. The motion was denied. The appeal was based solely on the grounds that the search and seizure contravened the Fourth Amendment and, therefore, the use of liquor in evidence violated the conviction.

The following facts are substantially undisputed. On 3 March 1947, investigators of the alcohol unit and a special investigator parked their car beside a highway in Northeastern Oklahoma. This point is about five miles from the Missouri-Oklahoma state line. Brinegar drove past headed west in his Ford coupe. One of the officers, who had arrested Brinegar five months earlier, recognized him and knew that he had a reputation for hauling liquor. They also recognized the Ford car as belonging to Brinegar. These officers later testified that the car appeared to be heavily loaded. Brinegar increased his speed and the officer gave chase at top speed for a mile. Finally, they overtook Brinegar, crowded in front of him, and pushed him off the road. As the agents got out of their car one of them spoke to Brinegar and asked him how much liquor he had. His comment was "not too much," but, after further questioning, he admitted that he had twelve cases in the car. One of the officers later testified that one of these was in the front seat and visible from outside the car, but the defendant claimed that it had been covered with a robe. The officers placed Brinegar under arrest and seized the liquor (Hall 1981a).

The district judge was of the opinion that the mere fact that the agents knew that the defendant was engaged in hauling whiskey, and the fact that the car "appeared" to be heavily loaded was not probable cause. Therefore, there was no probable cause when the agents began the chase. The district judge, however, held that the voluntary admission made by the defendant after his car had been stopped, constituted probable cause for a search regardless of the legality of the arrest and the detention of the defendant, and therefore, the evidence was admissible. The court of appeals, with one judge dissenting, took essentially the same view. The dissenting judge thought that the search was unlawful from its inception and, therefore, statements made during its course, could not justify the search.

The crucial question was: Was there probable cause at the time of Brinegar's arrest (and in the light of prior cases). The *Carroll* case was cited frequently during the *Brinegar* case.

Pintails feeding in a wetland.
Courtesy U.S. Fish and Wildlife Service, Carl Burger.

In discussing *Brinegar,* it was argued that the situation was very similar to *Carroll.* The four Supreme Court judges affirming the decision pointed out that the troublesome line posed by the facts in the *Carroll* case and the *Brinegar* case is one between mere suspicion and probable cause. That line must necessarily be drawn by an act of judgment, formed in the light of particular circumstances, and with account taken of all these circumstances. No problem of searching a home or other place of privacy was involved in either case. Both cases involve freedom to use the highway in swiftly moving vehicles, and to be unmolested by investigation and search during these movements (and in dealing with contraband). The citizen who has given no good cause for an officer to believe he is engaged in that sort of activity, they said, is entitled to proceed on his way without interference. But one who recently and repeatedly has given substantial ground for believing he is engaged in forbidden transportation in an area has no such immunity. If the officer sights him in that region and knows the situation about the man, he may make an interception.

This does not mean, as it might be assumed, that every traveler along a public highway may be stopped and searched at the officer's whim, caprice or mere suspicion. The question presented in *Carroll* lay on the border between suspicion and probable cause, but the Court carefully considered the problem and resolved it by concluding that the facts within the officer's knowledge amounted to more than mere suspicion. The judges believed this conclusion was right and that it was consistent with the Fourth Amendment, and the objection should have been overridden. Accordingly, the judgment was affirmed (Hall 1981a).

The opinion of the three dissenting judges in *Brinegar* is interesting. They dissented because they regarded the *Brinegar* decision as an extension of *Carroll* which, according to them, had already allowed enforcement officers to go too far in taking blanket authority in stopping and searching cars on suspicion. They point out they do not contend that officials may never stop cars on a highway. Such things as traffic

Cases 8–14

regulation, identification, traffic census, quarantine regulations, and many other causes give occasion to stop cars in circumstances which do not imply arrest or charge of crime, and to stop or to pursue a suspected car to its destination, or to keep it under observation. These things are not, in themselves, an arrest. But when a car is forced off a road, summoned to stop by a siren, and brought to a halt under such circumstances as are disclosed here, the three judges believe the officers were then in the position of one who has entered a home. They said, ''A search at its commencement must be valid and cannot be saved by what turns up later.'' Several cases were quoted to demonstrate this. The dissenting opinion continues: ''The findings of the two courts will make it clear that the search began and proceeded through critical and coercive phases without the justification of probable cause. What is later yielded cannot save it. 'I would reverse the judgment'.'' This opinion was written by Justice Jackson.

The dissenting judges point out the legality of such things as regulation of traffic, driver identification, traffic census, and quarantine regulations. We might reasonably assume that if an officer were hauled into court for stopping cars for a census of game harvest he should look to this particular case for some help; since it would appear that a game census and a traffic census have much in common. Something that is borne out here, and should be kept in mind, is that the reason Justice Jackson felt that certain practices are legal is the fact that the act of stopping a car does not imply arrest or the charge of a crime.

United States v. Matthews[8] demonstrated the value of officer personal knowledge. Military officers on base observed a civilian car with military license plates, but no military decal. When Matthews was stopped he had registration papers for a Chevrolet (the car was a Ford), and he had no log books. The Court determined there was probable cause (Hall 1981a).

The plain view doctrine states that objects in plain view of an officer who has a right to be there, are subject to seizure and can be used as evidence.[9] At night a flashlight may be used. The sense of smell may also be used. Presumably K-9 dogs would fall in this category. Officer expertise, as well as knowledge, can also be used to establish probable cause. Officers can use secondhand information that they deem reliable as if it were their own (Hall 1981a). In establishing the validity of informant's information, the Supreme Court listed two requirements:

(1) the underlying circumstances the informant used were true; and

(2) the reason for believing the source.[10]

In *Chambers v. Maroney,*[11] the U.S. Supreme Court ruled there was probable cause to arrest the occupants of a vehicle, who were suspected of earlier robbing a service station. The search of the vehicle at a later date was also held valid (Hall 1981a).

In *United States v. Menke,*[12] a federal appeals court ruled an unoccupied vehicle could be searched on probable cause and without a warrant. In *Scher v. United States,*[13] it was ruled that where officers have reasonable grounds for searching an automobile which they have been following, a search of the vehicle immediately after it has been driven into an open garage is valid. The existence of reasonable cause for searching an automobile does not, however, warrant the search of an occupant thereof, even though the contraband sought is of a character which might be concealed on the person.[14] This and other cases indicated a person may expect less privacy in an automobile than in a home. There are several reasons for this:

(1) a vehicle is for transportation;

(2) it rarely serves as a residence;

Cases 15–16

(3) people in a vehicle are open to public scrutiny; and

(4) vehicles are subject to licensing, etc., and may be stopped for inspection.

A motorhome may also be stopped under the vehicle exception rule.[15] None of this means that vehicles are excluded from the Fourth Amendment. Probable cause and exigent circumstances are requirements in all cases. Warrantless searches are subject to close court scrutiny, and an officer who conducts one assumes the risk of having the case overturned. A prudent officer will, when it is practical, first procure a warrant (Hall 1981b).

STOPPING A MOTOR VEHICLE

Once an officer has made the decision to stop a motor vehicle, the potentially dangerous action is just ahead. Most of the people stopped will do no harm, but an officer cannot assume the current one is in that category. Thomas and Boyer (1981) offer several suggestions for minimizing danger to the officer. Roll the driver's window down; you can hear better and it is safer. Have whatever you are going to take, ticket books, hat, flashlight, ready to exit the car as soon as you stop. Communicate with the dispatcher, giving license number, make of car, and location. Hang the microphone over the steering column for quick access. Keep your gun hand free. As you pull up behind the suspect's car stop at 15 to 20 feet, turn the front wheels 1½ to 3 feet to the left and stop. You may want to ask the driver to exit his car and bring you whatever material you want. Or, if you are uncomfortable with this procedure, walk toward the other car checking to see if the trunk is locked, if there is someone hidden in the back seat, and stop before you are even with the rear of the front door. Watch the hands of the driver and everyone else in the vehicle. Ask everyone to keep their hands in sight, for their safety as well as yours. Accept whatever material you have asked for in your off-hand.

Some people believe an officer should not turn his back on a stopped vehicle and its occupants. There is danger even while you are in your own vehicle if the suspect decides to shoot, especially if a shotgun or large rifle is used. If you believe you have stopped a dangerous felon, do not hesitate to call for backup. Your duty is to enforce the law, but that does not include undue and unnecessary risks (Thomas and Boyer 1981).

EXIGENT CIRCUMSTANCES

Exigent circumstances exist when delay in stopping and searching a motor vehicle would endanger lives or result in the loss of evidence. Under these circumstances (assuming probable cause has already been established) a warrantless search by police officers is justified. This was established in *Carroll*.[16] Arrest of the occupants was not effected until after the contraband had been found.

ROADBLOCKS

Roadblocks fall into three general categories:

(1) a barrier is thrown up and everyone is forced to stop;

(2) there is no barrier, but by a series of signs and lights everyone is asked to stop; and

(3) a select group (e.g., hunters) or random vehicles (e.g. one of five) is asked to stop.

Canada geese goslings.
*Courtesy U.S. Fish and
Wildlife Service,
DeWayne Anderson.*

Cases 17–18

The first type is used almost solely to stop dangerous felons who may try evasive action, to crash through the barrier, or resort to firearms. Pockrass (1984) points out there is more written on how to avoid roadblocks than there is on how to work them. This is true because kidnappers try most often to take wealthy or important people when they are in their automobiles. Therefore, much has been written for these people by their protectors. This technique for stopping motor vehicles is rarely, if ever, used by wildlife law enforcement officers. Stopping all vehicles without a roadblock is used by wildlife officers to some extent.

Roadblocks set by wildlife personnel are to collect management data and check licenses, fish, and game taken by hunters and anglers. Generally signs indicate only hunters or anglers should stop.

Roadblocks do not operate under the Fourth Amendment, since at the onset probable cause and exigent circumstances are not present. Roadblocks as administrative searches have the best chance of surviving court scrutiny, when carefully conceived and implemented. Here the courts impose no requirements of probable cause. Public welfare outweighs the invasion of individual privacy (Campane 1984). Written guidelines should be prepared by a qualified legal advisor. A complete written report should be made covering the facts surrounding the stopping of motorists, so that if a motorist is cited for criminal activity an officer will be able to give the reasons for his suspicion. It should be clear that the stopping of motorists has as its primary purpose the checking of licenses or game, and the stop was not a pretext for investigating other activities. Professionalism and fairness should be the hallmark of all roadblocks (Schofield 1980). At least two states, Oregon and South Dakota, have held hunting license check roadblocks to be reasonable: Oregon in *State v. Tourtillot,*[17] and South Dakota in *State v. Halverson.*[18]

Literature Cited

Campane, J. G. Jr. 1984. The constitutionality of drunk driver roadblocks. *FBI Law Enforcement Bull.* (Federal Bureau of Investigation, U.S. Dept. of Justice, Washington, D.C.) 53 (7): 24–31.

Hall, J. C. 1981a. The motor vehicle exception to the search warrant requirement (part 1). *FBI Law Enforcement Bull.* (Federal Bureau of Investigation, U.S. Dept. of Justice, Washington, D.C.) 50 (11): 24–31.

———. 1981b. The motor vehicle exception to the search warrant requirement (conclusion). *FBI Law Enforcement Bull.* (Federal Bureau of Investigation, U.S. Dept. of Justice, Washington, D.C.) 50 (12): 20–26.

Pockrass, R. M. 1984. Managing hazardous roadblocks. *FBI Law Enforcement Bull.* (Federal Bureau of Investigation, U.S. Dept. of Justice, Washington, D.C.) 53 (5): 20–23.

Schofield, D. L. 1980. The constitutionality of routine license check stops—a review of Delaware v. Prouse. *FBI Law Enforcement Bull.* (Federal Bureau of Investigation, U.S. Dept. of Justice, Washington, D.C.) 49 (1): 25–27.

Thomas, W. J., and F. W. Boyer. 1981. Police officers' survival during traffic stops. *FBI Law Enforcement Bull.* (Federal Bureau of Investigation, U.S. Dept. of Justice, Washington, D.C.) 50 (2): 1–6.

Notes

1. 267 U.S. 132 (1925).
2. *California v. Acevedo,* 111 S. Ct. 1982 (1991).
3. *Florida v. Jemino,* 111 S. Ct. 1801 (1991).
4. 442 U.S. 753 (1979); 29 CrL 3115 (1981).
5. 367 U.S. 643 (1949).
6. 367 U.S. 643 (1961).
7. *Brinegar v. United States,* 338 U.S. 160 (1949).
8. 615 F. 2d. 1279 (10th Cir. 1979).
9. *Harris v. United States,* 390 U.S. 234 (1968).
10. *Aquilar v. Texas,* 378 U.S. 108 (1964).
11. 399 U.S. 42, 50 (1970).
12. 468 2d. 20 (3rd Cir. 1972).
13. 305 U.S. 251 (1938).
14. *United States v. Di Re,* 332 U.S. 581 (1948).
15. *California v. Carney,* 105 S. Ct. 2066 (1985).
16. *Carroll v. United States,* 267 U.S. 132 (1925).
17. 618 P. 2d. 423 (Ore. 1980).
18. 277 N.W. 2d. 723 (S. Dak. 1979).

Recommended Reading

Anonymous. 1982. Motor vehicle identification. In *Law Enforcement Bible No. 2.* Ed. R. A. Scanlan. South Hackensack, New Jersey: Stoeger Publ. Co.

Brownlee, Gardner E. 1974. *Trial judge's guide: Objections to evidence.* National Judicial College.

Canudo, E. R. 1980. *Evidence.* New York: Gould.

Goodreau, G. E. Jr. 1982. Fingerprinting tools and techniques. In *Law Enforcement Bible No. 2.* Ed. R. A. Scanlan. South Hackensack, New Jersey: Stoeger Publ. Co.

LaPlante, J. A., and C. C. Tate. 1976. *Handbook of Connecticut evidence.* Little Brown and Co.

Lowli, G. C. 1978. *An introduction to the law of evidence.* West Publ. Co.

Scanlan, R. A. 1982. Doing it Right: How to Handle Evidence and Testify in Court. In *Law Enforcement Bible No. 2.* Ed. R. A. Scanlan. South Hackensack, New Jersey: Stoeger Publ. Co.

Sewell, J. D. and W. D. Beckerman. 1982. Stop that vehicle and stay alive. In *Law Enforcement Bible No. 2.* Ed. R. A. Scanlan, South Hackensack, New Jersey: Stoeger Publ. Co.

Multiple-Choice Questions

Mark only one answer.

1. The most dangerous moment for an officer when stopping a motor vehicle is:
   a. when the suspect's car is stopped and the officer leaves his car and approaches the other one
   b. when both cars are completely stopped
   c. when the officer makes the decision to stop the car
   d. when stopping unknown persons
   e. when the stop is not routine

2. The Carroll and Brinegar cases are:
   a. dissimilar
   b. in the same ballpark
   c. quite similar
   d. no two cases are similar
   e. all cases are similar

3. The Carroll Rule is also known as:
   a. the Brinegar rule
   b. the exception to probable cause rule
   c. the third rule of law
   d. the Fourth Amendment rule
   e. the vehicle exception rule

4. Which of the following manner of stopping vehicles does not imply arrest?
   a. regulation of traffic
   b. traffic census
   c. stopping for quarantine regulations
   d. driver identification
   e. all of the above

5. In *United States v. Menke*, the court ruled:
   a. an unoccupied vehicle can never be searched
   b. an unoccupied vehicle can be searched only with a warrant
   c. an unoccupied vehicle can be searched on probable cause alone
   d. an unoccupied vehicle can be searched without proper cause or a warrant
   e. an unoccupied vehicle can be searched if the owner is known to the officer

6. Once an officer has decided to stop a vehicle, he or she should:
   a. notify the dispatcher of his location, make of other vehicle, and license number
   b. roll the left front window down
   c. have ticket book, and whatever else the officer is going to take, ready to pick up
   d. hang the microphone over the steering column
   e. all of the above

7. The stopping of motor vehicles falls into how many categories?
   a. one
   b. two
   c. three
   d. four
   e. five

8. The Fifth Amendment prohibits unreasonable search and seizure and warrantless searches are per se unreasonable, but a wildlife officer can stop and search a car on:
   a. suspicion of guilt
   b. third party information
   c. orders from one's immediate superior
   d. probable cause
   e. the belief that the suspect's behavior fits the pattern of a violator

9. One of the reasons for the Carroll Rule is:
   a. it is sometimes not practical to obtain a search warrant for a vehicle because it can rapidly be moved away
   b. it is convenient for the officer
   c. it is the quickest way to get a job done
   d. warrants are sometimes used under these circumstances
   e. warrants are interchangeable with probable cause

10. The line an officer must draw between probable cause and suspicion is:
    a. never really clear
    b. always clear
    c. spelled out in plain detail by the law
    d. one that the officer must make in the light of evidence and on-the-spot judgment
    e. not troublesome

Chapter 11

Evidence

SCIENCE IN CRIME DETECTION

Every human society, even the least advanced, has been obligated to draw a distinction between lawful and unlawful conduct (Sannié 1954). As a result, the concept of crime and misdemeanor is inseparable from the concept of society itself; and the struggle against crime is of utmost importance. This hasn't always been apparent, partly because of the involvement and the resulting confusion caused by religious considerations. Religions have laid down moral laws which define right and wrong, and hold out prospective reward for virtue and punishment for evil. While these rules of social behavior may parallel legal codes, they are rarely the same.

It is axiomatic that in order to enforce a law the first step is to discover that the law has been broken; the second, to find the guilty party. In our society this involves the investigator, the prosecutor, the judge, and often the jury.

Social Values of Scientific Investigation

A judge or a jury has to ask the experts, ''Just how far can we rely on scientific evidence?'' Doubt is frequently raised due to disputes between various experts in the field or, in some cases, by an overly cautious attitude on their part. Scientific criminal investigation produces absolute proof only in special cases, such as fingerprinting. Forensic science contributes to court proceedings via circumstantial evidence which is often conclusive, but in most cases short of being absolute proof; however, one of its benefits is that in many cases limited circumstantial evidence produces a confession which otherwise might not have been forthcoming. The introduction of more accurate methods of detecting crime makes for greater certainty in the administration of justice.

Method of Judicial Inquiry

There are two methods used to gather evidence for judicial inquiry. Accusatory procedure, which is the older of the two, consists of bringing adversaries face-to-face. The defendant and the plaintiff debate the case orally and publicly. Accusatory procedure is still used in some places where trials are conducted by the ''Judgment of God.'' Most countries use the inquisitorial method for seeking out and detecting crime. The principles are the same in both procedures. The steps are:

(1) establish that a law has been broken; and

(2) suspect a certain person of breaking this law.

Truth is not discovered in criminal investigation—*it is proved.*

When collecting proof, a police officer relies on two complementary methods. The first is obtaining evidence from human witnesses. The second is obtaining evidence by studying the facts.

In the first, human testimony is obtained by examining either the accused or a witness or both. The cornerstone of many preliminary investigations is a report based on these examinations. This is essential to any investigation and explains why every law enforcement officer needs to be at least an amateur psychologist.

The shortcomings of this system should not be overlooked. It is obvious that an accused person who cannot speak the truth without condemning himself is, at best, in a hazardous position. Some of the weaknesses of human testimony are:

(1) unreliability of memory;

(2) unwillingness to tell the truth;

(3) a mental condition which does not permit a person to tell the truth; or

(4) one who is afraid to tell the truth.

There is a danger in using leading questions in examinations. In many cases the accused may even wrongfully admit a crime. The term "leading question" means that the question, in itself, suggests or points to the answer.

Case 1

Obtaining facts by the use of science was initiated in forensic (judicial) medicine. For example, using poisons as a method of homicide was first discovered by chemical analysis of the corpse. About 1880, a French investigator, Bertillon, concluded that every criminal leaves behind some tangible evidence of the crime.[1] This evidence, if correctly interpreted, may, independent of human testimony, supply proof or circumstantial evidence of the highest order. For the past fifty years there has been a growing tendency to substitute such evidence in place of human testimony.

Experience over eighty-five years in every major nation in the world has shown that fingerprinting is stable. It is predicated on the thesis that no two sets of fingerprints are identical. Experts throughout the world have checked, but have not discovered two identical prints. It is of interest to note that statisticians concluded this would be the case. Fingerprints appear before birth, roughly four months after conception, and persist after death until total destruction of the skin. An interesting sidelight—a few years ago it was discovered that in certain leprosy cases the ridge patterns on the fingers were partly or totally obliterated. This gave considerable concern until it was found that if the disease were treated, and improvement followed, the ridge patterns returned in exactly the same form as before the illness. It is impossible for the police to trace most individuals unless they have a set of ten prints. The FBI has a limited file of the more dangerous criminals. In this file, the latent print file, one finger can be identified.

Since no two people have identical fingerprints, they are evidence of the highest order. They are also a type of circumstantial evidence linking the accused to a particular crime by placing him or her at the scene of the crime. It is less well-recognized that the friction regions of the soles of the feet and palm prints of human beings are as individual and permanent as those of the fingers. In all three cases, the same identification techniques are used.

Preparation for Trial

It has been said that most lawsuits are won or lost before they are heard in court. A good lawyer prepares a case carefully, completing extensive research and investigation before going into court. The lawyer interviews witnesses, takes statements, and ascertains who is to be called by the other side and what questions are likely to be asked of them. The lawyer also decides whether or not expert witnesses will be called and discusses the case with his or her own witnesses from time to time, preparing them for court pitfalls, including questions which may be asked by opposing lawyers. The same general procedure should be followed by the conservation officer, as by a lawyer, to whatever extent the case merits.

The evidence (*top*) and the confiscated guns (*bottom*) (note the division tags).
Courtesy Utah Division of Wildlife Resources.

THE OFFICER AS AN INVESTIGATOR

One of the responsibilities of the wildlife officer is to collect and preserve enough evidence to justify taking the violator before the court with reasonable chances for a conviction. This preparation includes the identification, collection, and preservation of evidence so that its admissibility and effectiveness in court will serve the ends of justice. For the officer to do this, he or she must exercise the qualities of a trained investigator.

The officer's investigative role requires that he or she understand and know how to handle proof of a wildlife law violation. The officer must know, for example, what constitutes evidence; what evidence is admissible in court; when and in what manner to apprehend presumed law violators so as to avoid legal obstacles to conviction. The

Experts in the FBI laboratory can identify a specific shoe from an impression left at the scene of a crime.
Courtesy Federal Bureau of Investigation.

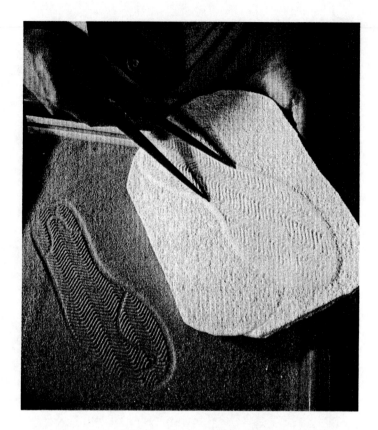

officer should also know how to develop information; how to recognize evidence in the field; how to collect and preserve evidence to safeguard its admissibility; how to obtain evidence from witnesses and from others who may be able to help; how to detect discrepancies, dishonesty, or general lack of good faith; when to call upon experts for help; how to testify, and other courtroom techniques.

NATURE AND KINDS OF EVIDENCE

Evidence is anything that can be shown or told in court to support the charge of a violation. In its nature it may be direct, circumstantial, or presumptive. In its several forms it may be called real evidence, documentary evidence, oral evidence, or opinion. Since the whole matter of evidence is too broad and technical to be covered in this text, the following treatment, while helpful to the student, to the officer in the field, and to the instructor of wildlife courses, is somewhat cursory.

Direct Evidence

Direct evidence tends directly to prove or disprove a fact in issue. It testifies to the very fact. For example, it is direct evidence if an officer testifies that he saw a hunter shoot and kill a deer, or a person in a boat throw trash into the lake.

Circumstantial Evidence

Indirect or circumstantial evidence tends directly to prove or disprove a fact or circumstance from which, either alone or in connection with other facts, a court may, according to the common experience of mankind, reasonably infer the existence or nonexistence of another fact which is in issue. It is established from cause to effect and leads to the fact that a violation was committed. For example, it is circumstantial evidence that a person killed a deer when it is found in his possession; although, if mere possession is illegal, the carcass is direct evidence.

Circumstantial evidence, even when believed, does not always conclusively establish a fact; but it may, with other known facts, prove the elements involved. It tends to show that under the established set of circumstances the accused may have, or at least was in a position to have, committed the violation. Circumstantial evidence is not necessarily inferior to direct evidence. In all cases, the plausibility of the statement and the competency of the witness must be weighed.

There is no general rule for contrasting the weight of circumstantial versus direct evidence. The assertion of an eye witness may be more convincing than contrary inferences that have been drawn from certain circumstances. Conversely, an inference drawn from one or more circumstances may be more convincing than a contrary assertion of an eye witness.

Presumptions

Presumptions form an important part of the rules of evidence in that they serve to relieve parties from the burden of presenting evidence to prove certain facts. A presumption may be defined as a rule of law that attaches probative value to specific facts not actually known but arising from their usual connection with other particular facts which are known. A presumption is a conclusion or deduction drawn from reasonable and logical inferences associated with usual probabilities and attended upon associated facts. Presumptions as facts are those based on experiences of mankind which has found them to be valid inferences which naturally arise in either common experiences or particular circumstances (Donigan and Fisher 1965).

The term presumption is applied to facts which courts are bound to assume in the absence of adequate evidence to the contrary. Examples of this are: an accused person is presumed innocent until guilt is proved beyond reasonable doubt; an accused is presumed to have been sane at the time of the offense charged, and to be sane at the time of the trial until a reasonable doubt of sanity at the time in question is raised by the evidence (U.S. Government 1975).

Other examples are: A sane person intends the natural and probable consequences of his act because it is recognized that people ordinarily intend to do what they do; a person found slain did not commit suicide, since self-destruction runs contrary to the instinct for self-preservation; a person was sane at the time in question since most people are sane; a person is of good character, sober, law abiding, and honest in his dealings because most people are of this character; public officials, bodies, and tribunals properly perform their duties since they are legally bound to do so.

There are a number of permissible inferences encountered during the course of a trial which are sometimes loosely referred to as presumptions that are not presumptions at all, but are merely well-recognized examples of the use of circumstantial evidence. The fact that evidence is introduced to show the nonexistence of a fact which might be inferred for proof of other facts, does not, if the evidence can reasonably be disbelieved, necessarily destroy the logical value of the inference, but rebuttal evidence must weigh against the inference. The same is true if the evidence is introduced to show the nonexistence of the facts upon which the inference is based. It may be inferred that a condition shown to have existed at one time continues to exist. For example, that a person's residence remains unchanged, or that immediately after a collision the lights on the vehicle were not burning, although in working order at that time, would support an inference that the lights had not been turned on at the time of the collision. Proof that a letter correctly addressed and properly stamped or franked was deposited in the mail will support an inference that it was delivered to the addressee; a similar inference is permissible in regard to telegrams. Possession of recently stolen property may support an inference that a person stole the property. The fact that one or more inferences contradict or are inconsistent with one or more other inferences does not necessarily neutralize or destroy the inferences on either side of the question. Relative weights of conflicting emphasis are assessed in accordance with the logical value of each and in light of attendant circumstances (U.S. Government 1975).

Real Evidence

One of the forms that direct or circumstantial evidence takes is that of real evidence. This includes physical objects, such as bodies of game animals, blood, weapons, empty shell cases, projectiles, glass fragments, and articles of clothing. However, to be admitted in court as real evidence, such items must be proved relevant to the issue involved.

If an item of real evidence which has been introduced into the case is not to be attached to the record because of the impossibility or impracticability of so doing, or for some other reason, the item should be clearly and accurately described by testimony, photographs, or other means.

Documentary Evidence

Demonstrative or visual evidence is called documentary evidence and includes, among others, writing, such as confessions, field notes, photographs, sketches, and fingerprints. Each is herein described briefly.

Writings. The word writing means every means of recording data upon any medium. This includes handwriting, typewriting, or other machine writing, printing and all documentary, pictorial, photographic, chemical, mechanical, or electronic recordings or other representations of facts, events, acts, transactions, communications, places, ideas, or other occurrences or things, whether expressed by words, letters, numbers, pictures, signs, symbols, marks, or chemical, mechanical, or electronic media, including all types of machine, electronic, or coded records, memoranda, or entries (U.S. Government 1975).

A writing is the best evidence of its own contents, and the original or its equal must be introduced to prove its contents. For example, when the law specifically makes it illegal to set traps at a certain time and place, then, a signed confession that the suspect then and there did knowingly and willfully set a line of traps with which the suspect took game constitutes documentary evidence that a violation was committed.

In order for this type of evidence to be admissible in court, however, it is necessary to present evidence authenticating the original document and then to introduce the original document in evidence. This means the wildlife officer must be able to prove that the suspect wrote the confession, and understood what he or she was signing, and that the suspect's signature is authentic.

A copy of a document, as complete as the original copy in all the essential respects, including relevant signatures, if any, or an identical copy made by a photographic process or other duplicating instrument is considered to be a duplicate of the original and equally as admissible as the original. If admissible writings have been lost or destroyed, or when for some other reason they cannot be produced, then the contents may be proved by an authenticated copy or by the testimony of a witness who has seen the writing.

Best and Secondary Evidence. As applied in the courts today, the Best Evidence Rule means that if the contents of a written document are sought to be shown in evidence, the original writing should be produced; a copy or other substitute can be resorted to only when the original is shown to be unavailable through no fraudulent act of the party seeking to introduce it (Donigan and Fisher 1965). The Best Evidence

Rule is not absolute and does not utterly exclude all secondary evidence. If it is established that production of the original writing is impossible or impracticable, secondary evidence may be resorted to. The original evidence may be lost, destroyed, outside the jurisdiction of the court, in the hands of a third person, or the opposing party, an inscription on a monument, part of a sign board, a sign on a building, or brand on cattle. These are examples of original evidence that can rarely be shown in court.

When it becomes necessary and proper that secondary evidence of a writing be used, the law requires the party to produce the best secondary evidence obtainable. A party will not be permitted to introduce inferior secondary evidence unless it can be shown that it is the best available.

Field Notes.   One of the primary sources of documentary evidence available to the wildlife officer is field notes. The officer's notebook should serve as a combination diary and casebook. Entries should show the area and the weather, including precipitation, percent of overcast, and direction and velocity of the wind.

Immediately after making an arrest, the officer should make a complete notebook entry, listing the contraband game and equipment together with all identifying marks placed on the contraband. Notes should include the full name and address of the accused, plus any distinguishing marks or characteristics that will aid in future identification. The day and time, place where the offense occurred, particular acts which constitute the violation, names and addresses of accomplices or witnesses, and the exact location of the officer in relation to where the accused was at the time the alleged offense was committed, should all appear in the notes. There should be no empty lines or space where material could be later added.

The extreme importance of initialing and dating all field notes at the actual time they are made was emphasized in a recent, widely publicized, criminal case. The defense attorney repeatedly questioned the laboratory technician, who was testifying for the state, as to whether he was briefed and as to what evidence he had looked for. Over and over again the defense asked to see the notes which had been made by the expert witness as he went along. As they were produced, the defense attorney noted that some were undated, and charged, ''So you are now simply relying on your memory, is that correct?''

Many cases are lost every year because of an officer's neglect or carelessness. The officer may have thought that evidence was conclusive when in fact it was not, the evidence having been insufficient or inadmissible. The officer may have counted on the defendant's oral plea of guilty only to learn at the trial that the defendant had decided to plead not guilty.

Notes should be accurate, complete, legible, and understandable. They should contain only authorized abbreviations, because it may later be necessary to introduce the notebook into court. Personal remarks or conclusions should not be recorded.

In discussing crime scene search and evidence integrity, Brunetti, Levine, and Banks (1977) offer five recommendations:

(1) the officer should use caution when searching the area so that the search is systematic and complete;

(2) take photographs, if possible;

(3) prepare notes or other records as items are collected;

(4) place distinctive marks or evidence tags on the objects collected;

(5) keep the chain of custody as short as possible.

Use of Photography/Video.   Photography has long been recognized as a basic law enforcement tool. Improvements in film quality, and in the diverse capability of "point-and-shoot" 35-mm cameras has increased the applications for which still photography is appropriate. However, no matter how sophisticated a 35-mm camera, it remains a still format.

Video cameras, and associated playback, editing, and viewing technology are practically icons of the late 1980s and early 1990s, as evidenced by the popularity of television shows using "home videos" for entertainment. Video cameras currently available range in price from a few hundred dollars to several thousand dollars. Video, like still photography previously, has a role in wildlife law enforcement which centers around its ability to make and preserve factual documentation, which records and preserves various aspects of evidence. As with still photography, use of video requires a certain level of skill and knowledge to produce useable film (admissible in court or simply to record events that an officer can refer to at a later date). Law enforcement personnel should not only standardize video recording procedures, but develop an understanding of the camera as a tool, and police and photographic techniques as well. If possible, some level of formal training, whether attendance at a school, experience with a trained person, or self-study, may be useful. Video footage of any aspect of wildlife law enforcement activity should be accompanied by complete and accurate notes; fortunately, it is possible in most instances (not including clandestine operations) to narrate a scene as the video portion is being recorded. Video of a crime scene or of an interaction between a law enforcement officer and a suspect should include sufficient footage of the surrounding area to facilitate identification of the site, presumably in legal proceedings. A timer may be installed in the camera to avoid the charge of editing.

Standardized use of video for wildlife law enforcement activities can be developed by individuals or agencies. A determination of which of the many video camera products to purchase is also required. The information below should be helpful in determining the video equipment which meets agency or individual needs.

Videotape recorders, or simple video cameras, have been used by the television broadcasting industry since the 1950s. Home video recorders, or VCRs, are the most commonly known type of recorders currently in use. Camcorders are a generic name for portable videotape recorders where the camera and the recorder are combined into a single unit, varying in size, quality of video produced, and price. The first commercially successful home videotape device did not appear until 1975. By the mid-1980s, two systems had emerged: BETA and VHS. These two formats are not interchangeable.

Current technology allows selection of the type, size, and cost of unit to meet specifications for use. Among the options, or variables, which can be selected singly or in groups are: lenses, format (High 8, etc.). Costs associated with acquiring a suitable VCR system for use in most law enforcement applications should be less than $2,500.

Other aspects of television and associated video production include the use of "low light" television recording, and use of infrared laser illumination devices with accompanying "night vision" goggles or other viewing apparatus. Many of these applications may be useful for monitoring large game animals during darkness as well. Local dealers should have specific information.

Illustrative Material.   Maps, sketches, diagrams, and other relevant drawings may be used to illustrate a scene or an event. These materials are admissible when they aid the court or jury's understanding and apply the facts to a particular case. Admissibility of illustrative material is conditioned upon a preliminary showing that it is reasonably accurate. Sketches may be either rough drafts or finished drawings. They may be the ones made at the scene of the violation or ones made later to scale. In any case, the scale should be approximately correct and the measurements reasonably accurate. The investigator should know the degree of accuracy of the sketches and figures and be willing to state it. In many cases a finished drawing is made later for courtroom presentation, but it is based on the rough sketch which is kept with the investigator's notes. These notes are available for court presentation if necessary.

Fingerprints. Another example of documentary evidence is fingerprinting. In the case of fingerprints on the handle of a knife found near the carcass of a deer, the only legal question involved may be the identification of the particular prints.

Oral Evidence. Oral evidence, proceeding from exclamation or from word of mouth, includes everything pertinent to the case, which the officer or other witness has seen or heard or otherwise observed through personal experience. Confessions, spontaneous exclamations, and dying declarations are examples of oral evidence.

Spontaneous Exclamation. This is an utterance concerning the circumstances of a startling experience made by a person while in a state of excitement, shock, or surprise. The utterance is caused by the person's participation in or observation of the event and warrants a reasonable inference that the utterance was made spontaneously and instinctively as an outcome of the event, as opposed to a statement made after deliberation or by design.

Spontaneous exclamations are admissible as evidence contrary to the general hearsay rule. They are admitted as evidence on the theory that a person who has just committed an act is more likely to tell the truth at that moment than later, after thinking it over. For example, if immediately after a man shoots a fellow hunter he says "My God it's a man," this undoubtedly would be a spontaneous utterance and admissible as part of the *res gestae*. Whereas, after he has recovered from the shock he may not be so truthful. A spontaneous utterance may be proved by testimony of a person who heard it made.

Dying Declarations. In homicide trials, dying declarations of alleged victims, including the identity of the person or persons who caused the injury, are admissible as evidence. The reason for accepting this type of evidence is that the statement would obviously otherwise be unavailable; also, presumably those who believe that they are about to die tell the truth.

In order to establish admissibility of this evidence, it must be demonstrated that the victim believed, either by his actual statement or by other facts or impressions, that death was imminent. The fact that the victim did or did not die is immaterial. On the other hand, the statement is not admissible if the victim had some hope of recovery but died shortly thereafter.[2]

Case 2

Dying declarations are not acceptable if they were obtained under duress, but they may be rightfully obtained in answer to leading questions or upon urgent solicitation. The declaration may be either oral or in writing and in favor of or against the accused. As an example, a declaration by a dying hunter to an officer or other witness that his companion shot him as a result of an argument over ownership of a deer may be used as admissible evidence in court to the prejudice of the second hunter.

Limitations of Oral Evidence. While this type of evidence is the most common, it may also be the least reliable. This is because of the possibility of honest mistakes, improper observations, bias, or other incompetencies.

Opinion

Another type of evidence is opinion. It is a general rule that the officer or other witness must state facts and not opinions or conclusions, but impressions under some circumstances, such as those based on personal observation, are required by the court. Examples of these are the distance from which a shot was fired at a game animal; the speed of an automobile; the identification of a voice as that of a man, woman, or child; and, in many cases, an opinion as to whether a given individual was behaving as if drunk, under the influence of drugs or ill.

To be admissible in court, opinions of a technical nature must be given by an expert witness. Such a witness is defined as one who is skilled in some art, trade, profession, or science, or a person who has knowledge and experience in relation to matters at hand which are generally not within the knowledge of people of common education and experience.

An expert witness may express an opinion on facts within his or her specialty without specifying data on which it is based. However, an admissible expert opinion is regarded as evidence only when it pertains to the matter on which the opinion relates.

Confessions

A confession is an acknowledgment of guilt. An admission, on the other hand, is a self-incriminatory statement falling short of an acknowledgment of guilt even though it was intended to exonerate its maker.

Case 3

Before a confession can be used against an accused, it must be shown to represent voluntary acknowledgment of guilt, or have been obtained under conditions or circumstances which could not reasonably be considered as rendering it untrustworthy. By the court's test of voluntariness, the type of force, threats, or promises, that would be considered sufficient to rule a test involuntary, would probably be sufficient to make the confession untrustworthy. In *McNabb v. United States,*[3] the Court held that not only must a confession be voluntary and trustworthy, but it must have been obtained by civilized interrogation (Inbau and Reid 1974).

SOURCES OF EVIDENCE

To be effective, the wildlife officer must develop a number of sources of information. The right sources may provide evidence essential to the case in addition to real evidence readily available at the scene of the violation.

The Complainant

The complainant is anyone who notifies a law enforcement officer or agency of an actual or suspected offense. The complainant is generally the witness or discoverer of the violation and the one who is willing to so state, rather than giving the information confidentially.

Witnesses

Witnesses usually are a source of information. An investigator should make every effort to locate known witnesses to a violation, and to obtain thorough and detailed interviews from as many of them as appear necessary.

Suspects

When properly questioned, suspects may be persuaded to give valuable information. Frequently this is the only source of information available to the officer.

Other Sources of Evidence

Other sources of information include records of telephone companies, credit bureaus, finance companies, banks, city directories, hospital records, transportation companies, insurance companies, light and power companies, newspapers, state record divisions, and post office departments.

INTERVIEWS AND INTERROGATIONS

A confession from the suspect, or information from material witnesses to a violation, when not forthcoming at the time of the violation, may be obtained through questioning. A successful investigator is effective in questioning the various persons involved, or those who may contribute to the solution of the violation.

Investigators generally divide questioning into two broad classifications:

(1) The interview uses the indirect approach to learn facts from persons who may have a knowledge of a violation but who are not themselves involved. It is a general discussion carried on in a conversational manner and permits the person to talk without being asked direct questions; and

(2) The interrogation, or direct approach, is used to obtain facts, admissions, or confessions from those who are directly implicated.

However, an interrogation is not necessarily confined to individuals suspected of committing the offense. It may include suspected accessories, or people who have a knowledge of the violation but are reluctant to admit it. These persons, with whom the indirect approach may have been unproductive, are restrained from giving information for such reasons as fear or dislike of the officer, concern of retribution, desire to protect someone else, or a general animosity of law enforcement agencies.

Usually, throughout the following sections, the terms interview, indirect approach, and direct approach, are used with reference to either the interview or to the interrogation. The context should make the meaning clear.

Methods of Procedure

Obviously, the most opportune time to conduct the interview is when a suspect is surprised in the act of committing a crime. An admission of guilt and a signed confession are probably the easiest to obtain under these circumstances. In other instances, the interview should be conducted as soon as possible after the violation has been committed.

The steps in an interview should be carefully planned in advance so that, if possible, all questions may be used directly or indirectly at one session. A favorable environment should be selected whenever possible, and adequate time allowed to complete a thorough interview.

The interviewee should be made to feel that he or she is not in any way being forced. While trying to be friendly and businesslike, the investigator attempts to get the person into a talkative mood; and whenever the conversation is wandering, the interviewer guides it back into more productive channels. Digression may be permitted when the interviewer feels that possible points will be touched upon that might otherwise be omitted. The interviewer must be circumspect in word and deed, trying to avoid a clash of personality, undue acts of familiarity, profanity, violent terms such as murder, steal, or thief, improbable stories, or distracting mannerisms. All of these are likely to hinder the success of the interview.

In the case of dishonest or otherwise unreliable interviewees, it may be desirable to start an interview with the indirect approach. This gives the interviewee an opportunity to trip himself or herself through admissions, inconsistencies, known misstatements, etc. The interview would then shift from the indirect to the direct approach. Talkative persons should be confined to the general subject. The discussion should be changed to a more direct approach when the suspect is obviously stalling for time.

Trained polygraph
operators are an
essential part of a
crime laboratory.
*Courtesy Federal Bureau
of Investigation.*

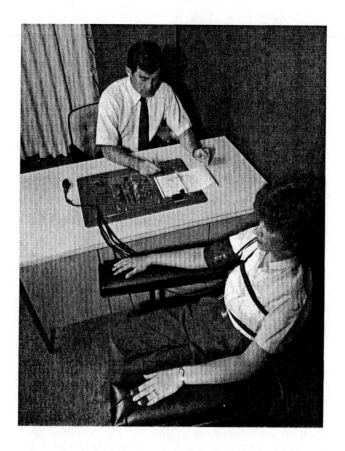

Precautions   Throughout the interview the interrogator must be careful to refrain from threats of
violence or promises of reward or leniency. If the suspect admits or implies guilt, the
investigator must not become condescending, emotional, or overanxious.

The fact that a confession was made during an investigation does not render the
statement inadmissible. The admissibility of a confession, when it is challenged, must
be established by an affirmative showing that it was voluntary. Voluntary admissions
of guilt, that is, without urging, interrogation, or request, are acceptable. The point in
question is whether or not the statement was voluntary or forced.

Silence on the part of the accused during the interview cannot be assumed to imply
guilt. Silence may appear to be evidence of guilt, for example, when a simple statement
of denial would tend to clear the issue. Even though it appears reasonable that the
accused is guilty, mere silence cannot, by itself, be taken as evidence unless it is
substantiated by other material indicating that the offense as accused has probably been
committed.

Since a confession not made in open court cannot be relied upon unless it is supported by other evidence, every effort should be made to obtain additional evidence of
guilt. Also, because defendants may change their minds and claim that confessions
were made under duress or as the result of promises or threats, or that they were
confused and did not realize what they were signing, officers should read back confessions to the defendants, in the presence of witnesses, and obtain the defendants'
signatures. Generally, no difficulty arises from a person making a confession if he or
she is taken immediately before the court.

Case 4

Case History. The situation is briefly this. The defendant, Henry P. Garin,[4] was accused of killing (taking) and possessing twenty ducks in one day (the legal possession limit was twenty) and the daily limit was ten. The prosecution proved, and the defendant admitted, that on 10 October 1941, Garin had twenty ducks, all killed in one day (this was not contested) in his possession. Since the daily limit was ten, the prosecution contended that: "We have proved there were twenty ducks in one day, in the possession of the defendant. There is no conflict as to that. And I say that it raises a presumption that shifts the burden to him of proving that he didn't kill them." At this point the court made a statement that it would like to see any law that the prosecution had to demonstrate this point. The defense had previously contested this point. It should be pointed out that the prosecution was apparently thinking of what it considered a similar circumstance—the presumption of guilt in a case of theft. If a person is found in possession of stolen goods, it raises the presumption that he, himself, has stolen these goods and that the burden of proving his innocence is shifted from the plaintiff to the defendant. However, as it later developed in this case, this is not a similar situation.

The case was built in this fashion. Three federal officers (U.S. Fish and Wildlife Service) surrounded and observed the duck club in question on 18 October. No shooting was observed by any of them and this was not denied by the defendant. The next day these three officers raided the club and found 286 ducks; 236 of these were seized. The U.S. attorney attempted to prove that because the legal limit was ten birds per day, and that since there were only thirteen men present at the club, that it was obvious that the defendant had in fact killed twenty birds in one day, because they had already proved that no birds had been killed the previous day. One of the officers had a conversation with the defendant in which he essentially admitted his guilt; however, when the prosecution attempted to enter this conversation into the court record it was objected to by the defense and sustained by the court that it was incompetent, irrelevant, and immaterial, because, since proper foundation had not been laid, *there was no proof of the corpus delicti* (that is the body of the crime). In other words, the defense contended that the prosecution had to first prove that a crime had been committed before he could introduce as evidence an admission of guilt by the defendant. This the court sustained as basic.

The defense pointed out that the charge was "did then and there take and possess" twenty ducks in violation of the law. And that "possessing twenty ducks in one day" was no violation, but rather, the only violation was if one had taken twenty ducks in one day. And that the prosecution had, in effect, redefined the law in such a way that it, according to his implication, meant the mere possession limit of twenty ducks was unlawful. At no time did the defense deny that the defendant had twenty ducks in his possession nor did defense deny that they had been taken in one day. Defense counsel pointed out that there was no evidence in the record that the defendant had shot (took) a single duck at any time during the two days. If the prosecution could have proved that there had been a violation, then the conversation between one of the officers and the defendant could have been admitted to the evidence and a conviction would probably have been sustained; however, this was not the case. The defense's argument was that no suspicion of an illegal act arises from the fact that a lawful act has been committed. The proof of possession of twenty ducks killed in one day (a legal act) offers no suspicion of an illegal act, the killing of twenty ducks in one day. This was sustained by the court.

SAFEGUARDING EVIDENCE

The identification and collection of evidence of wildlife violation is an exacting and arduous task. Professional training and field experience is often necessary before competency in these areas is attained.

Chain of Custody

Each person who handles a piece of evidence, from the time it is discovered until it is properly linked to the case, should so handle it that the continuous chain of custody will be provable. This includes a complete record as to who has had the material and where it has been stored. This record should include the date, name, time, place, and signature of each person. Each one releasing material should obtain a signature from the receiver. In the event it is sent through the mail, it should be registered. Technicians handling film should place an identifying mark on the edge of the negative and on the back of the final prints. Failure to keep intact this chain of custody may render evidence inadmissible.

Marking Evidence

Any article or thing that comes into an officer's hands and which afterwards he or she may be called upon to identify in a judicial proceeding, should be carefully marked by the officer. On receiving from any person an article or thing that may be used as evidence, the officer should first request such a person to mark it, then make a notation of both marks in a memorandum.

Placing a tag on an article or thing is not always sufficient as a measure of identification since it can be removed. A witness may not be able to prove to the satisfaction of the court the fact that a tag showed the article to be the article taken from the defendant or taken from the defendant's premises. Therefore, even though an article or thing is tagged, a private mark of identity should be put on it.

Evidence, wherever possible, should be marked for identification before it is placed in an envelope. After it is placed in an envelope, the envelope should be sealed in the presence of witnesses. The time, date, names of the witnesses, and the name of the person sealing the envelope should be placed on the flap-envelope junction. When the evidence contained in the envelope is presented in court, the seal should be broken in the presence of the court. After the hearing, the evidence should again be placed in the envelope, the envelope resealed and the time and date of sealing placed thereon. This procedure is ideal but may not be required in many courts.

The serial number, caliber and make of a revolver, pistol, or rifle should be noted as should any evidence of recent discharge. Empty cartridge shells should be marked with a scratch mark, and any letter or numbers appearing on them should be noted. Loaded rounds should be marked in a like manner, and the projectile marked with a scratch where it enters the case.

Discharged projectiles should be marked on their noses, and all such marks should be deep enough to prevent tampering. Any marks on them at the time of finding should be noted. Care should be taken to preserve marks made on the surface of the bullet made by its discharge through the barrel.

There are a number of special marking devices that are used in connection with game exhibits. These may include knife marks or notches on the hooves or antlers of deer and other mammals, extraction of a tooth of certain large animals, insertion of a metal tag or coin in the flesh of the animal, clipping the nails of game birds, or the clipping of a wing feather and the retention of the clipped feather. Any such marking should be carefully recorded by the officer in order to present a positive identification.

The FBI Mineralogy Unit can identify material from crime scenes.
Courtesy Federal Bureau of Investigation.

Preserving Evidence

It is important to physically preserve evidence connected with the violation for trial presentation in the same condition as when it was taken.

With equipment, this is no great problem; however, this is not always possible when fish and game animals are involved. Often, considerable time may elapse between the arrest and the actual trial of the case and, unless methods of preservation are available, it will not be possible to preserve the evidence intact. In such cases, photographing the evidence provides a visual record which can be presented to a jury, or a small piece of the carcass often will suffice.

AIDS TO AND RESTRICTIONS ON EVIDENCE

While gathering evidence to support the charge that a suspect has violated a wildlife law, the investigator should bear in mind some of the aids which the courts have provided. The investigator should also recognize the restrictions which, while safeguarding the rights of the suspect, make it more difficult to marshal sufficient evidence to support the charge.

Judicial Aids

Included among those aids which the officer may use in preparing a brief for the prosecuting attorney are: judicial notice, official records, and *prima facie* evidence; proof of character; character of the accused; *modus operandi;* and double jeopardy. Each one will be discussed in turn.

Judicial Notice. Certain kinds of facts need not be proved by the formal presentation of evidence, because the court is authorized to recognize their existence without such proof. This recognition is termed a judicial notice (U.S. Government 1975). The facts are assumed to be known for a certainty or, in other words, they are common knowledge. Examples of judicial notice are:

(1) the statutes of the several states or those of the federal government;

(2) the day of the week a given date of the month was on;

(3) the ordinary divisions of time into years, months, weeks, or other periods;

(4) general facts and laws of nature, including ordinary operations in effect and general facts of history;

(5) the political organization and the chief officials of the government of the United States, of its territories and possessions, of the District of Columbia, and of the several states of the United States;

(6) the signatures and duties of persons attesting official documents, or copies thereof, kept under the authority of any government agency of the United States;

(7) the treaties of the United States, and executive agreements between the United States and any foreign country;

(8) current political or *de facto* conditions of war and peace;

(9) organic and public laws, including regulations having the force of law;

(10) the seals of courts of record, and the seals of public offices and officers of the United States and its territories and possessions; and

(11) the public laws or regulations having the force of law in effect in any country or territory or political subdivision occupied by the armed forces of the United States.

Official Records.   An official record is a statement in writing made of a certain fact or event. For example, when an officer or other person, in the performance of an official duty imposed on him or her by law or customs, states facts or ascertains the truth of a matter, and the channels of information have been demonstrated to be trustworthy, this is deemed to be an official record. In this connection it may be presumed *prima facie* when:

(1) a person who has performed such a duty performed it properly;

(2) a foreign or domestic record which reflects a fact or is even required to be recorded by law, regulation, or custom was made by a person so required to make it, provided the record is authenticated as an original copy; and

(3) foreign or domestic records which are thus authenticated and shown to be on file in a public office and which reflect facts or events of a kind generally recorded by public officials, such as births, deaths, marriages, etc., were made by persons who in official duty by law or regulation are accustomed to record such facts and to know or ascertain through appropriate and trustworthy channels of information on the truth thereof.

Maps, photographs, X rays, sketches, and similar projections of localities, objects, persons, and other matters are admissible as official record when verified by any person, whether or not he made or took them, if he is personally acquainted with the locality, object, person, or any other thing thereby represented or pictured; and if he is able, from personal knowledge or observation, to state that they actually represented the appearance of the subject matter in question.

Prima Facie Evidence.   *Prima facie* means ''first face,'' ''on first view,'' ''on first appearance.'' If the defense were to move for dismissal on grounds that the prosecution

The FBI laboratory maintains numerous reference files, which are used in the analysis of questioned documents.
Courtesy Federal Bureau of Investigation.

has failed to present sufficient evidence to sustain the burden of proof, and the motion fails, then the prosecution is said to have made a *prima facie* case. It may well fail on rebuttal.

Statutes which provide that certain evidence is *prima facie* merely set up a rebuttable presumption. It may be that the conclusion required to be drawn from certain facts as set up by the statute or the rule of law amount to the guilt of the party or something substantially less than this. *Prima facie* evidence and presumptions are synonymous, and it should be clarified that the presumption is almost universally rebuttable.

Proof of Character. Whenever the character of a person is admissible in a case, the opinion of a witness as to that person's character may be received in evidence, if it is first shown that the witness has an acquaintance or relationship with the person in question so as to qualify him or her to form a reliable opinion. Another method of establishing character is through the general reputation which a person holds in a community in which he or she lives or pursues a business or profession. In this case, the testimony must come from a person who has lived in the community and has formed the opinion as a result of having been a member of the community. Admissible evidence in this case cannot be given by a person sent into a community with the specific job of investigating the character of the accused.

Character of the Accused. The general rule in evidence is that an accused may not have evidence of bad moral character introduced against him or her for the purpose of raising inference of guilt. In order to show a probability of innocence, an accused person may introduce evidence of good character as well as the fact that he or she is a moral, well-behaved, law-abiding citizen.

Evidence of particular traits of character are inadmissible as evidence unless there is a reasonable tendency to show that the particular trait indicates improbability that the offender had committed some particular offense. For example, it is assumed that a person who had never broken a fish and game law is less likely to have done so than a person who has a long list of proven fish and game violations. As a general rule, character traits which are not directly involved in the charge against the accused are used only for the purpose of raising an inference that the accused may have a disposition to act in a particular manner.

However, in some cases, evidence of other offenses may have substantial value as tending to prove something other than facts to be inferred about the character of the accused. Some of these examples are listed below.

1. The evidence tends to identify the accused as the perpetrator of an offense. An example of this is where two similar game violations have been committed in an area, and the accused has previously been convicted of one. It may reasonably be assumed that the accused is also involved in the second violation. Another example is where the accused leaves a piece of personal property at the scene of the offense. To connect the personal property with the accused is to connect him or her with the offense.

    A third example relates to the suspect's method of operating in committing a violation. Identification of a *modus operandi* may lead directly to the accused.

2. The evidence tends to provide a plan or design of the accused.

3. The evidence tends to prove guilty knowledge or intent, if such knowledge or intent is an element of the charged offense. For example, if a person charged with killing game on posted land claims not to know that it was illegal, he or she could be confronted with evidence from a previous accusation and conviction of a similar act. However, a previous assault charge cannot be used to indicate a tendency toward assault if the first assault was under different circumstance and presumably with different intent.

4. The evidence tends to prove a motive. For example, proof that a person was obligated to kill quail to later be sold to a certain party would be sufficient evidence of motive that this person would be interested in killing quail. Another example includes previous evidence of having falsified daily work sheets to cover up the fact that a person had not been at work at all on a particular day. This evidence tends to prove that the person was not where he or she had avowed to be, and that person could have been hunting quail as was charged.

This judicial aid provides the officer with yet another valuable tool. Connecting the suspect with the offense in any of the aforementioned ways permits the officer to go before the court with a much-strengthened hand.

Modus Operandi. One of the fundamental concepts in crime investigation is that certain criminals have methods of operation which are peculiar to their individual behavior. That is, they develop, perhaps subconsciously, patterns and habits which they tend to display each time they commit a crime.

The study and cataloging of operations have become so specialized that certain city police forces are at times able to predict approximately when and where a certain type of crime will be attempted; enabling officers to prevent the crimes. Criminals may have peculiar habits such as chewing gum, breaking up match sticks, or spitting; peculiarities of speech or eating; or nervous mannerisms. These habits may point to some individual or to a group of individuals.

An example of how *modus operandi* helps the wildlife officer apprehend a violator, and later supply the prosecution with admissible evidence, is seen in the case of illegal hunting. An illegal hunter may follow a pattern of driving along back roads at twilight shooting game from an automobile, then speed away with the intention of returning later to retrieve it. Having determined a behavior pattern, the officer can station himself or herself to surprise the offender in the act of violation. Later, the signed and dated entry of this case in the officer's field notes may be incorporated into the brief for the prosecuting attorney.

Judicial Restrictions   While judicial aids work in favor of the wildlife law enforcement officer, judicial restrictions may work in favor of the suspect.

In order that the law violator not be found innocent of the charge because of insufficient or inadmissible evidence, the officer should remain acutely aware of these restrictions on evidence: testimonial knowledge, hearsay evidence, rules of privilege, *corpus delicti,* competency of a witness, reasonable doubt, and sanity.

Testimonial Knowledge. Ordinarily, a witness may testify only to what is learned through the witness' own senses. That is, an officer may testify that he or she heard a shot in a thicket. Later, when investigating the thicket, the officer found the apparently freshly killed carcass of a deer, and the person now before the court is the one that was seen leaving the thicket. The officer cannot testify by his or her own knowledge that the person leaving the thicket was the one who killed the deer, although circumstantial evidence would justifiably permit this inference. Thus, a point of evidence on which the officer might have relied could be adjudged inadmissible.

There are many exceptions to this rule. A common one is that people may testify as to their own age, including the date of their birth, even though they obviously do not know this of their own knowledge.

Hearsay Evidence. Normally, witnesses may not give testimony which is not their own. In other words, they may not make statements which are not from their own observations, or which were made by someone else and repeated to them. However, all third party statements are not hearsay. It depends on the purpose of the testimony and why it is being introduced. For example, a person may testify as to a statement made by another individual, and it is fact and not hearsay that the conversation was carried on. Whether or not the third person was actually stating the truth, and whether the evidence is admissible as such, is another question. If the testimony is introduced only to show that there was a conversation about a particular subject carried on between the witness and a third party, then the testimony is admissible as such.

Case 5

There are numerous exceptions to the rule that hearsay evidence may not be introduced. One of these was covered in part in the Montana Supreme Court case of the *State v. Russell.*[5] Hearsay evidence was used, but it was brought out by preliminary questions that it was wholly immaterial to the case, and it was claimed that the defendant was not prejudiced thereby. The defendant was convicted by the justice of the peace court for unlawfully taking fish from a stream by the use of fishberries ground up with meat.

The case was appealed to the district court where Russell was again found guilty. From judgment and an order denying him a new trial, Russell appealed to the Supreme Court of Montana. The proceedings follow:

1. An attack was made upon the complaint by the defendant, but court held that the complaint was sufficient to charge the unlawful taking of fish from a stream of the state.

2. The defendant claimed that the trial court admitted certain hearsay evidence. A witness of the state had testified that he was prompted to make a report to the deputy game warden because the Wagner boys told him someone down the river was killing fish. This explanatory evidence was hearsay, but it was brought out by a preliminary question that it was wholly immaterial, and it was inconceivable that any substantial right of the defendant was prejudiced by it.

The supreme court held that the only fair deduction from the evidence produced by the state is that the defendant deposited in the Bitter Root River, in Ravalli County, fishberries ground up with meat which were eaten by the fish, with the result that they were stupefied and rendered easy prey; that while the fish were in this condition the defendant, by means of a landing net, took at least one fish from the river. Had the jury believed the evidence offered by the defense, a different verdict would have been

Cases 6–7

required. The only conclusion from the verdict is that no credence whatever was given to the story told by the defendant and his companions. There was evidence in the record to justify the verdict.

Rule of Privilege. Under certain conditions, the defendant has what is known as the rule of privilege, or privileged communication. The oldest of these is between attorney and client. It is the privilege of the client to forbid his or her attorney to testify in regard to any confidential communication made by the defendant to the attorney. If this were not the case, defendants would be afraid to go to lawyers and discuss their problems because of the possibility of having the information used against them.

Husband and wife are competent witnesses in favor of each other. They are also competent witnesses against each other. The general rule is that both are entitled to a privilege prohibiting the use of one of them against the other, although this may be waived by both spouses. This privilege does not exist when the husband or wife is injured by the offense with which the spouse is charged. Examples of this are assault on one spouse by another, bigamy, polygamy, unlawful cohabitation, the use of the wife for immoral purposes, and forgery by one spouse under certain conditions.

In many states a doctor cannot, over the objection of his or her patient, be permitted to testify regarding confidential information, unless the court rules that it is in the interest of justice. In addition, no person accused of a crime may be compelled to give evidence against himself or herself.[6] This is known as privilege against self-incrimination. A confession obtained under duress or promise of leniency may later be excluded from evidence on the basis that it violates the right against self-incrimination.

The officer should probably be more concerned with rules of privilege relating to husband and wife testimony and to self-incrimination, although on occasion the officer may deal with the others. It is well to remember that in court the officer cannot count on evidence which, at the scene of the violation, a man's wife may furnish or promise to furnish against her sportsman husband. Neither can an officer expect the accused persons to go before the court and voluntarily incriminate themselves, even though they might have promised at the time of the arrest that they would ''go to court and tell the truth about the whole thing.'' The wise officer forges a chain of evidence without these links.

Corpus Delicti. One of the important rules of evidence in criminal cases is that which requires proof of the *corpus delicti*. Literally defined this means ''the substance or foundation of a crime; the substantial fact that a crime has been committed.'' It has been incorrectly referred to as some visible piece of physical evidence. The basic concept of law is that the *corpus delicti* in a criminal case cannot be established by extra-judicial confession of the defendant alone. ''It is well-established law, not only in Indiana, but practically all jurisdictions where the common law prevails, that the state cannot prove the admission of a crime by the extra-judicial confession of a defendant alone. To hold otherwise runs counter to the generally accepted principles of the common law, that one may not be induced to convict himself. The crime or the *corpus delicti* must be established by some independent, additional corroborative evidence or probative value, aside from the confession alone.''[7]

Competency of a Witness. A competent witness is one who is legally qualified to be heard under oath before a judicial tribunal; one who has the requisite legal qualifications to give testimony in a court of justice. Competency is to be sharply distinguished from credibility (Donigan and Fisher 1965). The former has to do with one's personal qualifications to testify and must be determined by the court before the witness can give any testimony. On the other hand, the credibility of a witness, determined by opposing counsel, relates to that quality which renders the testimony worthy of belief. In earlier times it was held that in order to be a competent witness a person must ''Possess a conscience alive with true accountability to a higher power than human law in case of falsehood.'' Otherwise, it was felt the oath was not binding. It is not essential to the witness' competency that he or she be aware of God's existence or believe in a Supreme Being. It is now generally said to be sufficient qualification if a witness understands and undertakes the obligation of an oath. As a general rule, a person offered as a witness is presumed to be competent to testify unless the contrary is shown. Competency is the rule and incompetency is the exception. Ordinarily the burden of showing incompetence rests upon the party asserting it (Donigan and Fisher 1965).

There is no arbitrary minimum age limit below which a child is automatically disqualified as a competent witness. The competency of a child fourteen years old is generally presumed, but below that age a judicial inquiry into the child's mental capability usually is required and becomes more searching in proportion to the child's chronological immaturity. A witness, be it child or adult, must be capable of distinguishing between truth and falsehood. It is not required to be able to define the meaning of an oath, but rather that the witness appreciate the fact that as a witness he or she assumes a binding obligation to tell the truth, and that a violation of that obligation is subject to punishment by the court.

An insane person, or one who is otherwise mentally incompetent, is not necessarily incompetent as a witness even though he or she may have been committed to an institution for the insane. Rather, it is necessary that the person has sufficient mind to understand the nature and obligation of an oath and correctly to receive and impart impressions of the matters which he or she has seen or heard. A witness may be competent to testify although in some respects he or she is mentally unsound or has some mental impairment. However, where so impaired that a witness does not understand the obligation of an oath, or has no respect for the truth, the witness is not competent.

In the Wisconsin Supreme Court an eighty-six-year-old man under guardianship was allowed to testify as a prosecution witness in assault, despite inconsistencies and discrepancies in his story. The court said in part ''it was for the jury to accept such part of the contradictory testimony as it believed.'' The Wisconsin court further stated that ''witness may be competent to testify although in some respects he is mentally unsound or has some mental impairment, but where a witness is so impaired that he does not understand the obligations of an oath or has no respect for the truth, he is not competent.'' The court further states ''competency has two aspects:

(1) the mental capacity to understand the nature of the questions, and to form and communicate intelligent answers thereto; and

Case 8

(2) the moral responsibility to speak the truth, which is the essence of the nature and obligation of an oath.''[8]

Reasonable Doubt. In spite of all the efforts to gather evidence enough to convict a suspect whom the wildlife officer strongly believes to be guilty, it sometimes happens that the courts find the suspect innocent. The evidence just would not support the charge.

It is necessary that the court (or the jury) be satisfied beyond all reasonable doubt that the accused is guilty before a verdict of guilty may be brought in. This means not fanciful or ingenious doubts or unwarranted speculation, but rather an honest, substantial, conscientious doubt as suggested by the evidence, or lack of it, in the case. It should be a misgiving, generated by insufficient proof of guilt, and not a doubt suggested by the ingenuity of the counsel, nor a doubt borne of merciful inclination, or personal prejudice as in the case of an attractive woman. The rule of reasonable doubt extends to every element of an offense as charged. If a person appears to be guilty of a lesser charge, that person must be declared innocent of the current charge. This does not mean that every fact advanced by the prosecution must be proved beyond a reasonable doubt; it rather means that there is an overall sufficient amount of evidence to warrant a conviction, and that on the whole the court is satisfied beyond a reasonable doubt that the accused is guilty.

Sanity.   There are times when it is necessary or desirable to determine whether or not a suspect is competent to stand trial. In legal terms, "Whether or not the individual understands the charges against him and is able to assist counsel in his defense." This is the test used in most jurisdictions (Rappeport 1975). The question of competency can be raised by the judge or the prosecutor, as well as by the defense. Once the issue is raised, the individual must be evaluated. In some jurisdictions this is done by a psychiatrist appointed by the court; or it may be done in the county jail. If it is concluded the individual is not competent to stand trial, the psychiatrist makes a report to the court and the defendant is held at a hospital until he or she becomes competent to stand trial. The Supreme Court has stated that the hospital must report to the court within a short period of time on how long they feel it will be before the individual will become competent.[9] If the hospital does not feel that the defendant will ever become competent, then he or she must either be released or civilly committed. At this point the state must decide whether or not to drop charges.

Case 9

The American Law Institute Test of Responsibility set out in the proposed draft (1962) of the Modern Penal Code is used in twelve states and every federal jurisdiction except one. It states: "A person is not responsible for criminal conduct if at the time of such conduct as a result of mental disease or defect he lacks substantial capacity either to appreciate the criminality (wrongfulness) of his conduct or to conform his conduct to the requirements of law." This does not include abnormalities manifested by repeat criminals or otherwise antisocial conduct (Rappeport 1975).

As a general rule the accused is presumed initially to be sane and to have been sane at the time of the alleged offense. This presumption, as a matter of law, authorizes the court to assume that the accused is sane until, from the evidence, a reasonable doubt of sanity appears in the minds of the judge or jury.

One who is insane necessarily must have been incapable of having a criminal intent or of knowing he or she was committing a crime at the time of the act. Neither mental weakness nor a low order of intelligence constitutes a defense of insanity. A basic test is whether or not the defendant knew the difference between right and wrong at the time of the crime. A mere assertion of insanity is not necessarily sufficient to impose any burden of inquiry on the court. However, the actions and demeanor of the accused as observed by the court or the bare assertion from a reliable source that the accused is believed to lack mental capacity or was mentally irresponsible at the time of the crime may be sufficient reason for directing an inquiry.

There are times when it is necessary to consider the sanity of an individual who has been apprehended for a wildlife violation. The question of sanity may come either before or during the trial.

Mental diseases as such do not amount to insanity. Nor does insanity constitute proper defense for a person of low intellect who commits an offense under the delusion of righting a wrong, as long as the person knows at the time of the act that it was contrary to law, and the person was not acting under an irresistible impulse. On the other hand, an accused is not responsible for an act if, as a result of mental disease, he

Metallurgists test to determine the comparative potential strength and mechanical properties of metal specimens. *Courtesy Federal Bureau of Investigation.*

or she believed that the act was legally and morally correct. A mere defect of character, will power, or behavior is not grounds for declaring the suspect even temporarily insane.

However, even though the suspect may have been entirely sane at the time the alleged offense was committed, the suspect should not be brought to trial if considered mentally irresponsible at the time of the trial. This ruling rests upon the principle that the accused should possess sufficient mental capacity to understand the nature of the proceedings and to intelligently conduct or cooperate with his or her own defense.

A medical board having at least one member trained in psychiatry passes on the sanity of an individual, although lay witnesses may state their opinion of the individual's general mental condition, as judged by the bounds of common experience. If it is found that the accused is mentally responsible, then the trial proceeds as usual; however, if the accused is judged as having been mentally irresponsible for acts at the time of the alleged violation, the court will enter a finding of not guilty as to the charges and specifications. If there is a reasonable doubt as to the mental capacity of the accused at the time of trial, the court will adjourn until such time as the proper mental capacity of the individual has been established.

Literature Cited

Brunetti, O. A., K. F. Levine, and J. D. Banks. 1977. Laboratory aids to wildlife law enforcement—wildlife physical evidence. In *Proc. of the Forensic Sci. Symposium.* Calgary, Alberta: Alberta Recreation, Parks and Wildlife, Fish and Wildlife Div.

Donigan, R. L., and E. C. Fisher. 1965. *The evidence handbook.* Evanston, Ill.: The Traffic Institute, Northwestern Univ.

Inbau, F. E., and J. E. Reid. 1974. *Criminal interrogation and confessions.* Baltimore, Md.: The Williams and Wilkins Co.

Rappeport, J. R. 1975. Forensic science. *FBI Law Enforcement Bull.* (Federal Bureau of Investigation, U.S. Dept. of Justice, Washington, D.C.) 77 (7).

Sannié, C. 1954. The scientific detection of crime. Smithsonian Inst. Report Publication 4203.

U.S. Government. 1975. *Manual for courts-martial.* United States 1969 (Rev. Ed.). Executive Order 11476 (1969) and Executive Order 11835.

Notes

1. The Federal Bureau of Investigation adds ''or takes away.''
2. State laws regarding dying declarations may vary from this interpretation.
3. 318 U.S. 332 (1943).
4. *United States v. Garin,* No. 7920 (N.D. Calif. 1942).
5. 52 Mont. 583. 160 Pac. 655 (1916).

6. *Constitution of the United States,* Fifth Amendment.
7. *Hogan v. State,* 235 Ind. 271, 132 N.W. 2d 908 (1956).
8. *State v. Schweider,* 5 Wis. 2d 627, 94 N.W. 2d 154 (1959).
9. *Jackson v. Indiana,* 406 U.S. 715 (1972).

Recommended Reading

Baker, W. 1983. Poachers in paradise. *Outside* 8 (4).

Bessey, K. M. 1983. Analysis of the illegal harvest of White-tailed deer in Agro-Manitoba: Implications for program planning and management. Master's thesis, Univ. of Manitoba, Winnepeg.

Decker, D. J., T. L. Brown, and C. P. Dawson. 1980. Deer hunting violations and law enforcement in New York. In *Transactions,* Northeastern Society, Northeast Section, The Wildlife Society, 37: 113–128.

Farnsworth, C. L. 1980. A descriptive analysis of the extent of commercial poaching in the United States. Ph.D. diss., Sam Houston State Univ.

Fox, R. H. and C. L. Cunningham. Crime scene search and physical evidence handbook, Dept. of Justice. Washington, D.C.: U.S. Govt. Printing Office.

Lot, E. 1961. Evidence in wildlife law enforcement. In *Proceedings Southeastern Assn. of Game and Fish Comms.,* Rep. No. 15, 475–479.

McIntyre, T. 1992. Wildlife detectives. *Sports Afield* 208 (4): 23–26.

Parsons, B. 1957. Gathering and presenting evidence. In *Proceedings Southeastern Assn. of Game and Fish Comms.,* Rep. No. 11, 372–374.

Purol, D. A. 1982. Field estimating the legality of harvested deer. Michigan Dept. of Natural Resources, Law Enforcement Div., Rep. No. 4.

Steele, H. M. 1962. Photography and game and fish law enforcement. In *Proceedings Southeastern Assn. of Game and Fish Comms.,* Rep. No. 16, 468–470.

Strong, P. J. 1969. *Prima facie* evidence. In *Proceedings Southeastern Assn. of Game and Fish Comms.,* Rep. No. 23, 660–661.

U.S. Congress. House. Federal rules of evidence. Washington, D.C.: U.S. Govt. Printing Office.

Multiple-Choice Questions

Mark only one answer.

1. What does not constitute judicial notice?
   a. the laws of the various states
   b. the justices of the U.S. Supreme Court
   c. the day of the week in a given week and year
   d. nonpolitical organizations throughout the United States
   e. treaties of the United States
2. What is not an official record?
   a. an officer or other person in the performance of official duties states facts or ascertains truth of a matter
   b. maps, photographs, X rays, sketches, and so forth when verified by any person whether or not he made or took them if he is personally acquainted with them
   c. records of births, deaths, marriages
   d. that a foreign or domestic record reflects a fact or is even required to be recorded by law recognition, or custom
   e. a statement made by a witness not under oath
3. Chain of custody is important in the preservation of evidence. Which is not good practice?
   a. marking the rifle with a tag giving the officer's name and address
   b. marking an empty brass by a special mark as well as recording the caliber and make
   c. recording the make, caliber, type, and serial number of a rifle
   d. sealing an envelope of evidence in the presence of a witness and signing it and having the witness sign it across the flap-envelope junction

e. retrieved projectiles marked with a personalized scratch across the nose

4. Game animals may not be properly marked by:
   a. notches cut in the hooves of mammals
   b. a magic marker used on a particular place on the antlers
   c. clipping bird nails or feathers and retaining and recording the detached part
   d. a coin inserted in a slot in the skin with a record kept of identity and year
   e. extraction of a tooth in mammals

5. When collecting proof of an alleged crime the officer looks to (one is not reasonable):
   a. the verbal or written statement of a witness
   b. the verbal or written statement of the accused
   c. the possibility a witness may not be telling the truth
   d. the use of psychology to ferret out or force the truth from the witness
   e. the witness may be judged by the officer to be insane

6. Fingerprints are stable and valid because:
   a. no two sets of fingerprints are alike
   b. they appear before birth and persist for some time after death
   c. identical twins have identical fingerprints
   d. fingerprints that are destroyed by disease return to the original pattern when good health is restored
   e. fingerprints are valuable evidence because a person's prints at a given place prove he was there

7. An investigating officer should know how to handle proof that a crime was committed and:
   a. recognize evidence per se
   b. know the evidence is admissible in court as collected
   c. know how far facts can be rationalized when he or she knows the person accused is guilty
   d. know how the evidence should be handled to maintain a legal chain of custody
   e. know how to detect discrepancies in statements and physical evidence

8. What is evidence?
   a. anything that can be shown or told in court to support the charge of a violation
   b. evidence may be classified as direct, circumstantial, or presumptive
   c. in its several forms it may be called real evidence, documentary evidence, oral evidence, or opinion
   d. evidence is anything which tends to prove or disprove a fact at issue
   e. only direct evidence is acceptable in court

9. Circumstantial evidence:
   a. tends to prove or disprove not a fact of violation but a fact of circumstances from which either alone or in connection with other facts a person may reasonably infer the existence of another fact
   b. cannot stand alone in court
   c. is established from cause to effect
   d. does not always conclusively establish a fact but it may, with other known facts, prove the elements involved
   e. there is no general rule for contrasting the weight of circumstantial versus direct evidence

10. Writing:
    a. includes copies even when they are unauthenticated
    b. means every way of recording data upon any medium
    c. includes handwriting, typewriting, or other machine writing, printing, and all documents
    d. may be pictorial, photographic, chemical, mechanical, or electronic recordings
    e. is the best evidence of its own content and the original or its equal must be introduced to prove its content

11. Dying declarations:
    a. in homicide trials dying declarations of alleged victims, including the identity of the person or persons who caused the injury, are admissible as evidence
    b. the reason for accepting this type of evidence is that the statement would obviously otherwise be unavailable
    c. are not valid if the victim thought he was about to die but recovered
    d. the validity is based on the presumption that those who are about to die tell the truth
    e. the statement is not admissible if the victim had some hope of recovery but died shortly thereafter

12. Proof of character:
    a. the opinion of a witness as to that person's character may be received in evidence if it is first shown that the witness has an acquaintance or relationship with the person
    b. the general reputation of a person must come from someone who has actually lived in the community and formed the opinion as a result of having been a member thereof
    c. the general rule in evidence is that an accused may not have evidence of his bad character introduced against him for the purposes of raising an inference of guilt
    d. in entrapment cases proof of negative character may be used by the prosecution
    e. is a highly regarded and often used procedure by defenders

Chapter 12

Undercover Investigations in Wildlife Law Enforcement[1]

SETTING UP AND RUNNING AN UNDERCOVER INVESTIGATION

Case 1

In traditional undercover investigations an agent masquerades as a criminal and seeks to get involved in an already existing and ongoing illegal enterprise. Increasingly, however, in the FBI and other agencies, the criminal activity itself is bogus and the operation is called a sting. In such investigations the agents establish a criminal enterprise which is supposed to provide criminal opportunities and, thus, attract predisposed persons to engage in such activities.

The first question when setting up an undercover investigation is, Are there alternatives, or is this action the best, or possibly only, rational choice? Gathering information is an in-depth, detailed investigation that would solve most ordinary cases. In covert investigations it just presents the depth and scope of the problem. The suspects of the investigation are profiled, their actions described, and an approach suggested.

SOURCES OF INFORMATION

Persons or organizations that may provide information for undercover investigations include taxidermists, wholesalers and retailers of wildlife products and gear, persons living in areas where violations occur, school teachers, outdoor writers, hunting guides, locker plant operators, airline employees, and conservation organizations. Courts recognize and accept the use of confidential sources who supply information with no expectation of reimbursement. However, it is important that a background check be made on sources before the department commits itself. The possession of a criminal record can be raised in court at a later date and destroy the credibility of the witness as well as the information furnished. On the other hand, there are times when using a person with a criminal history is advantageous because he or she gains the confidence of the target. If a person is known to have violated the law, he or she may not be suspected of being undercover. Confidential sources should be advised of legal rights, of potential dangers, and the fact that they may, in spite of the agency's best efforts, be exposed to the public as "stool pigeons."

AGENT QUALIFICATIONS

An investigation may involve one or more agents who have the requisite personality and other traits that will fit into the activities of the alleged violators. For example, an agent pretending to be a big-game hunter looking for an illegal animal must be a competent hunter who understands the slang of the trade and knows about others who have taken animals in this class. Recruits should have a number of years of overt enforcement. The agents should know the laws, the areas, and the resources they are going to be working with. It is also necessary that undercover investigators be free from exposure identifying them as law enforcement officers. The agents should be able to be away from family and colleagues (except for a contact) for extended periods of time. They should be able to defend themselves physically (this may be viewed as a test of courage by the targeted persons or groups) and, perhaps above all else, be good actors. They should not forget for a moment that the job is dangerous and their cover can be blown without warning. They should have developed, along with the aid of colleagues, one or more scenarios for fallback if their covers are blown. However, it may be necessary to play it by ear.

Undercover investigators should be confident, even-tempered, and flexible so they can adapt to unpredictable situations. They should try to anticipate the questions that they will be asked, and prepare spur-of-the-moment answers. Covert investigators must be able to blend in easily with the lifestyle and behavior of the people they are targeting. They must be able to lie and not let it bother them. Also, they should be able to think on their feet and have good recall of previous undercover conversations. Psychological testing may help to determine if an agent can handle stresses involved with undercover investigations. If the investigation is long term, some testing during the process helps to assure that the agent is holding up. Once released from an undercover assignment, an agent should be deprogrammed to get back into the mainstream of law enforcement. A person in covert work too long tends to get away from normal lifestyles and behavior. Freedom from personal entanglements, domestic problems, and other things that may distract while undercover are important, as is physical stamina. Undercover agents take on entirely different identities. They are living three different lifestyles:

(1) undercover agent,

(2) possibly a family man or woman, and

(3) law enforcement officer.

An undercover officer is selected to fit the role—not the other way around. For example, someone who has fished and hunted for a long time could qualify in an undercover investigation as a hunting guide or as a hunter who wanted to collect a unique specimen. Generally speaking, a person should be of the same ethnic group as that being infiltrated.

Once investigators are selected for a specific assignment they should be given on-the-job training as well as classroom experience. These experiences should include specific deepcover activities, setting up a background covert identification, learning the laws that will be involved, and discussing the possibility of being charged with entrapment or outrageous government conduct. They should be both psychologically and physically ready to enter the investigation.

POTENTIAL TARGETS

Only people who are doing serious damage to the resource and who cannot be stopped by conventional methods should be considered as targets for undercover investigations. This includes commercial hunters, gross illegal takes by hunters, the taking of endangered species or limited resources. Undercover operations are expensive both in funding and manpower, yet the benefits far outweigh the costs when targeting primary killers and traffickers, rather than those who are incidentally involved. Some investigations may snowball, with end results that are not always satisfactory.

Examples of Targets

Potential undercover investigations include:

(1) A clan is suspected of killing and selling just about any wildlife species it can lay its hands on. Clans may be extremely dangerous when crossed;

(2) There is evidence of illegal dealing in animal skins by a well-known local company. The problem is that the reliable informants are afraid either of physical violence from company personnel or political retribution from their employers; and

(3) A well-known hunter is looking for a sheep. He lets it be known that cost is no object, and he knows there is an acceptable animal in one of the national parks in Alaska.

Failure to properly plan and execute a covert investigation can result in disastrous results, which includes losing in court, losing support of prosecutors and management for future operations, loss of the resource, and loss of human life.

NEGATIVE FACTORS IN INVESTIGATIONS

There is a downside to covert investigations that should be recognized and understood by everyone from the agency director to the field personnel. This aspect of FBI covert investigations, discussed in House Document 98–267, will be drawn on here. Undercover investigations carry with them the potential for serious damage to both society and individuals.

The first danger is to public institutions—especially in public corruption investigations, which may be intended to restore the public's faith and integrity of the institutions. Ill-conceived and poorly managed undercover investigations are likely to have the opposite effect.

Second, according to the committee report, the FBI's use of elaborate, lengthy, and deceptive practices, and the need to avoid discovery, have resulted in severe harm falling on totally innocent citizens; either through a failure to monitor the activities of the informants, or as the result of carelessness, neglect, or conscious design on the part of undercover agents.

Third, the house subcommittee, in its review of undercover cases, found that the techniques carry with them the potential for subjecting innocent citizens to persecution and even conviction. Because agents create the crime rather than detect it they hold the power to create the appearance of guilt. The subcommittee found that the discussions with targets were highly ambiguous, leaving considerable doubt as to whether there had been any meeting of the minds or that the subject understood the criminal activities that were being discussed. Since the general public tends to equate investigation with guilt, just being mentioned in an undercover investigation can lead to great personal and professional difficulties for targeted individuals.

THE DANGERS OF UNDERCOVER INVESTIGATIONS

The dangers of undercover investigations cannot be removed entirely no matter how well the investigation is planned or how experienced the operators. But there are ways to substantially reduce the dangers. Wildlife undercover investigations are more dangerous today than they were ten to fifteen years ago. This is because ''big money'' is involved now more than ever. The violators are more likely to be under the influence of drugs or alcohol. There are more foreign nationals who may have a proclivity toward violence or a perverse attitude toward wildlife laws. And the cost in money and loss of liberty for being caught and convicted has increased substantially in recent years (Wade 1990).

Undercover investigations should have a written operational plan. A manuscript need not be prepared for every undercover scenario. However, the details of the investigation should reflect the risks involved.

The most critical time for a covert investigator is the initial contact. In the first few seconds or minutes the targeted person or persons will make a tentative opinion of the investigator. They will decide whether or not they like and trust the investigator. Investigators should be mentally planning a way out in case the first or later contacts turn sour. Agents may want to develop confidence of targets by being with them off and on for some time before pushing an illegal transaction. It may take six months to a year before completing an illegal transaction in a major operation; just establishing presence and gaining confidence of individuals.

The second most critical time for the investigator is when the targeted individual is arrested. Ideally the covert investigator will not be around when the arrest occurs. One of the most dangerous actions an undercover officer can take is the arrest of a violator. At this stage in the investigation, the violator is sold on the officer's cover and may perceive an attempted arrest as an act of treachery. The suspect may resort to violence if he believes he has been tricked. If the undercover agent cannot be far removed from the arrest scene, he or she should maneuver into the best available defense position and prepare for any level of violence when the backup crew moves in to make the arrest. Another approach is to stage simultaneous arrests on the suspect and the covert agent. It may look better to the defendant if there is a fictitious arrest of the undercover agent. It is also a good way to get the agent out of possible danger.

Another danger of undercover work is complacency of investigators, who begin to miss small signs of approaching danger. People working long-term with a group are inclined to let down their guard, which is dangerous. Usually, when a suspect begins to doubt the investigator's true identity, there will be hints. The suspect may not discuss suspicions, but will be standoffish.

ARREST PROCEDURES

The time and date of arrest are set to maximize the number of defendants in a known location. Arresting suspects at home is not preferred because they are in their comfort zone. They know where everything is. They know what is going on. It is preferable to make an arrest away from a residence so agents are in control and there is little, if any, access to weapons or situations that can negatively impact the execution of the arrest warrant. Consideration should be given to persons who are not involved, such as children, customers, and bystanders. The personality of the arrestee should be known and considered. It is desirable when possible, to put the defendant in a position where he or she is least able to offer resistance. The arresting team should have sufficient personnel for the investigation and assume something not predetermined may occur and demand more force. The logistical team includes photographers, communication personnel, and clerical personnel to complete reports. The chain of command follows rules set before the operation is initiated. Generally, an undercover officer should have access to a communication device such as a beeper or a cellular phone. The officer should realize that sometimes communication devices fail or can't be used.

Whether or not to carry a weapon is a matter not uniformly agreed upon by undercover officers. Some feel uncomfortable with a weapon and believe it restricts them in what they can do and where they can go. When a weapon is carried it should be deeply hidden, or the officer should have a glib excuse ready for carrying it. Most undercover agents go without firearms unless they feel they can come up with a feasible reason why they are carrying them. There is no way to make an undercover investigation safe, but one that is well-planned and executed can lessen the dangers considerably (Wade 1990).

If a body wire is worn, the best evidence it can record is when a person says, ''Yes, this is what I did and it is illegal,'' or something to that effect. However, if there is any chance that the targeted persons will pat down the investigator, a wire should not be worn. People associated with drugs, illegal guns, counterfeit money, and organized crime may suspect everyone and insist on searching them for a wire or for a gun. The agent can always claim that a gun is for personal protection, but there is no reason for wearing a wire other than the obvious one. Supervisor-agent contact generally is essential on a daily basis. A monitored body wire is obviously the best contact. Alternatively, there are telephones and personal meetings. Cellular or radio phones are not satisfactory, except for brief messages, since they can be picked up by scanners.

THE RAID

A raid by law enforcement officers is a sudden assault on an area, building, or individual with the object of apprehending some person or persons, or of obtaining evidence or other material. Raids are generally planned, but on occasion may be spontaneous. Under no circumstances should a raid be executed unless there is no question as to its legality.

In the course of a wildlife officer's career he or she may be compelled by circumstances to conduct or participate in a raid. A knowledge of raid tactics will stand the officer in good stead. A successful raid is a masterfully executed culmination of a covert investigation. Agents go in, perform a safe operation, obtain the evidence and whatever documents are wanted.

PLANNING AND EXECUTING A RAID

Spontaneous raids should normally be made only when it appears that evidence may be lost or individuals may escape if there is a delay. Planned raids are much more effective, but to insure their success the planning must be carefully done. A number of important factors will be briefly considered.

1. Is the raid necessary to the performance of duty? Is the raid legal?

2. What is the composition of the raiding party? Will there be superiority of manpower? Is each member of the party an experienced officer, competent in the role he or she is to play? Is it understood who is in charge?

3. Is the party properly oriented? Does each member understand the plan of approach? Do the officers have a knowledge of the terrain, the number and location of buildings or other structures, the possible positions of lookouts and their scope of view, and the possible routes

Hunters may be out in all kinds of weather. *Courtesy Wyoming Game and Fish Department.*

along which the quarry may try to escape? In fish and game violations, particularly, the raiding party should consider the positions of wild or domestic animals as well as routes which can be used to avoid these and other obstacles.

4. Is the available equipment adequate for the task? Are the weapons optimum for the situation? Are signals agreed on?

5. What hazards may be expected? How many persons are being raided? What weapons may be expected to be available? What is the character and nature of each person? Will opposition probably continue to the bitter end, or will it cease at the first show of superior force? Is the possibility of trickery considered? What communication facilities may the suspects be able to use?

Depending on the circumstances, agents may consider initially going on low-profile contact. The raid team will be present, but they stand back. A couple of agents will go to the door, they will knock and say, ''Hey, we are here. We have to talk to you.'' A search warrant may not be needed. The agents will say, ''We would like to look for something.'' Agents may even take candy to the children. They may allow the family to continue with dinner.

Normally, a raid should be conducted when there is a minimum of opposition and interference from outsiders or from uninterested persons. The vast majority of raids start at 6:00 A.M. because most courts authorize search warrants to be executed between 6:00 A.M. and 6:00 P.M. On extensive, well-planned raids, one or more alternate plans should be drawn up in detail, in the event the original plan appears to be failing. For example, one plan may be followed until the suspect surrenders, and a second put in operation if the suspect chooses to flee or resist. A large raiding party should have maps prepared so that each individual will have no question as to his or her routes and duties. Night raiding parties should avoid wearing such shiny objects as luminous watch dials or anything that can reflect even a small amount of light.

As a final precaution, just before a raid is to begin, all equipment should be carefully checked to see it is in proper working order and no items have been omitted. These items should include handcuffs, which may be needed if stiff resistance is offered after the suspect is under temporary custody. If pictures are to be taken in total darkness, the proper infrared equipment should be included.

Operations involving several agencies are becoming more common since both information and funds can be pooled. In every case before a raid begins it is necessary that a predicate law has been broken. It can be either state, federal, Indian, or foreign. The Lacey Act is an umbrella statute. It is best known for the phrase that states if a state law is broken and the item in question is transported across a state line the Lacey Act is automatically violated. In 1988, Congress expanded the Lacey Act and gave the Fish and Wildlife Service the authority to enforce federal laws.

Two final points:

(1) since covert investigations are quite expensive, and law violations are numerous, operations must be on a priority basis, and

(2) it is necessary to have a green light from the prosecuting attorney and to keep him or her advised of progress and problems from the beginning. This helps deflate charges of entrapment and outrageous governmental conduct.

ENTRAPMENT AS A DEFENSE

In undercover cases, the defense of entrapment is often invoked. The best defense against entrapment is proof that the defendant was predisposed. A prosecutor's arsenal of predisposition evidence is vast and powerful. It includes both prior and subsequent activity, and it can include a defendant's ready acceptance of an agent's inducement. Where there is proof of predisposition the defense of entrapment is defeated. However, failure to prove predisposition turns the government's case into entrapment (Callahan 1984b).

Predisposition can be defined as a defendant's preexisting willingness to commit a crime whenever the opportunity is presented. The following instructions are often given to the jury in entrapment cases: "Where a person already has a readiness and a willingness to break the law, the mere fact that government agents provide what appears to be a favorable opportunity is not entrapment" (Callahan 1984a). A person can be predisposed to commit a crime even if he does not form an intent to commit the specific criminal act charged until solicited by a government agent. For example, a regular deer poacher might not be predisposed to poach on a particular evening until an agent originates the idea. The concept of predisposition can be traced to the first Supreme Court case which considered the issue of entrapment. In *Sorrells v. United States,*[2] an undercover agent made several requests of the defendant to provide him with illegal whiskey. The defendant finally agreed and provided the whiskey. At the trial the defendant asserted the defense of entrapment, but the trial judge ruled as a matter of law that entrapment was not present. A federal appellate court affirmed and the case went to the Supreme Court, where the decision was reversed. The majority opinion recognized the rights of the defendant to offer evidence that he had not committed the crime charged at the instigation of the government. The Court also observed that when the entrapment defense was raised, the prosecution should be able to have the chance to prove the defendant was not an innocent victim of police inducement but rather predisposed to commit the crime charged. The majority opinion in *Sorrells* has come to be labeled the "subjective view" of the entrapment defense because the focus is on the defendant's subjective state of mind and whether or not he was predisposed to

Case 2

Hunter pack trains are common in the high country out West.
Courtesy W. F. Sigler.

Cases 3–5

commit the crime. Three judges took the position that the entrapment defense should focus exclusively on the conduct of the government in the individual case. This view of entrapment has come to be called the "objective view." When this view is held, predisposition evidence is generally regarded as inadmissible. Since 1932, four Supreme Court cases have been decided in which a majority of the justices adopted the subjective view.[3] All of these cases approved of the admissibility of predisposition evidence to defeat the defense of entrapment. As a consequence federal circuits and a majority of the state courts generally follow the subjective view. However, at least eight states have adopted the objective view. They are Alaska, California, Hawaii, Iowa, Michigan, New Hampshire, North Dakota, and Pennsylvania (Callahan 1984a).

Because entrapment is an affirmative defense it must be raised by the defendant at the trial. The defendant bears the initial burden of producing evidence showing that the government initiated, suggested, or proposed the crime. Further, the defendant must produce some evidence that he was not predisposed to commit the crime.[4] If the defendant is successful in meeting this initial burden, the proof then shifts to the government to produce evidence that the defendant was predisposed beyond a reasonable doubt.[5] If the defendant succeeds in the entrapment defense he will be acquitted, because a failure to prove predisposition turns inducement into entrapment (Callahan 1984a).

Other crimes may be used by the prosecution to prove the defendant's participation in criminal activity other than the specific crime for which he has been tried. This is known as other-crime evidence. In general, rules which govern the admissibility of evidence prohibit the admission of other-crime evidence when it is offered to show that the defendant committed the act for which he is charged. Such evidence is considered prejudicial because it has a tendency to distract the jury. However, in entrapment cases the federal courts have routinely admitted such evidence for this purpose. This is because the question before the jury is not whether or not the defendant committed the acts charged. That is a foregone conclusion in most entrapment cases. In fact, in some federal circuits the defendant must admit to the acts charged in order to raise the entrapment defense. Thus, the principal issue for a jury is whether or not the defendant was predisposed to commit the crime charged and, therefore, other-crime evidence is directly relevant. The evidence is offered to prove intent, which is a specific exception to the general rule that other crimes are inadmissible as evidence (Callahan 1984a).

Cases 6–10

Predisposition may also be shown by evidence that the defendant has been previously convicted of similar offenses to the one charged. The U.S. Supreme Court implied this when it decided *Sherman v. United States.*[6] In *Sherman,* a government informant met Sherman at a doctor's office where both were being treated for narcotic addiction. Sherman turned down repeated requests from the informant to provide heroin. Finally, after the informant appealed to Sherman's sympathy based on his knowledge of narcotics, the defendant provided heroin. Sherman was indicted for three heroin sales and raised the entrapment defense at his trial. The entrapment issue was decided by the jury, which returned a verdict of guilty. A federal court of appeals affirmed. Sherman then appealed to the U.S. Supreme Court and argued that entrapment had been established as a matter of law. The Supreme Court agreed and reversed. The Court examined the sufficiency of evidence to establish predisposition and decided that the defendant's nine-year-old and five-year-old convictions for illegal possession were insufficient as a matter of law to establish predisposition. It should be noted that federal appellate courts have consistently refused to approve the admissibility of prior convictions to overcome an entrapment defense when the cases are very dissimilar. In *United States v. Pagan,*[7] the defendant was convicted of selling heroin. To overcome the entrapment defense the government introduced a prior conviction for interstate transportation of a stolen motor vehicle. On appeal, the court stated that prior convictions must involve offenses similar to those in question in order to constitute relevant rebuttal evidence (Callahan 1984a).

Prior arrests may also be used by the prosecutor to overcome the entrapment charge by the defendant. *Pulido v. United States*[8] provides an example. *Pulido* involved two defendants, one of whom was Luna. Luna was convicted of selling heroin, and during trial he asserted the entrapment defense. In attempting to show predisposition the government introduced evidence of a twelve-year-old arrest for narcotics violation. Luna appealed and argued that the trial judge erred in admitting this evidence to the jury. The court of appeals affirmed the conviction and held that prior arrest records are a permissible method of establishing predisposition as long as the evidence is relevant. Prior criminal activity is also regularly allowed by federal appellate courts in establishing predisposition. For example, *United States v. Salisbury,*[9] provides another illustration. Salisbury was tried and convicted of selling a load of stolen carpet to an undercover FBI agent. During the trial the defendant claimed entrapment and the government introduced testimony from the informant regarding a prior attempt by the defendant to sell him a truckload of stolen tires. On appeal, the court held that the prior criminal activity was relevant and its prejudicial nature did not substantially outweigh its evidentiary value. The court observed ''the evidence was squarely on point as to Salisbury's criminal predisposition. There was little in the way of unfair prejudice to counterbalance the probative value of the evidence'' (Callahan 1984a). The conviction was affirmed. Also in *United States v. French,*[10] the defendant was working behind the counter of a convenience store when he was approached by an undercover officer who asked him if he knew of anyone who was buying food stamps for money. Purchasing food stamps under these circumstances is a violation of federal law. The defendant stated that the owner did not like for him to do that but asked how many she had to sell. The undercover agent stated she had three books of stamps and he offered her $50.00 for them. She provided him with the books and received the money. Approximately five months later, the defendant once again purchased food stamps on two occasions from undercover officers. During trial the defendant claimed entrapment but the jury returned a guilty verdict which by implication means the jury found him predisposed. The court of appeals affirmed and stated that the defendant's initial contact with the undercover officer involved a question from the undercover agent to the

Cases 11–14

defendant concerning whether or not he knew if anyone was willing to purchase food stamps. The court considered the defendant's response to that question and concluded that it amounted to predisposition (Callahan 1984a).

Another method of proving a defendant's predisposition is by establishing that he became involved in similar activities after his participation in the offense for which he was charged. In *United States v. Mack*,[11] the defendant sold five grams of cocaine to an undercover agent on 10 November 1978. Mack met again with agents two months later and sold them two ounces of cocaine. He was later indicted for the 10 November sale. At the trial Mack claimed entrapment and the government rebutted with the defense of proving predisposition by showing that after his first indictment he again sold two ounces of cocaine. Mack was convicted and the court of appeals rejected his argument and held that the second sale was not only admissible to prove predisposition but was also relevant to the offense charged and was not substantially outweighed by the danger of undue prejudice. The court further observed, "despite Mack's protest to the contrary, the evidence showing that he had engaged in the December 16th drug sale, even though that sale occurred after the incident for which he was tried, did have a tendency to make it more probable that he was predisposed to commit the original offense charged, the sale of November 10th" (Callahan 1984b).

Another example of post-offense attempts at criminal activity is shown in *United States v. Moschiano*.[12] Moschiano was convicted for selling heroin to undercover DEA agents on 11 September 1980. During the trial he asserted the defense of entrapment in order to rebut the charge of predisposition. The government introduced evidence that on 5 December 1980, almost three months after the sale, Moschiano had asked another undercover agent to sell him fifty thousand Preludin tablets, a controlled substance, so he could sell them to truck drivers. The deal was never closed but the evidence remained. On appeal, Moschiano argued that such subsequent criminal activity should not be admissible to prove predisposition because it is irrelevant under the federal rules of evidence. The court of appeals disagreed and ruled that commissions of similar crimes after the indicted offense is often relevant with respect to predisposition and should be admissible as long as its relevancy is not specifically outweighed by the danger of undue prejudice (Callahan 1984b).

Ready acceptance, also proved to be a highly significant tool in proving a person's predisposition, lies in the defendant's ready and unhesitating acceptance of the government's offer to commit the offense.[13] The question of whether hearsay evidence should be admissible to prove predisposition has been answered negatively by several federal appellate courts. In *United States v. Webster*,[14] the defendant, Webster, became involved with a woman who was a government informant. Webster was convicted on hearsay evidence and appealed to a three-judge appellate panel which affirmed. Later, the fifth circuit vacated the opinion of the panel and reversed. The fifth circuit noted that earlier cases in the circuit had approved a rule which allowed hearsay evidence to prove predisposition, but the court observed that such a rule enables a jury to consider unsworn, unverified statements of unidentified government informants whose credibility is not subject to effective testing before the jury and whose motivations may be less than honorable. The court noted that it was hard pressed to conceive of a situation where the disparity between the relevance of certain evidence and the prejudicial effect could be greater (Callahan 1984b).

It has been noted that the defense of entrapment is often raised by defendants in covert action cases. As was pointed out earlier, a proof that a defendant is predisposed will often overcome the entrapment defense in spite of government inducements. However, a failure to turn a government inducement into entrapment raises the critical issue that the entrapment equation is predisposition.

Cases 15–18

In February 1984 a fifty-six-year-old man, with no record of law violations, ordered and received from an adult bookstore two magazines containing photographs of nude teenage boys (*Jacobson v. United States*).[15] Subsequent to this, Congress passed the Child Protection Act of 1984 which made it illegal to receive such material through the mail. Over the next two and one-half years government investigators, through five fictitious organizations and bogus pen pals, repeatedly contacted the defendant by mail, exploring his attitude toward child pornography. Twenty-six months after the mailing to the defendant commenced, government investigators sent him a brochure advertising photographs of young boys engaged in sex. At this time the defendant placed an order that was never filled. Later a catalog was sent to him and he ordered a magazine containing child pornography. After a controlled delivery the defendant was arrested and charged with receiving child pornography through the mail in violation of federal law. He defended himself by claiming the government's conduct was outrageous, the government needed reasonable suspicion before it could legally begin an investigation, and he had been entrapped by the government investigative techniques (Kukura 1993). The lower federal courts rejected these defenses, but in a five-to-four decision the Supreme Court reversed his conviction based solely on entrapment.[16]

In *Jacobson*, the Supreme Court held that law enforcement officers may not originate a criminal design, and plant in an innocent person's mind the disposition to commit a criminal act and then induce the commission of the crime so that the government may prosecute. The first question was, did the government induce the defendant to commit the crime, and, second, assuming the government improperly induced the defendant to commit the crime, was the defendant nevertheless predisposed to commit a criminal act prior to first being approached by agents (Kukura 1993).

Since the government did not dispute that it induced the defendant to order the pornography, the sole issue before the Court was whether or not the government had proved beyond a reasonable doubt that Jacobson was predisposed to order illegal pornography. The Court rejected as insufficient evidence the government's claim that the mailing to the defendant of legal material before the child pornography law was passed showed evidence of predisposition. The Court said, ''there is a common understanding that most people obey the law even when they disapprove of it'' (Kukura 1993).

Federal courts have held that no federal constitutional requirement for any level of suspicion is necessary to initiate undercover operations. The issue was whether or not the government needed reasonable suspicion to approach the defendant. In *Jacobson* this was resolved in the government's favor by lower courts and the Supreme Court refused to overturn that holding. These decisions rejected the claim the government needs a preexisting basis for suspected criminal activity before targeting an individual in an undercover operation (Kukura 1993).

The federal defense of entrapment requires a defendant first establish he was induced to commit a crime, then the burden shifts to the government to prove the defendant was nevertheless predisposed (*United States v. Van Slyke*).[17] Inducement generally requires more than merely establishing that an officer approached and requested a defendant to engage in criminal activity. Evidence that the government engaged in persuasion, threats, coercive tactics, and harassment may amount to inducement. Nevertheless most courts require the defendant to demonstrate that the government conduct created a substantial risk that an undisposed person or otherwise law-abiding citizen would commit the offense (Kukura 1993).

For example, in *United States v. Skarie*,[18] a government informant moved in with the defendant and asked her to put him in touch with someone who could sell him and his friends drugs. She declined but the informant continued pressure. At one point he impaled one of her chickens on a stick and later stated that what happened to her chicken could happen to people as well. The defendant subsequently took the informant to the source of drugs, who later brought approximately three pounds to the defendant's house. At this point the police arrested the defendant. The U.S. Circuit Court of Appeals for the Ninth Circuit found in *Skarie* the government induced the defendant to break the law because the informant initiated the idea of a drug sale then pressured the defendant repeatedly and threatened her. In considering all factors the court found that no reasonable jury would find the defendant was predisposed to sell drugs independent of insistence and threatening action of the informant (Kukura 1993).

NONCITED CASES

Illegal Deer and Bear Operations

One summer evening in 1988 a cream-colored Chevrolet pickup truck drove slowly by a Korean herb shop in a quiet section of downtown Manhattan, New York. After circling the block several times the pickup parked down the street from the shop and two men alighted from the pickup. Both were in their mid-thirties with full beards and long hair. One of the men had his hair pulled back in a pony tail. They both wore blue jeans, sweatshirts, work boots, and baseball caps. They walked up to the front of the shop where there were about five Korean men dressed in business suits. One of the bearded men pulled something out of a bag he was carrying and the Korean nodded, smiled, and invited him inside. When asked how much they had, one of the men replied "forty-two." The Korean offered to pay five dollars per gram and a handshake sealed the bargain. The illegal items being sold were gallbladders of wild black bears, and the two men in the pickup were not narcotics dealers but undercover environmental conservation investigators who were deeply involved in a three-year operation (Brewer 1990).

This particular investigation had been initiated at a meeting between Massachusetts and Connecticut officers. It was stated that a Connecticut meat dealer was supplying venison to a restaurant in western Massachusetts. The Connecticut officers believed there was a widespread abuse of permits issued in their state granting landowners and their designees the authority to kill deer which were damaging agricultural crops. This privilege was being used illegally. The investigative team was composed of agents from the two states and the U.S. Fish and Wildlife Service.

Two agents were given fictitious identifications and then proceeded to establish their new identities near where the alleged violations were taking place. The agents were also given resumes on the background of possible suspects. One operated a fishing lure business, one a slaughterhouse, and a third a smokehouse. In early September the agents drove to the home of one of the suspects where they stopped and asked directions on how to get to a lake that had particularly good fishing. The result of this conversation was a fishing trip with two of the suspects. Contact had been made.

Over the next year the agents gave the suspects what appeared to be illegally killed deer to butcher and smoke. No questions were asked but remarks from the suspects indicated they believed the deer were killed illegally. The agents were able to document that Connecticut had a major problem with its deer damage permit system. Landowners could do about whatever they wanted with the deer killed except sell them. The agents quickly discovered that deer taken with these damage permits were being sold. Deer were killed elsewhere, marked with tags provided by the landowner, and sworn to having been killed on his place (Brewer 1990).

Up to this point the agents had concentrated entirely on Connecticut's deer damage permit problems. However, they were beginning to receive information that a market existed for black bear parts, particularly the gallbladder. The size of the market was believed to be widespread, and commercialization of black bear was probably having a heavy impact on the resource. The decision was made to shift some of the emphasis to the illegal taking and trafficking in black bear. One of the agents learned that a Chinese woman had requested her business card be passed around for anyone who might be selling deer parts or bear gallbladders. The black bear gallbladder is prized by Koreans and Chinese as a medicine. It is also alleged to be an aphrodisiac. The person killing the bear would normally get $50 to $250 for each gallbladder, which would then bring $500 to $600 in the Orient. An entire bear carcass, including the gallbladder, could bring from $300 to $500. The agents decided they would pretend to be sellers and try to get invited on bear hunts by some of the suspects. In September 1987, one of the agents met a suspect who operated a slaughterhouse and sold him four black bear gallbladders. Later that same day a Korean woman and her father met the investigators and purchased four gallbladders, one entire deer carcass, and a set of deer antlers in the velvet. The woman then informed him that she wanted to purchase bear paws and more gallbladders. From this time on the buying and selling of bear and deer began to escalate. The agents' transactions involved the gallbladders from 374 black bears, 25 bear carcasses, and the heads, hides, and paws from another 15 bears. In one transaction alone the investigators bought 218 gallbladders for $5,000 and resold them for $11,000. However, in spite of this, the Korean and Chinese dealers were difficult to deal with. They did not trust anyone and dealers would taste and chew the gallbladders to assure that they were black bear and not pig (Brewer 1990).

Once the operation was rolling, it became a year-round seven-day-a-week detail for the agents. They would receive calls at all hours of the night to go on ''hunting'' trips. They would then have to hurriedly find ways to call their supervisors from telephone booths. Particularly hard to understand were the motives of some of the suspects. They were all respected, middle-class people, including loggers, butchers, gas station operators, and so forth. When these suspects went on a hunting spree they talked about beating the system, beating the game warden. At times they wanted to see how many species they could illegally kill. They did it for fun, the thrill of watching something die. For the agents these were trying times; they were never completely trusted, and if their cover was blown they would be in serious trouble (Brewer 1990).

At the end of 1988 it was decided to terminate the investigation. It was causing considerable stress to the two agents and there appeared to be an abundance of evidence that could be used in court. Many supervisors believe that an undercover investigation for one team of agents generally should not be allowed to go much beyond two years. In January 1989 the agents were pulled from their undercover role, and eighteen teams consisting of eighty-four officers armed with warrants began to arrest the suspects and execute searches in four different states. A total of 962 charges were filed against twenty-eight individuals (Brewer 1990).

Illegal Activities in Yellowstone National Park

The greater Yellowstone ecosystem in Wyoming, Montana, and Idaho, a four-million-acre wilderness, offers one of the largest concentrations of game animals to poachers of any place in the world. Some of these people are thrill hunters who want to take shots at animals to see them die, but many others are taking trophies for payoffs as high as $20,000 for a record bull elk, $50,000 for a full-curl bighorn sheep head, $3,000 for the paw of a grizzly, and $1,000 for a bald eagle. The risk of getting caught is negligible and the penalties usually lenient in spite of the amended Lacey Act. There are far too few team agents to do more than scratch the surface in this vast wilderness area. Violators use helicopters and monitor police radios. A three-year sting operation netting three dozen suspected poachers in 1984 is the exception rather than the rule. An increased number of enforcement personnel could save some of the big game and other animals, but perhaps more important is public involvement.

In a typical operation agents go undercover pretending they are outfitters. They set up a bogus big-game booking service and solicit clients. These agents then go to sportsmen's shows and conventions, advertise in magazines, and generally publicize their competence as guides in getting trophy-size, big-game animals. At fairs and other public meetings they have booths and circulate photos and brochures telling of their success. They may even set up legal hunts just to make themselves look legitimate for someone who is suspicious. The agents may, for two to three years, cut themselves off from their families and the outside world to infiltrate organized poaching rings. Undercover agents must be imaginative, independent, and able to maintain their cool under adverse and sometimes life-threatening conditions. It has been said that market hunters are as dangerous as bank robbers. They are well-armed with modern weapons, and they are willing to shoot if they believe they are threatened.

Literature Cited

Brewer, W. 1990. Undercover. *The Conservationist* 44 (4).

Callahan, M. 1984a. Predisposition and the entrapment defense, part 1. *FBI Law Enforcement Bull.* (Federal Bureau of Investigation, U.S. Dept. of Justice, Washington, D.C.) 53 (8): 26–31.

———. 1984b. Predisposition and the entrapment defense (conclusion). *FBI Law Enforcement Bull.* (Federal Bureau of Investigation, U.S. Dept. of Justice, Washington, D.C.) 53 (9): 26–31.

Kukura, T. V. 1993. Undercover investigations and the entrapment defense. *FBI Law Enforcement Bull.* (Federal Bureau of Investigation, U.S. Dept. of Justice, Washington, D.C.) 62 (4): 27–32.

Wade, G. E. 1990. Undercover violence. *FBI Law Enforcement Bull.* (Federal Bureau of Investigation, U.S. Dept. of Justice, Washington, D.C.) 59 (4): 15–19.

Notes

1. Extensive editing and material were furnished by Terry Grosz and Neill Hartman, Asst. Regional Director Law Enforcement and Deputy Asst. Regional Director Law Enforcement, respectively, Rocky Mountain and Prairie States Region of the U.S. Fish and Wildlife Service.
2. 287 U.S. 435 (1932).
3. *Sorrells v. United States,* 287 U.S. 435 (1932); *Sherman v. United States,* 356 U.S. 369 (1958); *United States v. Russell,* 411 U.S. 423 (1973); *Hampton v. United States,* 425 U.S. 484 (1976).
4. *United States v. Watson,* 489 F. 2d 504 (3d Cir. 1973); *United States v. Riley,* 363 F. 2d 955 (2d Cir. 1966).
5. *United States v. Garrett,* 716 F. 2d 257 (5th Cir. 1983), cert. denied. 104 S. Ct. 1910 (1984); *United States v. Walker,* 720 F. 2d 1527 (11th Cir. 1983); *United States v. Silvestri,* 719 F. 2d 577 (2d Cir. 1983); *United States v. Burkley,* 591 F. 2d 903 (D.C. Cir. 1978), cert. denied, 440 U.S. 966 (1979).
6. 356 U.S. 369 (1958).
7. 721 F. 2d 24 (2d Cir. 1983).
8. 425 F. 2d 1391 (9th Cir. 1970).
9. 662 F. 2d 738 (11th Cir. 1981), cert. denied, 457 U.S. 1107 (1982). See also, *United States v. Segovia,* 576 F. 2d 251 (9th Cir. 1978); *United States v. Biggins,* 551 F. 2d 64 (5th Cir. 1977).
10. 683 F. 2d 1189 (8th Cir. 1982), cert. denied, 459 U.S. 972 (1982).
11. 643 F. 2d 1119 (5th Cir. 1981); see also, *United States v. Burkley,* 591 F. 2d 903 (D.C. Cir. 1978); *United States v. Brown,* 567 F. 2d 119 (D.C. Cir. 1977); *United States v. Rodriguez,* 474 F. 2d 587 (5th Cir. 1973).
12. 695 F. 2d 236 (7th Cir. 1982). See also, *United States v. Jannotti,* 673 F. 2d 605 (3rd Cir. 1982).
13. *United States v. Jannotti,* 673 F. 2d 605 (3rd Cir. 1982); *United States v. Myers,* 692 F. 2d 823 (2d Cir. 1982), cert. denied, 103 S. Ct. 2438 (1983); *United States v. Anderton,* 679 F. 2d 1199 (5th Cir. 1982); *United States v. Fleischmann,* 684 F. 2d 1329 (9th Cir. 1982), cert. denied, 459 U.S. 1004 (1982); *United States v. French,* 683 F. 2d 1189 (11th Cir. 1982); *United States v. Rogers,* 639 F. 2d 438 (8th Cir. 1981).
14. 649 F. 2d 346 (5th Cir. 1981).

15. 112 S. Ct. 1535 (1992).

16. The Supreme Court declined to address the issues of outrageous government conduct or the need for reasonable suspicion.

17. 976 F. 2d 1159 (8th Cir. 1992).

18. 971 F. 2d 317 (9th Cir. 1992).

Additional Case Notes

United States v. Ivey, 949 F. 2d 759 (1991).

Appellants Ivey and Wallace were indicted and convicted on one count of conspiracy to bring bobcat hides illegally into the United States from Mexico, and several counts of smuggling, contrary to the law. Both appellants asserted the jury verdict should be reversed due to defects in the indictment and the jury charge as well for insufficiency of evidence. Ivey also appealed based on a claim of entrapment and outrageous government conduct. The court of appeals affirmed the district court's ruling.

The facts of the case were not in dispute. In 1986, the U.S. Fish and Wildlife Service received information that large numbers of bobcat hides were entering the country from Mexico in violation of the United States law. The government initiated an undercover investigation to identify those individuals involved in the illegal fur trade. As part of the investigation, the government opened a store named Van Horn Fur House. Agents of the Fish and Widlife Service established what is known as a ''reverse sting.'' They purchased pelts believed to have been trapped in Mexico and smuggled into the United States, and then resold those pelts to other fur buyers. The investigation revealed the likelihood that Ivey and Wallace were involved in the trafficking of illegally smuggled furs. They had established an extensive network of trappers and other individuals to procure and smuggle Mexican bobcat furs into the United States. In order to conceal their illegal action they established a pool of ''trappers,'' some legitimate and some who had never trapped. These people signed tags certifying the bobcat pelts were taken in accordance with state and federal law. The trappers were paid for their signatures. At the trial, government witnesses testified they received smuggled furs from Mexico and sold them to the defendants. The witnesses also testified that the defendants knew the furs came from Mexico. Ivey and Wallace claimed that many of these witnesses also trapped furs in the United States, and it was impossible to tell where a bobcat had been trapped merely by looking at its pelt.

These accounts included a conspiracy to receive, conceal, buy, sell, and facilitate the transportation, concealment, and sale, after importation of bobcat hides, knowing that said hides had been brought into the United States illegally from Mexico. The other counts involved smuggling, receiving, concealing, buying, selling, and so forth. The only difference between counts four, five, and seven was that each count alleged a violation on a different date. A jury found Ivey and Wallace both guilty on counts one, five, and seven of the indictment, and, in addition, found Wallace guilty on count four. Ivey and Wallace alleged that the indictment was defective for several reasons. First, they claimed that counts four, five, and seven should have been dismissed for failure to state a criminal offense. The Court said that the appellants' arguments were unconvincing. Their argument, the Court said, would undermine the purpose of the Endangered Species Act (see *Tennessee Valley Authority v. Hill*).[1] The Court also stated that in a complaint an indictment or conspiracy is flawed when it fails to allege the elements of the substantive crime ''ignores literally decades of black-letter law to the contrary . . . It is well settled that an indictment for conspiracy to commit an offense—in which the conspiracy is the gist of the crime—it is not necessary to allege with technical precision all the elements essential to the commission of the offense . . .'' The fundamental purpose of an indictment, the Court said, is to inform the defendant of the charges against him so he may prepare an adequate defense. Ivey and Wallace asserted the trial judge's jury instructions were defective in two ways. First, they failed to specify the elements of the crime and, second, failed to instruct as to the elements of specific intent. They also challenged the sufficiency of the evidence to support their conviction.

Ivey, individually, alleged his conviction under count five should have been reversed because of the entrapment or, in the alternative, outrageous government conduct. Ivey had requested a jury instruction on the defense of entrapment, but the trial judge did not grant this request. The trial court acted on the ground that the defense of entrapment was unavailable to a defendant as long as the defendant denied committing the act which constituted the crime. The court obviously was depending on *United States v. Henry,*[2] but the law had been changed at the time the defendants were tried in 1990. The Supreme Court held in *Mathews v. United States*[3] that the entrapment defense was available to a defendant even though he or she denied committing the act upon which the criminal charge was based. The court of appeals concluded that the reason the trial judge used *Henry* in refusing to give an instruction on entrapment was incorrect, but there remained the question of whether or not there was error in failing to give the instruction. In all entrapment cases the threshold question is the defendant's predisposition to commit the crime. Before the defense of entrapment can be raised there must be a showing of government inducement and a lack of predisposition to engage in criminal conduct on the part of the defendant (*United States v. Cantu*).[4] The Court said further that the evidence in the case amply supported the conclusion Ivey was predisposed to purchase smuggled bobcat hides. Thus, although the trial judge gave an incorrect reason for not instructing the jury on entrapment, no instruction was necessary.

1. 437 U.S. 153, 184, 98 S. Ct. 2279, 2297, 57 L. Ed. 2d 117 (1978).
2. 749 F. 2d 203, 205 (5th Cir. 1984).
3. 485 U.S. 58, 108 S. Ct. 883, 99 L. Ed. 2d 54 (1988).
4. 876 F. 2d 1134, 1137 (5th Cir. 1989).

United States v. Venter, No. 91–50084, United States Court of Appeals for the Ninth Circuit, 10 January 1992.

Wallace Charles Venter, a South African safari leader, brought nine Hartmann Mountain Zebra skins from Johannesburg, and shipped them to the United States. Hartmann Mountain Zebras are classified as a threatened species under the Endangered Species Act and are, therefore, prohibited for commercial use. Before the Fish and Wildlife Service released the skins to Venter he was asked to sign a notarized affidavit that all the skins were for personal rather than commercial use. Venter did so. Notwithstanding his disclaimer, he advertised the skins for sale in a local paper. This prompted the Fish and Wildlife Service to set up an undercover meeting with him. During this meeting Venter agreed to sell the nine skins for $1,500 each. When Venter placed the $1,500 check for the first skin in his shirt pocket, the Fish and Widlife Service agent arrested him. A grand jury indicted Venter on three counts of criminal behavior: (1) knowingly importing wildlife which he knew were possessed for sale in violation of the Endangered Species Act, (2) knowingly selling wildlife which he knew was in violation of the Endangered Species Act, and (3) knowingly and willfully making a false statement and representation as to the material facts. The jury acquitted Venter on counts one and two, but returned a guilty verdict on count three. Venter's court claim was that no rational juror could have acquitted him on counts one and two and convicted him on count three. The court said this claim is entirely without merit since the Supreme Court has clearly held that acquittal on one count cannot be used to impugn conviction on another account (*Dunn v. United States, United States v. Powell*).[1]

The court said that the defendant is only entitled to have his theory of defense in the jury instructions if it is supported by law and has some basis in evidence. Finally, Venter contested the sentence he had received and the court said that since Venter was no longer in detention, the sentencing issue was moot. ''A case is moot if the court can no longer grant effective relief.'' (*United States v. State of Oregon*).[2]

1. 284 U.S. 390, 393 (1932); 469 U.S. 57, 63–65 (1965).
2. 718 F. 2d 299, 302 (9th Cir. 1983).

United States of America v. Drake, 655 F. 2d 1025 (1981).

This is a case where two undercover wildlife officers put a down payment on two flamingos and then went to town to get a search warrant. The officers returned the next day to the defendant's residence with the warrant and the money. They helped capture the birds, paid for them, put them in their truck and then arrested the defendant. The officers then talked to a wildlife agent who released information to the press. Drake claimed that the wildlife officers did not have a search warrant the first day, and they prejudiced his right to a fair trial by releasing information to the press. The defendant also claimed that the arrest warrant should have been used immediately. The District Court of Colorado ordered suppression of the evidence of the two seized flamingos from the defendant and ordered their return.

The Tenth Circuit Court of Appeals reversed the ruling of the district court. The court of appeals said that although undercover agents of the Fish and Wildlife Service violated district court rules and policies of the Justice Department when they issued a press release and pretrial publicity following the arrest of the defendant, that the defendant's motion to dismiss indictment should not have been granted since there was no actual prejudice to a fair trial for the defendant. The issue of the appeal by the United States was whether or not the trial judge erred in (1) dismissing the indictment as a sanction for violation of pretrial publicity guidelines, (2) suppressing use of the flamingos as evidence on the ground that wildlife officers did not serve an arrest warrant they held until after completing the purchase and taking possession of the birds, and (3) ordering the return of the flamingos, asserted to be contraband, to the defendant. Although Drake relied on the due process clause of the Fifth Amendment and other grounds in support of his motion to dismiss, the trial court relied on its supervisory powers to correct prosecutorial misconduct, dismissed the action "for failure to follow the rules of this court and for failure to follow the policies of the Justice Department with regard to issuing news to the press." The district court said there is no requirement that an arrest warrant be executed immediately after its issuance; rather, the general rule is that, while execution should not be unreasonably delayed, law enforcement officers have a reasonable time in which to execute a warrant and they need not arrest a suspect at the first opportunity. They relied on *United States v. Joines,*[1] where delay under certain circumstances has been held fatal, and government agents purposely delayed execution in order to gain a tactical advantage they otherwise would not have had. The legal question, according to the circuit court, is whether such a delay is, in and of itself, impermissible, and the court held that it was not, when the evidence thus obtained could have been lawfully secured without the arrest warrant.

1. 258 F. 2d 471, 472 (3d Cir.), cert. denied, 358, U.S. 880, 79 S. Ct. 118, 3 L. Ed. 2d 109 (1958).

Recommended Reading

Adler, J., and M. Hager. 1992. How much is a species worth? *Natl. Wildlife* 30(3).

Begley, S., and M. Hager. 1981. The "snake scam" sting. *Newsweek* 98(4).

McKiernan, B. 1984. Montana sting. *Montana Outdoors* 15(3).

Milius, S., and D. Johnson. 1992. Where would they be without the law? *Natl. Wildlife* 30(3).

Post, C. 1982. Anatomy of a "sting." *South Dakota Cons. Digest* 49:1.

Reisner, M. 1991. *Game wars.* Viking Press.

Speart, J. 1992. A poacher's worst nightmare. *Natl. Wildlife* 30(3).

Starr, D. 1983. Paw scam. *Omni* 6:3.

Wilkinson, T. 1988. Yellowstone's poaching war. *Defenders* 63(3).

Williams, T. 1991. Open season on endangered species. *Audubon* 93(1).

Williams, T. 1992. Insight: Canned hunts. *Audubon,* Jan.–Feb.

Multiple-Choice Questions

Mark only one answer.
1. Which group of violators, in a short time, do the most damage to the resource?
   a. thrill killers
   b. habitual poachers
   c. guides who violate the law
   d. rich people who want a trophy regardless of cost
   e. spur-of-the-moment violators

2. Qualifications for a covert investigation agent are:
   a. wildlife law enforcement experience
   b. understanding the fine points of law and the scheduled operation
   c. ability to establish rapport with hunters and anglers
   d. being deadly with a pistol
   e. having no prior press exposure

3. Potential covert investigation targets are generally not:
   a. people doing serious damage to the resource
   b. people that cannot be stopped by conventional methods
   c. invariably hardened criminals
   d. hard to catch groups that are well organized
   e. wholesalers that mix legitimate and illegitimate business

4. Examples of targets for covert investigation are not:
   a. a clan known to be chronic poachers
   b. illegal dealers in bear gallbladders
   c. a wealthy hunter wanting to fill his bighorn sheep quota regardless of cost or the law
   d. a guide whose hunters invariably get their trophy no matter how inept they are
   e. a farmer with 10 children who is suspected of feeding them on mostly venison

5. Qualification for a covert investigation agent is not:
   a. self-confidence
   b. even tempered and no loss of control at any time
   c. can act wisely on the spur-of-the-moment
   d. cannot bring himself to lie
   e. can often anticipate problems and questions

6. The first and most important question asked in setting up a covert investigation is:
   a. are there alternatives or is this the best and only way to go
   b. who are the suspects
   c. are the suspects profiled
   d. is the operation too dangerous
   e. are all plans in place

7. Qualifications for a covert investigation agent are:
   a. a personality for this devious type of investigation
   b. personally known to the suspects
   c. being able to be away from family and friends for extended periods
   d. healthy and physically fit
   e. be a reasonably good actor

8. Peak levels of danger that a covert investigation agent may face include:
   a. a blown cover by a newspaper article or some other means
   b. the moment of initial contact
   c. when the suspect is arrested—assuming the agent is still present
   d. becoming complacent and letting down one's guard
   e. by and large there is very little danger in these investigations

9. Questions the leader of a raiding party asks oneself before commencement are:
   a. is the party totally oriented
   b. is equipment adequate for the task
   c. is everyone's life insurance active
   d. is the raid really necessary
   e. what hazards are to be faced and how well will they be executed

10. People who may provide information for covert investigations do not include:
    a. the violators themselves
    b. taxidermists
    c. wholesalers in wildlife supply
    d. hunters
    e. locker plant operators

Chapter 13

Forensics

Case 1

Forensics—pertaining to, connected with, or used in courts of law (Webster).[1] In wildlife law enforcement, forensic means applying natural resource knowledge to issues involved in civil and criminal law.

Science, technology, and law are rooted in mankind's never-ending thirst for knowledge. Despite the concurrent origins of science, technology, and law, their practitioners have, until recently, been virtually ignorant of each other's accomplishments. The scientists' disdain of politics, and the politicians' aversion to the lab only widen the gap of communication and understanding. These respective illiteracies can be a serious handicap. The role of science today affects government and people more than ever in social, economic, and cultural events. Fortunately, in the last fifteen years the legislative branches have added trained scientists to their staffs (Gibbons 1990). The problem may persist, but not to the degree it has in the past.

To increase effectiveness, every wildlife law enforcement officer should study a forensics manual before going afield. Field evidence initially will be used to establish probable cause. However, the evidence may be presented in court along with expert opinions of laboratory personnel who have examined the material. Therefore, strict care and detail must be followed in gathering field evidence. As a general rule one cannot collect too much evidence. Records should be kept of how many times each piece of evidence has changed hands, the results, and, whenever possible, accuracy (Adrian 1992b). The chain of evidence should be kept as short as possible and hand delivered. This is particularly important because opposing attorneys will exploit any weakness shown in the chain. Every individual who has had custody of evidence may have to appear in court to testify that the evidence recorded in the field is the same as that examined in the laboratory. Anyone who has had sole possession of the evidence may be required to so testify (Adrian and Moore 1992a).

Investigative techniques for wildlife crimes are similar in many respects to standard crime scene investigations. Smith (1992) lists several recommendations. The first officer on the scene should:

(1) take precautions to protect and mark the immediate crime area;

(2) do an initial walk-through determination of the scene and gather evidence to help plan the investigation;

(3) locate and preserve perishable evidence;

(4) assign one person to lead the investigation and another to gather and preserve evidence;

(5) see that vehicles are searched systematically and carefully; and

(6) document evidence by notes, photographs, and sketches or the use of a camcorder.

A 35-mm camera with color film is used in most cases. Use natural light outside and off-camera flash inside. Take several shots of each situation. Use a maximum depth of

State and local crime laboratory examiners receive hands-on training in serology at the Forensics Science Research and Training Center.
Courtesy Federal Bureau of Investigation.

field whenever possible. In many instances a camcorder may be used in addition to still pictures.

Social security numbers are available to law enforcement officers for use in identifying possible out-of-state hunters using resident licenses (Hoilien 1992b).

THE USE OF HEMOGLOBIN FOR IDENTIFICATION OF UNGULATES

Hemoglobin of elk, moose, pronghorn antelope, mule deer, and desert bighorn sheep is identifiable by analytical polyacrylamide gel isoelectric focusing. Moose and elk gel patterns are similar but not identical. With the exception of wild and domestic sheep, banding positions and densities are species specific. This is a reliable, consistent, rapid, and inexpensive technique that can be used to identify animals from fresh, frozen, and air-dried blood samples (Bunch et al. 1976). Hemoglobin patterns of frozen samples differ from fresh samples in that they have two additional bands which also are species specific. Banding positions and densities are characteristic for each wild animal. With the exception of domestic and wild sheep, genotype determinations based on hemoglobin patterns are definitive. Fresh hemolysates of wild and domestic sheep are different. Greater differences cannot be resolved by freezing or drying the hemolysates. Hemoglobin patterns of dried and frozen hemolysates of the same origin are identical. This technique is capable of resolving hemoglobin differences with a high degree of accuracy (Bunch et al. 1976).

AUTOMOBILE TIRE AND SHOE IMPRESSIONS

The term alibi means literally, ''I was someplace else.'' If someone can prove this, then they could not have been at the scene of the crime. And if it can be proved that a person was at the scene of a crime, they may have committed it. Additional evidence is needed, however, to prove they did. Since arrival at a particular point is usually a function of driving (generally an automobile) and walking, tire and shoe impressions enter the picture. No two tire impressions or shoe impressions are exactly alike, assuming there has been enough wear to give them individual characteristics. No two tires or shoes have identical experience and are, therefore, unique. New tires and new shoes may or may not have individual characteristics (Bodziak 1984).

People who break the law frequently wear gloves to avoid leaving fingerprints, and they may don masks to avoid eyewitness identification. However, these same people are rarely aware of, or make an attempt to conceal tire and shoe impressions. Shoe, and even tire, impressions on a hard surface may be difficult to detect. It is important that the scene of the violation be roped off and curious onlookers be kept outside the area.

Hard surfaces may have valuable dust or residue impressions which are not easily seen under normal lighting conditions. To locate these impressions, turn off all lights, then direct a strong beam of light from a low angle across the surface being searched. The strong beam of a flashlight lying on the floor is adequate. A very light coating of dust provides a good medium for holding shoe impressions. Dust impressions should not be dusted with latent fingerprint powder. However, as a last resort it is better than nothing.

Several, rather than just one, impressions should be taken for both tires and shoes. This involves two steps. The impressions should first be photographed, and then a casting made.

The photographs will be enlarged, so a camera with the largest negative format available should be selected. Use a fine-grained, slow-speed, black-and-white film. The camera should be mounted on a tripod and positioned over the center of the impression with the lens parallel to the surface. The flash held away from the camera provides an oblique light source. This flash should be held close to the surface and at least three feet away from the impressions. A ruler alongside the impression provides a proper scale when the photograph is enlarged.

Prior to casting an impression, it may be necessary to remove leaves and other debris which have fallen into it. However, no attempt should be made to remove debris if that is part of the impression or if there is any possibility of destroying some of it. A form should be put around the impression to give the cast extra strength. Class I dental stone and plaster of Paris are two forms of gypsum that can be used in impressions. Class I dental stone is stronger, easier to use, more durable, and superior to plaster of Paris. It is available from most local dental supply houses. It is important to use class I since other types of dental casting contain material that causes the gypsum to shrink. Approximately three pounds of dental stone are needed for a cast of each shoe impression. The mixture should be of the consistency of thin pancake batter. If the mixture is too thick, do not add water, discard it. The cast may be removed from the area in approximately thirty minutes. No attempt should be made to clean it. It should then be allowed to air dry for twenty-four to forty-eight hours. It should never be placed in an airtight container (Bodziak 1984). If plaster of Paris is used, approximately five pounds will be needed for each footwear impression.

Impressions in the snow can be cast using the same technique. However, a small amount of snow or ice should be mixed in with the stone or plaster to keep the temperature low. The mixture should be slightly more viscous than that used in warmer weather. A product known as snowprint wax helps in casting snow impressions.

If the photo or casting cannot be submitted to a laboratory because of size or some other reason, the impression should be lifted with a commercially available footprint lift.

In wildlife cases, the size more than individual characteristics are of value (Bodziak 1984). Individual characteristics are for species identification. Animal tracks may be photographed and a cast made.

FINGERPRINTS

All latent fingerprints submitted to a laboratory should, if at all possible, be accompanied by the fingerprint record of the suspects. United States courts accept twelve or more points of similarity between fingerprints as a match. In the case of unusual ridge characteristics, as few as six or seven points are considered sufficient. Single fingerprints of a limited number of hardened criminals and of criminals who specialize in certain crimes are indexed in the Identification Division of the Federal Bureau of Investigation.

It is essential that the officer have a thorough knowledge of the basic principles of fingerprinting if prints are to be properly taken and prepared for transmittal to a law enforcement agency, such as the Federal Bureau of Investigation. Fingerprint identification is based on the fact that no two fingerprints are identical and that they may not be obliterated permanently by the destruction of surface tissue. Some places to search for prints in the crime vicinity include: entrances and exits; doorknobs or frameworks; where there is dust; and any shiny, smooth surface, such as window glass.

FIREARMS AND AMMUNITION

Firearms identification is the science of identifying projectiles, cartridges, and weapons. Few field officers are capable of making these identifications; and if they were, they probably could not qualify as an expert in court. It is desirable to send these items to the state crime laboratory, or to the Federal Bureau of Investigation.

The cartridge case, as well as the projectile, has marks peculiar to the gun from which it was fired. Projectiles recovered from game animal carcasses may be used to compare with projectiles fired from guns of suspects.

The suspect's gun should be handled carefully so that any fingerprints will not be obliterated.

Firearms, projectiles, empty cartridge cases, and live ammunition should be transmitted after wrapping them individually and securely in a box to prevent scratching or shifting. Up to five rounds of ammunition should be included, if available. As much information as is available concerning the make of weapon, caliber, and serial number should be included. Weapons should be tagged, and projectiles, ammunition, and cartridge cases should be individually marked.

Live ammunition cannot be sent through the mail. If at all possible, weapons should be transmitted unloaded. If it is impossible to unload them or it is highly desirable that they be left loaded, then a letter of transmittal should give this information, and the guns should be properly labeled. The shipping box should be appropriately marked on the outside.

It is desirable to have a sketch, or preferably a photograph, of the area, showing the location of the projectile, cartridge, and weapon.

CARTRIDGE CASES AS CRIME EVIDENCE

Often, a fired cartridge case is the principal clue in a crime involving use of firearms. "The bullet may have whistled off into the night. The gun and criminal may have disappeared. An ejected case or cases left behind can be the starting point of the solution to the crime" (Berg 1969). Cartridge case identification has played a distinct role in firearms identification in the United States for well over half a century.

In 1907, U.S. infantrymen were involved in a riot at Brownsville, Texas. Shots were fired. The question naturally arose, Who fired them? Cases were recovered at the scene,

Case 2

along with bullets, and a number of suspected rifles. The accumulated evidence was sent to the ordnance staff at the Frankfort Arsenal, Philadelphia, for examination.

Cartridge case identification also figured into the famous Sacco-Vanzetti case in 1920 (*Commonwealth v. Sacco*).[2] One of the recovered .32-caliber cartridges was identified as having been fired from Sacco's .32-caliber pocket model automatic pistol.

In 1925, the adaption of the comparison microscope for both cartridge cases and bullet study put firearms identification on a firm footing from which it has continued to develop and be refined.

On examination of the markings on the cartridge case head and primer made by the firing pin, extractor, ejector, magazine, and breech face, and a comparison of these to known standards, it is sometimes possible to determine the make and model of gun from which the round was fired. Some markings are common to many guns, and no useful determination is possible; but when two fired cartridge cases are available, it is usually possible to determine if they were fired from the same gun. The identification of cartridge cases is partly based on analysis of the class and individual characteristics of the fired cases. The class characteristics are those peculiar to all guns of a certain make or model, or to all guns in a certain group or class. For example, semi-automatic guns leave one type of marking, revolvers leave another type.

Other class characteristics include the location of the extractor and ejector markings. The extractor usually marks the rim of the cartridge as it withdraws the cartridge from the chamber, and as the extractor snaps over and engages the rim.

Individual characteristics are those peculiar to only one gun, and are imparted to a cartridge case by the internal pressure in firing. Firing forces the hard case against the breech face and the sides of the barrel chamber. Outline of the firing pin mark, as well as other breech face openings, can frequently be seen on the cartridge.

Bullets recovered during a necropsy should be washed in cold water to remove blood, hair, and other debris. If blood remains on the projectile it can cause damage to the fine striations, sometimes to the point that the match is not possible (Adrian and Moore 1992).

In centerfire cartridges there is always the possibility that the cartridge has been reloaded once or several times. It may have even been used in a different make or model of gun (Berg 1969).

Yet, keeping in mind that there are usually exceptions to all rules, cartridge case identification has earned legal and scientific status.

GLASS FRAGMENTS

All suspicious pieces of glass at the scene of a violation should be retained. Laboratory examination may indicate the type of glass, the manufacturer, where the glass came from, and, in the event it was broken by a blow from a stick or bullet, the direction and angle of impact. If there is a series of holes in a sheet of glass, the sequence in which they were made may also be established. Automobiles or trucks traveling narrow, brushy lands may have headlights, taillights, or running lights broken, and these fragments of glass help to identify the vehicle.

SOIL SAMPLES

Soil from the suspect's shoes or from the tires of a vehicle may be compared with soil samples removed from the scene of the violation. Soil examination may indicate geological origin, area of origin, and similarity between the soil from the suspect's person or vehicle and that from the violation area. Soil containers normally should not be less than a pint in size, and preferably larger. The container should be sealed with tape, labeled on the outside, and transmitted by registered mail, along with complete information as to the area sampled, the depth at which the soil was taken, whether or not it was the A, B, or C soil horizon, and the amount of litter on the ground.

TIME OF DEATH

The ability to estimate the time of death (TOD) in fish or game species is a valuable skill to wildlife law enforcement personnel. A hunter may have killed a deer early on opening day—or the night before. Where bag limits are double the daily limit, hunters may claim, after the first day, to have killed part of the bag on the previous day when, in fact, they have killed them all on the current day. The two most reliable methods of determining TOD is temperature and electrical stimulus. Chemical and physical changes in the eye and rigor mortis sequence also provide valuable information. Since TOD techniques can rarely be performed in the laboratory, field officers must learn to perform them. The officers may have to show that they have used these techniques before, and they understand them (Oates 1992).

Temperature

After death, body temperatures approach ambient temperatures through loss of heat by radiation, convection, and conduction. Cooling rates depend primarily upon:

(1) initial body temperature,

(2) air temperature,

(3) body surface area and weight, and

(4) how the carcass was handled after death.

Initial body temperatures are generally the same for a given species. However, deviations from the norm can occur in diseased or stressed animals. Body surface area remains constant unless an animal is field dressed. The weather is among the more variable parameters used in determining time of death. Air temperature can vary substantially, and the windchill factor and humidity are also important factors (Oates 1992).

Electrical Stimulus

Response to electrical stimulus is chemically related to the availability of adenosine triphosphate (ATP) in muscles. After death ATP breaks down, and responses to electrical stimulus gradually decrease until nonexistent. Physiological variables which affect rigor can also affect response to electrical stimulus. Rested, healthy, well-nourished animals that die quietly will respond to electrical stimulus longer than an animal which has undergone a severe death struggle or is fatigued or stressed prior to death. Inability to respond to electrical stimulus may be hastened by quick death under extreme exertion, tension, or brain damage. The value of electrical stimulus tests is that it allows an estimate of death for the first few hours (four or less) after death, something that is not always possible when using a thermometer. The responses go from very good, that is, gross muscular contractions, to no contractions.

Rigor Mortis

Rigor mortis is a post-mortem state of rigidity which develops in muscle tissues when their supply of ATP and phosphocreatine have been depleted. This depletion comes about by a chemical reaction and is dependent upon both physiological parameters and ambient temperature. The presence of rigor is normally determined by manipulating selected joints. A fairly reliable sequence has been derived for humans and deer. The sequence incidentally has been observed to be reversed twenty-four to forty-eight hours after death in a deer and thirty to eighty hours in humans. This loss of rigor is generally credited to putrefaction. Laboratory devices are available to test rigor but are impractical for field use. Therefore, field people should flex the joints and rely on visual estimates such as rigor, partial rigor, or no rigor. Such things as handling, freezing temperatures, and gunshot wounds can cause aberrations in an estimated TOD (Oates 1992).

Physical Changes in the Eye

Among the changes that occur following the death of an animal are the loss of eye clarity, luminosity and shape, color changes in the pupil, and constriction of the pupil. At death the cornea and eye fluids are very transparent and the surface of the eye is tight and smooth. The transparency and luminosity decrease as the intraocular fluids become cloudy. The shape of the eye changes as the internal fluid pressure diminishes, putrefaction occurs, and the volume of the liquid is reduced (Oates 1992).

Chemical Changes in the Vitreous Humor of the Eye

Vitreous humor is a somewhat viscous liquid (especially in deer) found in the posterior chamber of the eye. The anterior chamber contains the aqueous humor (a less volatile liquid). Following death of animals, potassium levels increase in the vitreous humor due to the release of potassium as the retinal cells break down (Oates 1992).

Deer

The normal body temperature of a deer is approximately 102° F. However, temperatures as low as 100° F have been observed, and stressed or diseased animals may have temperatures higher than 102° F, perhaps as much as 107° F. To use existing temperature data one should know the ambient temperature for the first twelve hours after death, and the weight of the deer.

Electrical stimuli locations used for deer are eye, muzzle, ear, front leg above the elbow, back below the vertebrae, tail, inner exposed flank, and the tongue.

Rigor is detected by gently attempting to flex selected joints on their normal plane of movement. When flexing for rigor, both sides of the deer should be used and the most advanced stage of rigor selected for the best estimates. Rigor can proceed much faster than expected when influenced by high ambient temperatures, by quick death under extreme stress or exertion, and by brain damage. Freezing can be mistaken for rigor. It is suggested that one check the dew claws, ears, and tail for evidence of freezing since they are not appreciably affected by rigor. It should be noted that once rigor is broken it normally does not reappear. Rigor can cause pupil constriction of the eye, which can be measured with a pair of dividers (Oates 1992).

Mallard

The eyes of a dead mallard change in appearance in a definite pattern. The characteristics most useful are color and shape. At death, and for about two hours thereafter, mallard eyes maintain the normal appearance of a living bird. As moisture evaporates from the eye, turgor is lost, small wrinkles appear, and indentation begins to form in the eye. Changes in the pupil color are about five to eight hours after death. The pupil becomes turquoise-cobalt with a reflective or glossy appearance from inside the eye. No further change in pupil color occurs. However, with further moisture loss the corneal surface over the pupil generally becomes sunken and flat. Between seven and twelve hours after death, the iris becomes blackish-brown to aquamarine-violet at about the same time and has a clear watery look due to the fluid. After an indefinite period, the cornea over the iris sinks to the level of the pupil, and the entire eye appears flat. This is most characteristic fourteen to twenty-four hours after death. The iris loses its watery appearance at this time and becomes a dull or opaque ultramarine-violet. It may also appear medium gray or blackish-brown (Morrow and Glover 1967).

Pheasant	The normal body temperature of pheasants is 104° F. The thigh and breast muscles are considered the most reliable for temperature readings. Temperature loss of the thigh may be as much as 15° to 20° F in the first hour and most temperature loss is generally in the first two hours. The maximum reaction time to electrical stimulus for pheasants is about two hours. In many cases it may last no longer than one and one-half hours (Oates 1992).
Cottontail Rabbit	As in other animals, the most important factor in estimating TOD is the surrounding temperature, body weight, body covering, carcass handling, and general weather conditions. Rabbit carcasses stacked together lose temperature less rapidly than those lying out separately. Muscles respond to electrical stimulus for about three hours after death (Oates 1992).
Differentiation of Mule Deer and White-tailed Deer	Where mule deer and white-tail range overlap it is sometimes of value to be able to determine the species with rather limited information. One outstanding feature of mule deer is their very large ears. The white-tail's most distinctive feature is the large tail. Antlers also differentiate the two species. In the white-tail the points on each antler arise from a single main beam, much as the points on a garden rake arise from the iron crosspiece. The antlers of the mule deer form pretty much of a letter Y, and the upper ends fork to form two smaller Ys. The ends of these also fork. The white-tail's winter coat has a buff cast while the mule deer is a plain gray. Both species have white bellies, however, the mule deer brisket is a rich brown and it has a brow with a distinctive dark gray patch. The white-tail has a loping gait but the mule deer bounces along stiff-leggedly, striking all four feet at once. Most species have two sets of specialized skin glands. The preorbital tear glands which lie just in front of the eyes lubricate and clean the eyes. This gland in a mule deer is much larger than it is in a white-tail, measuring more than an inch. The metatarsal glands occupy an elongated area on the outside of the shanks. This area, which shows as a part in the hair, is four to five inches long in mule deer, but rarely more than an inch in white-tails (Oates and Walker 1992).

WATERFOWL IDENTIFICATION BY USE OF BREAST BONE

There are four types of ducks commonly observed in the United States. With few exceptions these are the puddle ducks, diving ducks, mergansers, and sea ducks. Visual physical differences may be used most of the time to differentiate one genera from another. However, to segregate species within a genus actual measurements are usually necessary. The puddle ducks have three genera: *Anas, Aix,* and *Dendrocygnia.* The first genus includes mallards, black ducks, pintails, and others. The second genus includes the wood duck, and the third includes black-bellied and fulvious tree ducks. The manubrial spine on the breast bone is one of the key characters used in identifying most genera of puddle ducks. *Anas* is represented by a long spine. *Aix* has either a very short or no spine, and *Dendrocygnia* no spine (Adrian and Oates 1992).

There are three genera of diving ducks that will be discussed. The first genus, *Aythya,* contains the canvasback, red head, lesser and greater scaup, and the ring-necked duck. The second genus, *Bucephala,* contains buffle head and golden eye, and the third genera, *Oxyura,* contains the ruddy duck. The genus *Aythya* has a shallow dip in the coracoidal sulcus, which is the key feature. There are two short manubrial spines which can generally be felt. This may be difficult, and occasionally one side is missing. The sternal notch is fairly long and normally not fused on the end. In the genus *Bucephala,* the manubrial spine is normally absent. There is usually a shallow dip in the coracoidal sulcus, which may not be smooth. The sternal notch is usually fused on the end. The genus which has only the ruddy duck, *Oxyura,* has a sternum that is very stocky. The broad spine forks at the end, and the pneumatic fossa is fused. The lateral process is very short and thick.

There are two genera in the mergansers. *Mergus* includes both the common and the red-breasted merganser, and *Lophodytes* is the hooded merganser. The merganser (*Mergus*) is characterized by no manubrial spine and a very definite dip in the dorsal lip of the coracoidal sulcus. As a general rule the merganser displays the largest pneumatic fossa of all ducks. The genus *Lophodytes* does not have a deep dip in the coracoidal sulcus, but the pneumatic fossa is normally large. The coastal ducks classified as sea ducks have five genera. This includes *Somateria,* which has the common and king eider (Adrian and Oates 1992).

Sex Identification of Field-dressed Pheasants

Since most states have laws that allow the killing of cocks only, and many hunters field dress their birds, it is desirable to be able to sex these dressed birds. Sex identification is not difficult if the ovaries or gonads remain attached. However, they often are removed in dressed birds. Adult roosters are much larger than hens, and they are much more colorful. Based on carcass size alone hens can easily be mistaken for young roosters. By including the cartilage in sternal measurements, young roosters at fourteen weeks may already have larger sterna than adult hens. To facilitate field use of data for legal records, a simple measuring device was constructed by the Nebraska Game and Parks Commission. This device includes instructions for use and specifically marked areas indicating range of overlap between sexes. These measurements include distance between the two end points on the crest of the sternum and distance from the high part of the crest perpendicular to the ridge that projects out from the keel. It also includes the distance between the tip of the sternum and the base of the notch formed by the ziphoid process (Oates et al. 1985).

Matching Carcasses

Whenever there is doubt that a given head belongs to a carcass, the easiest method for matching the two of them is to radiograph both the head/neck and the neck/shoulder pieces of the carcass. It is difficult to separate heads from carcasses in exactly the same location unless a knife has been used to separate the head and the first cervical vertebrae. In field situations one should look for saw marks versus cut marks versus axe marks on both the head/neck side and on the neck/body side. Special note should be made of cut angles and missing or extra pieces of bone or flesh (Adrian 1992a).

Matching hides to carcasses is quite subjective. It involves checking the hide size against carcass size, cut marks on the hide against cut marks on the carcass, and matching the amount of skin left on the carcass as opposed to the head. Hides shrink considerably as they dry. It is also a good idea to check for wounds on both the hide and the carcass to see if they match. On checking capes of skinned heads the best area to look at is the area on the skull where there is cartilage, since cartilage is easily cut during the skinning process. Matching of viscera to carcass looks almost impossible but can be done in many cases. This involves laying the gut pile out and checking all cut marks and matching them to those on the carcass. If there are legs with the gut pile, match the size and cut marks on the legs to those on the carcass. Pay particular attention to the cut edges on the diaphragm, and match them in both size and angle of cut. Finally, matching a large quantity of meat in order to determine the questions of how many animals are involved may look impossible but often is not. Identical parts should be laid separately by size and sex, such as front versus hind quarters and left versus right quarters (Adrian 1992a).

Fatal Wounds

Wildlife officers are often confronted with determining how an animal died. This most often occurs during hunting season but may happen any time of the year. The weapons most often used are archery, muzzleloading rifles, high-powered rifles, and shotguns.

Contact firearms wounds are created when the muzzle of the weapon is held close enough to the skin so that the soot from the burned powder is deposited on the skin. The soot may or may not have been wiped off. Intermediate range gunshot wounds are characterized by ''powder tattooing.'' The powder tatoo is caused by impact of unburned powder granules on the skin. Reddish-brown to orange-red points or dots are on the skin if the wound was created before death. In wounds after death these marks are gray or yellowish in color. With handguns, powder tattooing begins with a muzzle-to-skin distance of about ten millimeters.

Spire-point bullets create small entrance-inverted wounds and, generally, larger exit wounds. Round or flat nose projectiles generally produce larger entrance and exit wounds than the spitz-type bullets. Shotgun wounds from multiple projectiles can be one large, gaping hole, or a series of small holes if the range was more than a few yards. Shotgun slug wounds resemble bullet holes but are usually much larger. Arrows typically make puncture wounds with hemorrhage around the edge. Arrows with broadhead points create puncture wounds with lateral cuts. If arrow wounds are found that do not have hemorrhage along the edge, the wound is probably one that occurred after the animal was dead. Some people attempt to conceal that an animal was shot with firearms during the archery season by pushing an arrow through the bullet wound. Wounds of this type show hemorrhage around the center but not on the lateral cuts. Generally radiographs show bullet fragments and broken bones in these animals (Spraker and Davies 1992).

Metal Detectors in Wildlife Forensics

Metal detectors are used in recovering bullets from snow-covered fields, finding cartridges, locating projectiles in animals' carcasses, and in distinguishing steel shot from lead shot. Determining whether pellets are lead or steel may be done when shot is in the bird carcass, or, in the case of hand reloads, when shot is in the cartridge.

Spent cartridge cases at the crime scene may be helpful, but spent projectiles are solid evidence. To determine where to search for a projectile in the field, it is necessary to have a general idea of the point from which the gun was fired, and the approximate angle and distance of the projectile before it struck the ground. Deflected bullets are more difficult to find, but should be looked for nevertheless (McGill and Will 1992).

Blood and Tissue Identification

Early reports of electrophoresis were based on the observation of suspended clay particles migrating through an electric field toward the positive electrode. Today, zone electrophoresis is used to identify both birds and mammals from either blood or tissue. One of the earliest breakthroughs in wildlife law enforcement forensics occurred in California when a person was convicted with evidence based wholly on the precipitin test and its ability to differentiate venison from beef, mutton, and bear meat.

The single most important factor affecting protein patterns is antiserum, which is difficult to standardize since it is a biological product. Antiserums can differ not only among different species, but also within one species. Littermate rabbits given identical injections may give antisera with quite different characteristics.

Freeze-drying is probably the best nonchemical means of sample preservation, although freezing or refrigeration are more commonly used. Blood samples may come from a variety of sources such as knives, burlap bags, shirts, trunk linings, soils, spare tires, and newspapers. If sample size permits, triplicates are usually tested when the material is to be used in wildlife law enforcement cases. A latex suspension sensitized

Blood analysis is one of the many chores of a forensics laboratory. *Courtesy Federal Bureau of Investigation.*

with deer antiserum has been prepared, placed on plastic cards, dried, and packaged for field use by the Nebraska Game and Parks Commission (Oates et al. 1983).

Bones are also used in species identification. Bone development can be used to differentiate between juveniles and adults. And it can serve to verify tissue identification (Oates and Weigel 1976).

Mammal Identification with Dorsal Guard Hairs

To aid in ecological and food habit study, as well as law enforcement investigations, by providing key descriptions and keys to dorsal guard hairs, Wyoming has identified the hair of ninety-one species of mammals (Moore et al. 1974).

DNA (DEOXYRIBONUCLEIC ACID)

The ability to identify a species of animal or an individual animal—currently one of the most difficult problems—is often the difference between winning and losing a case. Individual identification of human victims is possible in crime labs, and wild animal identification will be possible in the near future. The U.S. Fish and Wildlife Service, in June 1989, established the world's first fish and wildlife forensic laboratory at Ashland, Oregon. Its director, Ken Goddard, and staff are busy collecting background material and applying known techniques to wild animals. Goddard believes that through DNA analysis, the lab director and his crew will eventually not only be able to sex an animal, but to identify its species and, ultimately, each individual's genetic blueprint (Knickerbocker 1990). The results of DNA examination can effectively rebut alibied defenses, corroborate the accuracy of what otherwise might be questionable eyewitness events, and correspondingly produce more guilty pleas. It can also exonerate the innocent. For the procedure to be effective in criminal or civil justice systems expert opinion and conclusions based on DNA identification must be admissible in prosecutions. Comparisons and conclusions based on DNA technique are considered by some scientists to be strong, if not overwhelming, proof. Prosecutors, investigators,

Cases 3–4

and forensic scientists should anticipate strong defense objections, for the next several years at least, to the admission of such testimony. All examined specimens must be obtained in compliance with constitutional standards and maintained in a manner that precludes contamination, sometimes difficult with wildlife samples.

A strict chain of custody for authentication and identification must be maintained. Some courts are presently viewing forensic science as a relatively novel application of scientific techniques and procedures. The law enforcement community should be prepared to satisfy specific admissibility requirements not normally associated with the introduction with other types of expert testimony. Meeting admissibility requirements falls almost entirely upon the ability of the expert witness to convince the courts through his or her testimony that these requirements have been fulfilled. It is also incumbent upon the expert witness to convince the court that the procedures were conducted in a reliable manner.

When assessing the admissibility of this "novel" scientific evidence, some courts may apply the traditional evidence test of relevancy. Under this test, scientific evidence is admissible if the testifying expert witness is duly qualified, the expert's opinion is relevant and will assist in finding the facts, and the testimony is not so prejudicial as to outweigh its probative value. For example, in *United States v. Baller,*[3] the U.S. Court of Appeals for the Fourth District applied the test of admissibility to testimony relating to a then new technique of voice prints. Most jurisdictions, however, apply the more stringent *Frye* standard when judging the admissibility of evidence from a relatively new scientific procedure. In *Frye v. United States,*[4] the courts also required the theory underlying the technique, as well as the technique itself, be generally accepted, or commonly recognized, by scientists in the relevant scientific community. In *Frye,* the U.S. Court of Appeals for the District of Columbia Circuit reviewed the admissibility of evidence based on a then relatively unknown polygraph technique and ruled as follows:

> Just when a scientific principle or discovery crosses the line between experimental and demonstrable stages is difficult to define. Somewhere in this twilight zone the evidential force of the principle must be recognized, and while the Court will go a long way in admitting expert testimony deduced from a well-recognized scientific principle or discovery, the thing from which a deduction is made must be sufficiently established to have gained general acceptance in the particular field in which it belongs.

What this says briefly is that the evidence must meet the *Frye* standard by establishing that the technique and the principles behind it are generally accepted in the relevant scientific community. This will not consider the entire spectrum of scientists, but only those whose specific background and training are sufficient to allow them to comprehend and understand the scientific principles involved, and form their judgment from it. These scientists will most often be those who have had direct experience with the procedure, or at least will be familiar with its use. There is not a great deal of guidance in court rules as to what evidence is generally sufficient in *Frye.* However, the standard does not require unanimity of agreement in application of the techniques. Rather, it requires a substantial section of the scientific community but not a universal one. The fact that there may be divergent testimony between expert witnesses does not necessarily mean that the DNA technique is not generally accepted. Scientists expect some degree of scientific divergence of view in expert witnesses. To repeat an earlier statement, prosecutors attempting to meet the rules of DNA acceptance must rely almost exclusively upon the expert witness to persuade the court that the DNA testing procedure used is acceptable to the appropriate scientific community. This may no longer be true in ten years. Prosecutors will also have to rely on these same witnesses to convince the court and jury that the procedure was effectively and properly applied. Courts which have accepted the expert testimony of scientists from the field of molecular biology and genetics have consistently agreed upon the principle underlying the DNA techniques. It is commonly recognized in the scientific community that cell nuclei contain DNA, and the structure of DNA is different in each individual except identical twins. From the expert testimony the courts have also recognized that certain DNA testing protocol are generally accepted as producing reliable and accurate results that satisfy both the *Frye* and "relevancy standards" of admission (Fiatal 1990).

Case 5

In *State v. Schwartz,*[5] the Supreme Court of Minnesota acknowledged the scientific acceptance of DNA testing but cautioned that the admissibility hinged on the laboratory's compliance with appropriate standards and controls. In summary, properly employed DNA identical procedures have been judicially acknowledged as meeting admissibility standards. The law enforcement community can, in most cases, confidently employ it in criminal investigations and prosecutions. However, the officer should be aware that a strong defense objection may be raised. The verdict may ultimately depend on the ability of the expert witness to convince the court of the scientific validity of the DNA process used in any particular conclusion or identification.

CRITICISM OF DNA TECHNIQUES

DNA is the molecular basis for heredity. Scientists say that each individual, except identical twins, has a unique sequence of DNA. Presumably this applies to twins of wild animals. Every cell of the human body contains a complete copy of that individual's DNA. That is the genetic grouperant of organic life. Most of DNA that all humans share is identical. Only about one percent of the genetic code between any two individuals has any difference. This small fraction is where science is able to differentiate between two individuals in DNA tests chemically extracted from a small number of cells; for example, a drop of blood, a semen smear, a drop of saliva, or a few hair roots. The long chains of the DNA molecule are treated with special enzymes that cut the molecule at specific points and permit the patterns to be examined. This is a five-step process and uses at least four different probes. Experts claim that the pattern of numbers is unique and the odds of two individuals having the same match are millions to one. These tests were pioneered by Cellmark (Rockville, Maryland) and Life Codes (Valhalla, New York) between 1982 and 1986. In 1986, the Federal Bureau of Investigation developed its own system (Garfinkel 1991).

The DNA technique is not without its detractors. Recently in Massachusetts the State Supreme Judicial Court ordered a new trial for a man who had been convicted with evidence using DNA. The court said that the scientific community does not have widely accepted uniform methods for testing and interpreting DNA evidence. In Arizona, a superior court judge ruled that the results of DNA testing could not be introduced into a trial because the apparent scientific accuracy of the test might convince a jury of the defendant's guilt. This even though the scientific community itself has questions about the underlying test. Again a prominent biology professor said that they, meaning a part of the scientific community, have rushed to the courts with their conclusions instead of doing the intensive, hard groundwork they need to make a system good. He also says that geneticists really don't know how many people have the genetic markers being used in the test, and there is quite a bit of variation from subgroup to subgroup in the American population. Civil Libertarians are more worried because of the large scale DNA testing. They believe it is likely that this technology will lead to a national identification system (Garfinkel 1991).

Cases 6–10

FBI scientists and others disagree with the criticism leveled at the DNA technique. They say that the population data demonstrates there is very little difference, for example, between Caucasian subgroups and very little difference between the black subgroups. Between blacks and Caucasians there are some differences, but they are small. Increasingly the scientific establishment seems to be accepting the legitimacy of the DNA tests. For example, the U.S. Congress Office of Technology Assessment states that forensic use of DNA tests are both reliable and valid when properly performed and analyzed by skilled personnel. More than a dozen police laboratories around the United States use DNA. The FBI has trained over 240 crime laboratory technicians from eighty different agencies. The FBI is also laying the groundwork for a national data bank of DNA information collected from crime scenes and convicted felons. Michael Baird, Director of Research Facilities at Life Codes, states that having testified as an expert in over fifty cases he believes that the introduction and the acceptance of this technique has been very good in today's courts. California and Virginia have started taking blood samples from convicts before they are released. Eleven states have the requirement that DNA typing be done on convicted offenders. Scientists believe that future tests may be able to analyze DNA for predisposition to disease or even abnormal behavior characteristics. In summary, the DNA technique appears to be well on its way to acceptance in spite of its detractors. The expert witness, however, should not lose sight of the detractors for even a moment when he or she is on a witness stand (Garfinkel 1991).

JUDICIAL ACCEPTANCE OF DNA PROFILING

Forensic DNA profiling has been under intense judicial scrutiny by courts for two years (1988–1990). However, an overwhelming majority of the courts have admitted forensic DNA evidence after reviewing it under varying standards. The courts have recognized in numerous decisions that a genetic profile developed by an individual is reliable, probative, and objective.[6] However, despite many favorable decisions, DNA evidence may be challenged and must continue to undergo a pretrial review until a court of appeals in the jurisdiction in which the evidence is offered addresses the question of whether or not DNA evidence is acceptable. Traditionally, two standards have been used to admit novel scientific evidence in U.S. courts. The courts have adopted either the *Frye* standard or the "relevancy standard" when deciding novel scientific evidence such as DNA profiling.[7] According to the courts, the role under *Frye* is properly limited to an assessment of the extent in which the scientific community has embraced the technique as a whole. The analysis performed in any particular case is not generally at issue in a *Frye* hearing. Many courts have turned to the "relevancy standard" as a basis for determining whether or not a court will accept evidence from new scientific techniques. The "relevancy standard" is based on the Federal Rules of Evidence and directs the court to consider the relevancy, the potential for unfair prejudice, and the reliability of the testimony.[8] For example, evidence may be rejected under the "relevancy standard" if the jury is asked to accept the expert's opinion on faith alone.[9] Recently a New York trial court expanded the traditional approaches during its review of DNA evidence. After determining that forensic DNA profiles met *Frye* standards, the court established a precedent for admissibility of DNA profiling evidence, not just to determine whether the DNA profiling technique is generally accepted, but also to determine whether or not the technique was properly applied to a specific case.[10] The court became convinced that the private laboratory did not properly apply the accepted technique for DNA profiling and excluded the evidence of a match from use at the

Crime lab evidence
analysis is rarely
routine.
*Courtesy Federal Bureau
of Investigation.*

Cases 11–13

trial. Forensic DNA profiling has been reviewed extensively by courts, and the number of favorable decisions is encouraging. Recently, however, a single state appellate court balked at recognizing DNA profiling but left the door open for future admission. In *Commonwealth v. Curnin*,[11] the Supreme Judicial Court of Massachusetts reversed the trial court's submission of DNA evidence presented by a private laboratory. The court stated that the offer of population statistics which conveys to the jury how common or rare the reported DNA profile is was not supported by testimony by an expert on population genetics. (This may be even more difficult concerning wildlife species.) The court felt that in the absence of such testimony the prosecution could not demonstrate the general acceptance of the statistical approach of private laboratories. The court, however, did leave the door open for consideration of future evidence (Sylvester and Stafford 1991).

No federal appellate court decisions currently address whether forensic DNA profiling is judicially acceptable (in 1991). However, two of the more significant challenges to forensic use of DNA profiling have been heard by two U.S. district courts. In *United States v. Jakobetz*,[12] the suspect was charged with kidnapping in the U.S. District Court in Vermont and later releasing the victim in New York. The defense in *Jakobetz* raised a substantial challenge through the admissibility of forensic DNA evidence attacking the reliability of the FBI laboratory procedure as well as use of population statistics in interpretation of the match (Sylvester and Stafford 1991). One of the most hotly contested DNA admissibility hearings occurred in *United States v. Yee*.[13] The victim, Yee, was shot fourteen times at close range in his own van. He had apparently been mistaken by his assailants as the leader of a rival gang. The U.S. magistrate issued an opinion recommending that the FBI's DNA test results be admitted based on his decision of the requirements of the *Frye* standard. The magistrate found that there was general acceptance in the pertinent scientific community of the procedure developed and implemented by the FBI sources and it was reliable (Sylvester and Stafford 1991).

In general, defense challenges of admissibility are based on three factors:

(1) bias;

(2) matching; and

(3) population statistics.

A few defense experts contend that forensic testing is biased against the subject, since the examiner is aware of which samples the contributor expects to match. Experts for the defense still challenge the ability of forensic DNA laboratories to reliably determine a match given the deteriorated or degraded condition of most forensic samples. A third defense objection is population statistics. The principal purpose of this attack is modeled after a match has been established. Because the current application of technology does not yet exclude one profile from that of every other person in the world, DNA profiling laboratories sample a portion of the population to determine if rare or common certain DNA profiles occur in the population. A few scientists have testified that the FBI has not sufficiently addressed the differences among the ethnic subpopulations within a race. Therefore, they cannot properly assess the resultant effects upon the statistical calculations provided.

It should be pointed out that while DNA profiling is fast getting acceptance by the courts, investigators should be mindful that forensic DNA evidence does not yet positively identify the depositor of a biological sample. It is but one factor of identification and cannot be relied upon alone to support a determination of guilt or innocence (Sylvester and Stafford 1991).

According to McElfresh (1993), DNA-based identification is a sound science and it is firmly established in the legal system. DNA identification technology has had, and survived, its midlife crisis quite nicely.

Literature Cited

Adrian, W. J. 1992a. Matching carcasses. In *Wildlife Forensic Field Manual,* ed. W. J. Adrian. Assn. of Midwest Fish and Game Law Enforcement Officers, 126–131.

————. 1992b. Introduction-court records. In *Wildlife Forensic Field Manual,* ed. W. J. Adrian. Assn. of Midwest Fish and Game Law Enforcement Officers, 2.

Adrian, W. J., and D. Oates. 1992. Sex and species determination in birds. In *Wildlife Forensic Field Manual,* ed. W. J. Adrian. Assn. of Midwest Fish and Game Law Enforcement Officers, 49–81.

Adrian, W. J., and T. D. Moore. 1992. Officers laboratory forensic services, evidence submission. In *Wildlife Forensic Field Manual,* ed. W. J. Adrian. Assn. of Midwest Fish and Game Law Enforcement Officers, 3–23.

Berg, S. O. 1969. Cartridge cases as crime evidence. *The American Rifleman* 112: 2.

Bodziak, W. J. 1984. Shoe and tire impression evidence. *FBI Law Enforcement Bull.* (Federal Bureau of Investigation, U.S. Dept. of Justice, Washington, D.C.) 53 (7): 2–12.

Bunch, T. D., R. W. Meadows, W. C. Foote, L. N. Egbert, and J. J. Spillett. 1976. Identification of ungulate hemoglobins for law enforcement. *Journal of Wildlife Mgmt.* 40 (3): 517–522.

Fiatal, R. A. 1990. DNA testing and the Frye standard. *FBI Law Enforcement Bull.* (Federal Bureau of Investigation, U.S. Dept. of Justice, Washington, D.C.) 59 (6): 2–12.

Garfinkel, S. L. 1991. The Jury is Out on DNA. *The Christian Science Monitor.* Mar. 27, 1991, 12–13.

Gibbons, J. H. 1990. Science, technology and the law in the third century of the Constitution. *Natl. Forum* 71 (4): 13–15.

Hoilien, G. I. 1992a. Imprint casting: "Snow-print-wax" an aid for casting impressions in snow.
———. 1992b. Social Security number area allocation by state. In *Wildlife Forensic Field Manual,* ed. W. J. Adrian. Assn. of Midwest Fish and Game Law Enforcement Officers, 29–31.

Knickerbocker, B. 1990. On the Trail of Wildlife Crime. *The Christian Science Monitor,* Sept. 11, 1990.

McElfresh, K. C., D. Vining-Forde, and J. Balaze. 1993. DNA-based identity testing in forensic science. *Bioscience* 43(3).

McGill and Will. 1992. Metal Detectors for Wildlife Forensics. In *Wildlife Forensic Field Manual,* ed. W. J. Adrian. Assn. of Midwest Fish and Game Law Enforcement Officers, 139–141.

Moore, T. D., L. E. Spence, and C. E. Dugnolle. 1974. Identification of dorsal guard hairs of some mammals of Wyoming. Wyo. Game and Fish Dept.

Morrow, T. L., and F. A. Glover. 1967. Experimental studies on post-mortem changes in mallards. Washington, D.C.: Bureau of Sport Fisheries and Wildlife, Special Scientific Publication Report Wildlife No. 134.

Oates, D. 1992. Time of Death. In *Wildlife Forensic Field Manual,* ed. W. J. Adrian. Assn. of Midwest Fish and Game Law Enforcement Officers, 82–111.

Oates, D., and D. N. Walker. 1992. Mule deer vs. white-tailed deer. In *Wildlife Forensic Field Manual,* ed. W. J. Adrian. Assn. of Midwest Fish and Game Law Enforcement Officers, 152–154.

Oates, D. W., and D. L. Weigel. 1976. Blood and tissue identification of selected birds and mammals, part 2. Nebr. Game and Parks Comm.

Oates, D. W., C. A. Jochum, K. A. Pearson, and C. A. Hoilien. 1983. Field technique for identification of deer blood. Technical note. *Journal of Forensic Science.*

Oates, D. W., G. L. Hoilien, and R. M. Lawlor. 1985. Sex identification of field-dressed ring-necked pheasants. *Wildlife Society Bull.* 13: 64–67.

Smith, B. 1992. Field Forensic Techniques Wildlife Crime Scene Investigations. In *Wildlife Forensic Field Manual,* ed. W. J. Adrian. Assn. of Midwest Fish and Game Law Enforcement Officers, 24–27.

Spraker, T. R., and R. B. Davies. 1992. Differentiation of Wounds in Carcasses. In *Wildlife Forensic Field Manual,* ed. W. J. Adrian. Assn. of Midwest Fish and Game Law Enforcement Officers, 131–138.

Sylvester, J. T., and J. H. Stafford. 1991. Judicial acceptance of DNA profiling. *FBI Law Enforcement Bull.* (Federal Bureau of Investigation, U.S. Dept. of Justice, Washington, D.C.) 60 (7): 26–32.

Notes

1. The examples here are used for humans because they are much more advanced legally than are wildlife cases. The same was true in the case of the Carroll Rule that established the right of officers to stop motor vehicles on probable cause only.
2. 225 Mass. 369 (1920).
3. 519 F. 2d of 463 (4th Cir. 1975).
4. 293 F. 1013 (D.C. Cir. 1923).
5. 447 N.W. 2d 422, 428 (Minn. Sup. Ct. 1989).
6. *State v. Wimberly,* 467 N.W. 2d 499 (S.D. 1991); *State v. Smith,* 807 P. 2d 144 (Kan. 1991); *State v. Pennington,* 327 N.C. 89, 393 S.E. 2d 847 (1990); *Caldwell v. State,* 260 GA 278, 393 S.E. 2d 436 (1990); *State v. Ford,* 392 S.E. 2d 781 (S.C. 1990); *Spencer v. Commonwealth,* 240 Va. 78, 393 S.E. 2d 609 (1990) (Spencer IV) (PCR); *Spencer v.*
Commonwealth, 238 Va. 563, 385 S.E. 2d 850 (1989).
7. *Frye v. United States,* 293 F. 1013, 1014 (D.C. Cir. 1923).
8. *United States v. Williams,* 583 F. 2d 1194, 1198 (2d Cir. 1978). Cert. denied, 439 U.S. 1117, (1979); *United States v. Jakobetz,* 747 F. Supp. 250, 254–255 (D. Vt. 1990).
9. *United States v. Downing,* 753 F. 2d 1224, 1234 (3rd Cir. 1985).
10. *People v. Castro,* 144 Misc. 2d 956, 545 N.Y.S. 2d 985 (Sup. Ct. 1989).
11. 409 Mass. 218. 565 N.E. 2d 440 (1991).
12. 747 F. Supp. 250, 254–55 (D. Vt. 1990).
13. 134 F.R.D. 161 (N.D. Ohio 1991).

Recommended Reading

Caplan, M. H., and J. H. Anderson. 1984. Forensics when science bears witness. Washington, D.C.: U.S. Dept. of Justice, U.S. Govt. Printing Office. 31 pages.

Federal Bureau of Investigation. Handbook of forensic science. Washington, D.C.: U.S. Dept. of Justice, U.S. Govt. Printing Office. 119 pages.

Glover, R. L. 1981. Detecting lead in ''arrow'' wounds in deer using rhodizonic acid. *Wildlife Society Bull.* 9: 216–219.

Glover, R. L., and L. J. Korschgen. 1980. Evaluation of two methods of preparing antisera for precipitant tests. *Wildlife Society Bull.* 8.

Johnson, B. C., L. A. McQuire, and D. R. Anderson. 1980. Determining time of death in mule deer by potassium levels in the vitreous humor. *Wildlife Society Bull.* 8: 249–252.

Moore, T. Forensic Science a reference list. Wyo. Game and Fish Dept. 36 pages.

Oates, D. W., and D. E. Borland. 1988. Waterfowl breastbone identification (audio and visual). Neb. Game and Parks Comm.

Oates, D. W., C. W. Brown, and D. L. Weigel. 1974. Blood and tissue identification of selected birds and mammals, part 1. Neb. Game and Parks Comm.

Oates, D. W., J. E. Seeb, L. N. Wishard, and P. Abersold. 1983. Biochemical identification of North American waterfowl. Neb. Technical Service No. 13. Neb. Game and Parks Comm.

Peterson, J. L., S. Mihajloeic, and M. Gilliland. 1984. Forensic evidence and the police. The effects of scientific evidence on criminal investigations. Washington, D.C.: U.S. Dept. of Justice, U.S. Govt. Printing Office. 234 pages.

Pex, J. O., K. D. Mineely, and F. C. Andrews. Time of death estimation in Blacktail deer by temperature and aqueous humor glucose. *Journal of Forensic Sciences* 594–600.

Steinberg, M., Y. Leist, and M. Tassa. 1983. A new field kit for bullet identification. *Journal of Forensic Sciences* 169–175.

Steltner, L. H. 1979. To catch a thief. *Colorado Outdoors* 28 (3).

Stone, W. B., S. A. Butkas, C. F. Kellogg, and D. E. Kebabjian. 1978. A lead detection kit for determining bullet wounds. *N.Y. Fish and Game Journal* 25.

Wilson, M. (Ed.). 1977. Bibliography of forensic science in wildlife law enforcement. Alberta Recreation, Parks and Wildlife.

Wolf, A., and C. Gremillion-Smith. 1983. Using vitreous humor to determine time of death: Problems and a review. *The Wildlife Society Bull.* 11 (1): 52–55.

Woodburn, J. A. 1972. Planning and funding a comprehensive law enforcement officer training program. In *Proceedings Southeastern Assn. of Game and Fish Comms.,* Vol. 26.

Multiple-Choice Questions

Mark only one answer.

1. DNA experts believe in the reasonably near future this technique will be able to:
   a. identify animal species
   b. determine sex
   c. develop a genetic blueprint
   d. identify the locale from which the animal was taken
   e. all of the above

2. Cock pheasants even when field dressed can be identified by:
   a. more colorful plumage as adults
   b. adults are much larger than hens
   c. by the fact that cocks are always larger than the hens
   d. by measurements taken on the sternum
   e. by the beaks

3. Mule deer can be separated from white-tailed deer by:
   a. having much larger ears
   b. by antlers that roughly form the letter Y on one or more levels
   c. by a brisket that is rich brown
   d. by having a smaller metatarsal gland
   e. by having a much larger preorbital tear gland—measuring an inch or more in adults

4. The normal body temperature in Fahrenheit of live pheasants is:
   a. 98°
   b. 100°
   c. 102°
   d. 104°
   e. 106°

5. The time of death in a mallard may be determined by:
   a. a continuous change in the color of the pupil from the time of death forward
   b. the eye maintains a normal appearance for about two hours after death
   c. from five to eight hours after death there are changes in the color of the pupil
   d. the iris become blackish-brown to aquamarine-violet at about seven to twelve hours after death
   e. the iris loses its watery appearance between fourteen and twenty-four hours after death

6. After death the body of an animal approaches ambient temperatures through loss of heat by radiation, convection, and conduction. Cooling rates depend upon:
   a. initial body temperature
   b. air temperature
   c. body surface area and weight
   d. how the carcass was handled after death
   e. all of the above

7. The parts of a rifle that leave distinctive marks on the projectile are:
   a. rifling
   b. firing pin
   c. extractor
   d. ejector
   e. breech face

8. A laboratory analysis of a piece of glass from an automobile light may reveal:
   a. the manufacturer
   b. where the glass came from
   c. the direction and angle of impact if the light was broken by a stick or bullet
   d. how it was broken
   e. all of the above

Chapter 14

Conservation Officer Law Enforcement Training

C. L. Garey, former Assistant Chief
of Law Enforcement, Idaho Fish
and Game Department

FIREARMS SELECTION AND TRAINING

Since the majority of wildlife conservation agencies now require their law enforcement personnel to attain commissioned peace officer status, the selection process for properly arming and training conservation officers differs little from that of any other police agency in the United States.

Duty firearms find their way into the holsters and patrol vehicles of conservation officers nationwide under two basic premises: department purchase and issuance, and personal acquisition under guidelines set up by the department. Some agencies take this process one step further by issuing duty firearms, yet allowing officers to use personal weapons within established policy guidelines.

The following information is useful for officers who are allowed to purchase and carry their weapon of choice. It also offers guidelines for agency issue of weapons, creates law officer awareness of decisions involved in the selection of duty firearms, and discusses training methods. Unfortunately, the bureaucratic process can result in issuance of firearms that do not meet the need. Officers should be aware of recent trends and be proactive within their agencies to ensure that proper firearm selection and training needs are met.

Sidearm Selection

Law enforcement agencies nationwide are clamoring to adopt semiautomatic pistols as the duty sidearm of choice. While most law enforcement officers still carry revolvers, the majority would not seriously consider replacing them with anything but a semiautomatic, despite the fact that revolvers can still adequately fulfill the need.

The switch to semiautomatic duty sidearms is based on a variety of reasons, some with practical rationale and some without. One premise is that the increased firepower of semiautomatic pistols enables the officer to keep pace with weapons commonly carried by the criminal element. The logic in this assumption is shaky, and will be discussed later in the text. There is basis in fact that the ease of manipulation, loading, and trigger control can be important reasons for adopting the semiautomatic pistol.

Starting with the center farthest left and moving clockwise, the sidearms are:

A. Taurus .38 special double action revolver 2½'' bbl.
B. Colt .45 semiautomatic pistol, stainless steel.
C. Browning 9-mm semiautomatic pistol, high power.
D. Ruger P89 9-mm semiautomatic pistol, double action.
E. Smith and Wesson .45 Colt double action revolver 8½'' bbl.
F. Glock Model 17, 9-mm semiautomatic pistol.

Courtesy Ted Hansen.

Semiautomatic Pistol Choices

Several European and American manufacturers supply the majority of sidearms for police duty use. A discussion of strengths and weaknesses of each brand is beyond the scope of this treatise. It behooves officers and agencies to glean all available testing and evaluation data on sidearms, and conduct extensive internal testing prior to brand selection. All semiautomatic pistols are not equal. Some brands are more reliable in mechanical operation, frequency of repair, and human engineering.

Magazine capacity is an issue only in grip size. A single column magazine with eight or nine rounds allows a grip configuration that fits the average officer better than a staggered magazine that holds twelve to eighteen rounds.

The most important criterion in selecting a semiautomatic duty pistol is the trigger system. There are several types of firing systems on the market, each with its own strengths and weaknesses. Originally, the single-action semiautomatic pistol was the only weapon of choice, with the Colt .45 semiautomatic being the most common. The simple, rugged mechanism and the consistent four-to-six pound trigger release endear the Colt to many knowledgeable law officers. But, the manual safety and the required cocked-and-locked carry mode cause most police agencies not to adopt this pistol, not to mention the extra training required to handle the pistol safely.

The advent of double-action semiautomatics heralded the beginnings of widespread semiautomatic pistol use by police agencies. Traditionally, the first trigger pull of the double-action semiautomatic was like that of a double-action revolver—long and heavy (eight to fifteen pounds); with subsequent shots being single-action. Manufacturers now offer double-action-only semiautomatic pistols, which duplicate a fifteen-shot double-action revolver. The double-action-only feature helps to ease the transition from using revolvers to using semiautomatic pistols. There are also trigger-cocked models such as the Glock ''safe action,'' which is neither double nor single action, and the manually cocked models, Heckler and Koch P-7 series. Several pistols can be set in either the double-action or single-action mode. Many action variations are available, but the models with a consistent trigger pull and simple design features should be the weapons of choice.

The inherent accuracy of issue semiautomatic duty pistols is a point of discussion, sometimes beyond practical consideration. Any sidearm that will hold a group of four inches at twenty-five yards will handle any practical shooting situation an officer may encounter.

Sidearm Caliber/ Ammunition Selection

Officers who carry revolvers are evenly split between the .38 special and the .357 Magnum. Agencies that issue big-bore revolvers represent a small minority.

In semiautomatic pistols, the most popular caliber is the 9-mm Parabellum. Many agencies have been led to believe that this cartridge is a substantial ballistic improvement over the .38 special, but it is not. The standard 9-mm loading is roughly equivalent to a +P .38 special load. The availability of 9-mm +P and +P+ loads gives the edge to the 9-mm, but there are .38 special loads in this category also. The two rounds are comparable in their terminal effectiveness. The minimal recoil generated by both rounds allows the shooter to concentrate on making better hits.

The second step up is the .357 Magnum, which is chambered primarily for double-action revolvers. The bullet diameter is the same as the .38 special (.357 inches) and slightly more than the 9-mm (.354 inches). With proper loads, the powerful .357 Magnum easily outdistances the other two rounds. It propels bullets at a higher velocity, which also increases the recoil and muzzle blast. For officers carrying .357 Magnums, extensive training with full-power ammunition is recommended.

The third step up in caliber selection for semiautomatic pistols is a relative newcomer—the .40 Smith and Wesson round. This round offers an increase in bullet diameter (.400 inches), and an increase in bullet weight (155–180 grains) compared to the .38 special and 9-mm (with respective .354 and .357 bullet diameters and 95–158 grains bullet weights). The .40 Smith and Wesson cartridge has the same overall length as the 9-mm Parabellum round, which is the key to its overnight success; it can be chambered in existing semiautomatic 9-mm pistols. This cartridge is being widely accepted by police agencies since it offers ballistics that approach the .45 auto, yet is available in a 9-mm frame size. Many departments that have contemplated switching to a 9-mm semiautomatic, have chosen the .40 Smith and Wesson round instead.

The 10-mm round preceded the .40 Smith and Wesson offering and is a true magnum semiautomatic loading that approaches the ballistics of the .41 magnum revolver round in a four-inch barrel. The 10-mm requires a large pistol frame to accommodate the cartridge, and the recoil can result in poor shot placement. The FBI adoption of a reduced-velocity 10-mm (.400 bullet diameter) load ultimately led to the birth of the .40 Smith and Wesson round. The .40 Smith and Wesson equals the ballistics of the FBI 10-mm load while offering the advantage of a smaller pistol in size and weight. With the current trend in emasculating the ballistics of the 10-mm round, and the advent of the .40 Smith and Wesson, it is safe to assume that the 10-mm will not see widespread use among police agencies.

The .45 ACP semiautomatic round is still popular and considered by some to be the best choice for law enforcement duty use. This round is being chambered in a variety of semiautomatic pistols and revolvers. With proper ammunition, it is a proven ''fight stopper'' with an easy-to-control recoil.

No discussion on the selection of duty cartridges would be complete without a short discourse on the relative ''power'' of pistol rounds. Everyone has heard the old adage "you can hit a man in the finger with a .45 automatic bullet and knock him off his feet." Any serious student of ballistics and actual shootings involving humans can only chuckle at this misconception. The terms firepower and bullet energy should be reserved for discussions involving an F-16 fighter or the battleship New Jersey. A pistol is a weapon of convenience for an officer. Its portability does not allow for a power factor that will reliably stop a human adversary one hundred percent of the time. A bullet velocity of approximately 2,000 feet per second is required to initiate hydrostatic shock and the low-velocity range (800–1,200 feet per second) of duty pistol cartridges results in no more than inflated "puncture wounds," without massive tissue destruction or immediate disruption of physiological processes. For this reason, an accurate first shot and the capability of delivering one or two extra rounds quickly and accurately are of great importance, as is adequate penetration.

Recent FBI ballistic wound studies confirm the known fact that increased bullet diameter results in better terminal ballistics when coupled with a minimum of twelve inches of tissue and bone penetration.

Standard Pistol Sights

Most law enforcement semiautomatic duty pistols are supplied with high-profile sights with the rear sight adjustable for windage only. Compared to a fully adjustable rear sight, with its attendant screws and springs, this type of system is more rugged and will stand up to the rigors of field duty.

Fixed-sight pistols will adequately handle all practical shooting situations, and officers will not be constantly adjusting the sight.

Most manufacturers of semiautomatic duty pistols offer a three-dot sighting system for higher visibility under less than desirable conditions. The usual combination is a white dot on either side of the rear sight notch and a matching dot on top of the front sight post. Proper sighting puts the three dots in a horizontal line. Many firearms instructors feel that three-dot sights are unnecessary, time consuming, and, therefore, detrimental to good shooting. If three-dot sights help officers place good hits on intended targets quickly, then they should be carefully considered as an option.

Self-Luminous Pistol Sights

Another pistol sight option is the three-dot self-luminous system. A low-level radio-active element, which emits a highly visible glow under low, ambient, light conditions, is contained in small capsules or encased in epoxy and placed in the rear and front sights in the standard three-dot pattern. The color of the container dictates the color of light emitted; usually green. Each element is outlined with white paint for daylight sighting. Since most law enforcement shooting situations occur during the hours of darkness, self-luminous sights are becoming standard issue in many agencies.

Special issues need to be noted for self-luminous sights: The enhanced visibility allows quick, precise sight alignment, enabling the shooter to fire highly accurate shots under lighting conditions which may not readily allow identification of the adversary and other persons in the potential line of fire. The high visibility of the sight may also result in the shooter taking excessive time for sight alignment when only brief confirmation is needed for a center-of-mass shot on the adversary. Self-luminous sights are also visible to persons other than the user, compromising the officer's location. In low-light conditions, the glow of the sights, when brought to eye level, can result in an inability to see the faint image of the target.

Learning proper "instinctive" front-sight skills with standard sights negates the need for self-luminous sights. Once again, choices must be made, and training must address the needs to ensure officer survival.

Duty Holsters   Duty holsters are usually issued, giving officers little choice. In general, a duty holster should retain the pistol in a safe, secure manner while allowing immediate deployment.

An alarming percentage of law officers are killed or wounded with their duty weapon by an adversary. During routine license checks, investigations, and in many arrest situations, officers work alone but are confronted by more than one adversary. Several special holster features make it harder for adversaries to remove an officer's weapon in these situations: "decoy" safety snaps, multiple safety straps, and built-in retention devices that require forward or rearward movement for pistol release. The drawback is that these features also increase the degree of difficulty required for the officer to draw the weapon. Most high-level security holsters require extensive range training followed by individual officer practice of drawing the weapon from any body position.

Agencies should consider authorizing officers to carry a backup gun. A small, five-shot revolver or small-frame semiautomatic pistol tucked away on the officer's person provides insurance in a gun take away situation.

Shotguns   Many agencies issue their officers "riot" shotguns, or allow officers to carry personal shotguns on duty.

Pump shotguns are the most popular due to their durability, simplicity, and most importantly, reliability. There are some semiautomatic shotguns that are worthy of law enforcement use, but others are overly complicated in loading, unloading, and firing sequences. Officers should also avoid the "confused actions" that allow a choice between pump and semiautomatic function; this flexibility sounds good in theory, but falls apart under the stress of practical application.

The appropriateness of shotguns in police work is controversial. But if one considers that there are at least fifty-four varieties of birdshot, twenty-four varieties of buckshot, and fifteen varieties of rifled slug, not to mention rubber pellets, gas cartridges, signal devices, and flechettes, the versatility of the 12-gauge police shotgun cannot be questioned. It is merely a question of proper training and application.

When using a shotgun it is important to know the capability of the ammunition. The term "scatter gun" is applicable only at extended ranges. From zero to 10 yards, with shot charges, the shotgun is essentially aimed like a rifle. Pellet patterns will generally be ten inches or less at the 10-yard mark, necessitating a precision approach to shot placement. A good rule of thumb is one inch of pattern spread for each yard increase in range. This equates to an optimum effective range of 10 to 30 yards, depending on the brand of ammunition and the chokebore of the shotgun. When the range extends from 30 to 50 yards, buckshot loads scatter. Rifle slugs can be used at any range, and a good match of slug and chokebore constriction allows accuracy at 50 to 100 yards.

Officers should test their guns for consistency of aim by studying the pattern/group at several distances. Attention should also be directed to the penetration qualities of buckshot and slugs. Buckshot can easily be deflected by light foilage and fails miserably in penetration of auto glass and car bodies. The same can be said of many types of rifled slugs.

Training sessions should require officers to change from buckshot to slugs and back to buckshot again to prepare for scenarios that actually could occur.

Shotgun magazine capacity is highly overrated, to wit the eight-, nine-, and ten-shot magazine extensions available. Field use seldom involves more than two or three rounds being fired.

Typically, shotguns have the same open sights as rifles. These sights are too high in profile, fragile, not easily adjustable, and too time consuming to line up to be effective in police work. Rifle sights are not needed for buckshot use, and credible aim can be achieved with slugs at 50 to 100 yards with only a bead sight. The "ghost ring" aperture sights now available are the best choice if rifle-like sights are required. The ghost ring is a large, rear, peep sight and a post front sight that allow rapid target acquisition. These sights are sturdy and will take much abuse and still maintain sight zero.

Many police agencies are discontinuing shotgun use in favor of semiautomatic carbines and, in some cases, submachine guns. With proper training and understanding of the shotgun's versatility, it will cover close-range situations with more finality than a handgun and will substitute for a carbine out to 100 yards—depending on the type of rifle slugs used, choke constriction, and the shooter's ability.

No gun is perfect. An accurate shot fired from the best gun may not stop an adversary. Always be prepared to shoot again, and have an alternate course of action. Still, within its range and performance capabilities, the shotgun is a valuable piece of equipment to have in a patrol vehicle.

Patrol Rifles

The use of patrol rifles in law enforcement is growing rapidly. A good rifle is a valuable tool for wildlife law enforcement. Since this discussion is about patrol rifles and not sniper rifles, caliber and action selection are less rigid.

A wide variety of patrol rifles can accomplish wildlife law enforcement tasks. The officer should assess the need and choose an appropriate rifle. A military grade semiautomatic that would be acceptable for a SWAT team member, would not necessarily be the weapon of choice for a conservation officer or rural deputy sheriff. A conservation officer's needs may range from shooting crippled big-game animals to using lethal force on adversaries at close or long range.

Patrol rifle criteria include function, reliability, ease of manipulation, practical accuracy, simple adjustable sights that allow rapid target acquisition, chambering for cartridges in the power range of the .30–30 to .308, magazine capacity of five rounds or more, a good trigger pull, portability, and lightweight. Practical accuracy equals a three- to four-inch group at 100 yards with iron sights. This standard may seem excessive, but it allows head shots at that range, and still allows good torso hits on human adversaries at 200 yards or a little more.

Sighting system selection is important to balance the speed needed to place a shot at 5 or 10 yards, yet allow the needed precision to print practical groups at 100 to 200 yards. Usually this necessitates the use of aperture or peep sights, preferably with a ghost ring or very large hole in the aperture rear sight. Optical sights are acceptable only if in low magnification powers that result in a wide field of view. Most patrol rifles are applied at handgun ranges, and a low magnification scope ($1\times$–$2.5\times$) will suffice and still allow the practical precision needed at 200 yards. "Scout scopes" that mount forward of the rifle receiver are also acceptable. Both the low-power and scout scopes allow shooting with both eyes open to retain the advantage of binocular and peripheral vision. The disadvantage of optics is the possibility of damage, and the effect of the elements on optical clarity: snow or water accumulation on the lenses, and lenses fogging when the rifle is taken from a warm vehicle into cold air.

With the premise that *any* rifle of adequate power is suitable during a fight, then cartridge chambering selection for a patrol rifle is not an issue. But experience has proven otherwise. A .30–30 lever-action rifle may have made a big difference in the FBI shooting in Florida a few years ago. An acceptable minimum-power patrol rifle can include several carbines chambered for pistol cartridges such as the .357 Magnum or .44 Magnum. Pistol cartridges fired from a pump- or lever-action carbine gain substantial velocity over that attained in pistol-length barrels and can stretch the cartridges' effectiveness to 150 yards.

Intermediate rifle rounds include the .223 Remington, the 7.62 × 39 Soviet, and the .30–30 Winchester. The .223 and 7.62 × 39 are currently available in the Ruger Mini-14 and Mini-30 semiautomatic carbines and in the Colt AR-15 semiautomatic rifles. The 7.62 × 39 round is comparable to the .30–30 in ballistics and is gaining popularity in law enforcement circles.

The .308 is reaching the practical limit in power as a lightweight, semiautomatic, bolt-action carbine. Officers should consider using a bolt-action .308 since it is easy to assemble, has a six-pound carbine, and has little recoil. A bolt action is of no disadvantage in a fight if proper tactics are employed. *Aimed* fire is the main consideration.

Sixteen-round pistols and thirty-round rifles are the norm for magazine capacity. This is more than ample as officers usually are not fighting off masses of enemy troops. The patrol rifle will, if used properly, be fired only one, two, or three times to settle the problem, and speed reloading will not be a major consideration. Shotguns with a five-shot capacity have served law enforcement officers well for years.

Another point to ponder is portability and weight of a patrol rifle. Some of the military-grade rifles can weigh up to ten pounds or more when loaded with twenty to thirty rounds. When considering carrying a heavy rifle several miles over rough terrain, the rifle is usually left in the patrol vehicle. Since the primary reason for carrying the rifle to a potential threatening situation is to increase hitting and stopping power, it makes sense to go as light as possible. In addition, the lever-action, bolt-action, and to a degree, the Ruger Mini-14 type rifles are all "acceptable" weapons to the public. They do not attract unnecessary attention as do the military-grade rifles and do not come with the attached "assault rifle" stigma.

Rifle action and caliber selection for duty usage necessitates careful consideration, testing, and evaluation. The first step is defining the practical field applications, then, selecting the rifle to meet the demands. A .223-caliber rifle, for example, may meet the needs of some, but officers that routinely have to dispatch wounded or crippled elk may find this caliber lacking.

Ammunition selection is subject to the same criteria as pistols and shotguns. Law enforcement agencies now typically issue jacketed softpoint and hollowpoint rounds for duty use. The same rationale that mandates these bullets for handguns applies for rifles, only more so: extensive tissue destruction to end the fight, without over penetration that may endanger innocent bystanders. The use of full-metal-jacket rifle bullets should be carefully considered. Some full-metal-jacket bullets will fragment upon impact with automobile windshields, resulting in little or no effect upon a target three feet behind the glass. Conversely, the same fully jacketed bullet may completely penetrate a fully exposed adversary with little or no initial effect on his fighting ability. Jacketed softpoint rifle bullets, depending on construction design, seem to offer the best compromise for stopping ability and penetration of an adversary's cover. Extensive testing is the only way to determine the proper issue rifle round to ensure that expected ballistic performance and safety factors are properly addressed.

Firearms Training

Firearms training is rapidly changing from practical police competition (PPC), which was once accepted as the state-of-the art for police officer firearms qualification and training. PPC courses allowed generous time limits to complete each shooting phase, stressed shooting small groups, and involved no tactical movement on the officer's part other than drawing the weapon, shooting, reholstering, and polishing brass at leisure. Officers were losing fights because they applied target-shooting principles to actual shooting situations. Officers often failed to use cover and concealment, failed to shoot and move, failed to look around for potential multiple adversaries, and failed to look around while reloading. There are documented cases of officers being killed or wounded in gunfights because they had taken the time to put empty cartridge cases in a pants pocket, as they had done countless times at the shooting range. Top competition shooters who could shoot groups that were one ragged hole, lost fights due to slow reaction time and failure to have a mental survival plan. Fortunately, the modern technique of police firearms training is gaining a strong foothold, with few agencies using PPC training.

The modern technique involves a training triad of weapon skills, tactics, and mental preparation for combat. The percentages may vary, but once the fundamentals of marksmanship are accomplished, training should involve seventy-five percent mental preparation, fifteen percent weapon skills, and ten percent tactical considerations.

In general, mental preparation training involves careful study of actual gunfight factors. These include target mobility, actual distances within which gunfights occur, duration of the fight, lighting conditions, target response, and after-action responses. Officers should also learn the strengths and weaknesses of their sidearm. Proper training can compensate for most weaknesses.

Officers should be aware of factors that may determine whether they live or die during an armed conflict, including mental preparation, skills, tactics, serviceable and functional equipment, opponent's skill level, and physical fitness of the officer.

The ultimate goal of mental preparation is awareness that an armed encounter can occur at any time. Training can replace helpless feelings with confidence and cool resolve.

There are several reputable firearms training consultants who offer this modern technique. Agencies should contract with these people to properly train their firearms instructors.

Apart from the primary training goal of weapon skill, tactics, and mental preparation, a firearms training program should include training and certification in low ambient light shooting, shoot-don't shoot decisions, and moving target practice. It should also include practice in tactical advance and withdrawing drills, utilizing available cover, cover drills, hostage drills, drills involving lateral, forward, and backward movement while engaging targets, injury drills, and weapon retention. Training in these areas is virtually mandated by the current atmosphere of legal liability that surrounds the use of a weapon for the protection of an officer or innocent bystander.

All agencies should properly document their firearms training efforts. Current philosophy indicates that if training is not documented, it never happened. Training programs should include sign-off sheets for every phase:

(1) the purpose of the instruction;

(2) that the instruction was received and understood;

(3) remedial training required; and

(4) officers meeting the minimum qualification/training standards required by the agency.

Minimum standards vary by agency. Much information, based on court decisions, is available that outlines the amount and frequency of training required for a successful program.

Firearms training, if properly conducted, takes time and effort. Many administrators have a hard time justifying the expense of a validated training program when there has never been a shooting incident in the entire history of the agency. There are several reasons for implementing a firearms training program, including increased officer morale and safety, to lessening of liability concerns. The frequency of officer-involved shootings is not lessening; it will be only a matter of time before most agencies will have to face the issue. When this time comes, only proper firearms training and documentation will ensure that officers are prepared, and that liability concerns for the agency have been addressed to the best degree possible.

POST-SHOOTING AND TRAUMA POLICY

Since officer-involved shootings are not widespread, many agencies have no framework set up for dealing with the aftermath of a shooting situation. Agencies often have proper reporting procedures should an officer fire a weapon for lethal force, but ignore the intricacies involved in the protection of officers and family members during and after such a traumatic event.

It is important for agencies to adopt a policy outlining steps that must be taken after an officer-involved shooting, and supervisory responsibilities to ensure that an officer and family members are immediately put under protection. From the first moments after a shooting situation to the time the officer returns to duty should be carefully planned and put into a policy format. This addresses the care needs required, ensures that the investigation is completed without interference from media and other outside interests. The same rationale applies should the officer become the victim of a shooting.

DEFENSIVE TACTICS AND INTERMEDIATE-WEAPON TRAINING

Most law enforcement agencies provide some type of defensive tactics training involving wristlocks, takedowns, and restraint holds. Instruction with a baton (PR-24, straight, or more commonly, a collapsible model) and a defensive aerosol spray may also be part of the program. The purpose of this training is to address officer self-defense and subject control needs that fall between verbal commands and gun force.

Many training courses rely upon modified martial arts techniques that may overwhelm officers with their complexity, and may take years to properly master. If officers have to think about a complex defense technique under stress, they may fail to quickly control the subject, resulting in needless injuries to the offender or the officer.

A defensive tactics program should incorporate only a few strikes and restraint holds that are easily remembered under stressful conditions and will result in quick control of the aggressive subject. Collapsible baton training, speed handcuffing, and defensive aerosol spray training are also included to ensure that the officer has means to properly respond to any type of aggression.

A defensive tactics program should address not only the practical application of personal and intermediate weapons, but how their use fits into the force continuum, which ranges from verbal commands to firearm use. The program should include the agencies' required justifications for each level of force inflicted.

Mandating officers to wear collapsible batons, defensive aerosol sprays, and handcuffs is an option that agencies may need to consider. Officer safety and liability involving excessive force should help decide the proper course of action; optional carry should not be a consideration. It would be a tragedy for an officer to resort to fatal use of a firearm to stop an adversary who could have been controlled with an aerosol spray or baton.

RECRUIT SELECTION AND TRAINING

Initial Selection

New recruit selection varies among agencies depending upon many factors, including state hiring guidelines. Initial procedures usually involve a combination of written and oral evaluations to gain placement on a hiring register. In some cases, the selection process ends at this point, with the highest scoring recruits being offered employment. However, many agencies now have internal assessment centers designed to test the recruit's knowledge, skills, and abilities in the area of fish and wildlife identification, public speaking, report writing, and interpersonal relations. The results of these assessment centers allow the agency to grade the recruits in the practical areas of agency concern rather than relying solely on a score based upon stated qualification and abilities.

With the liability issues of negligent hire and retention and the increasing costs of training law enforcement personnel, agencies have virtually been mandated to use additional screening procedures to ensure that only top-quality recruits are employed. These procedures include extensive criminal history background investigation, psychological profiles, polygraph examinations, medical examinations, and validated physical fitness and agility tests. Until recently, agencies could conduct these evaluations prior to an offer of employment. Under present federal hiring practice guidelines, a condition of employment must be satisfied prior to subjecting recruits to any of the previously mentioned screening procedures, except the polygraph. Should the recruit fail to satisfy the screening procedures, the only option for the agency would be termination under probationary guidelines.

Field Training Officer (FTO) Evaluation Program

After all employment screening procedures are satisfied, the next step is recruit training in a validated FTO program. These programs have been in use by police departments for years, and are now becoming the norm with fish and wildlife law enforcement agencies.

FTO programs address and alleviate concerns regarding negligent retention, negligent assignment and investment, inadequate and/or insufficient training, and supervision and career development. Objectives include resolving problems involving law suits, quality training for career development, instilling officer self-confidence, and establishing proper officer safety habits. Should the probationary officer fail to attain an acceptable level of performance, employment termination would occur during the FTO training phase.

Typically, FTO programs are twelve to sixteen weeks in length and require the recruit to work with certified veteran field training officers. The recruit usually works with three FTOs during the training period and is evaluated daily on appearance, attitude, knowledge, performance, and relationships. All acceptable and unacceptable performance is carefully documented, as is needed remedial training.

Proper implementation of an FTO program requires extensive training of all agency personnel involved in the program. There are recognized private companies that specialize in law enforcement FTO training, and it is paramount that agencies seek this training for development of field training programs.

FTO programs are primarily designed to be implemented prior to an officer's attending a police academy. Establishment of a sound FTO program will ensure that the new recruit is prepared when assigned to a permanent field patrol assignment. The days of issuing new recruits badges and citation books and turning them loose on the unsuspecting public should now be of only historical significance.

Career Training   Once the initial screening, assessments, FTO training, and police academy requirements are satisfied, agencies must focus on a long-term career training program for commissioned officers. Unfortunately, it is common for agencies to find that the only documented training that an officer received was ten years prior at the police academy!

Agencies should consider conducting a training-needs assessment and structuring future training contingent with officer career development, agency requirements, and legal/liability concerns. Training in fish and wildlife responsibilities can be established on an as-needed basis, but training in officer safety, search and seizure, and laws of arrest must be much more frequent due to ever-changing procedures and court decisions. Courses in laws, regulations, and policies must be freshly documented, even if there have been no changes. As redundant as this may seem, the current legal atmosphere dictates this approach.

Career development for officers aspiring to supervisory positions should also be available. Many agencies take the extra effort to mandate the hours and training required to be eligible for promotional examinations. This approach clearly outlines what training requirements are necessary for promotional advancement. It also ensures that new supervisors will start with an understanding of requirements and responsibilities, rather than relying on after-the-fact training methods, which are often inadequate and superfluous.

Recommended Reading

Adams, McTernan, and Remsberg. 1982. *Street survival.* Northbrook, Ill.: Calibre Press, Inc.

Ayoob, M. 1984. *Stressfire.* Concord, N.H.: Police Bookshelf.

Barber, D. L. 1986. *Law enforcement firearms training techniques tactics for police and security.* Garden Grove, Calif.: D. L. Barber Ventures.

Clede, B. 1985. *Police handgun manual.* Harrisburg, Pa.: Stackpole Books.

———. 1985. *Police shotgun manual.* Harrisburg, Pa.: Stackpole Books.

Cooper, J. 1989. *Principles of personal defense.* Boulder, Colo.: Paladin Press.

FBI Wound ballistic evaluation. Boulder, Colo: Paladin Press.

Harbaugh, C. R. 1985. The investigation of shootings by police officers. Gaithersburg, Md.: Intl. Assn. of Chiefs of Police.

Lindell, J. W. 1981. *Handgun retention system and training text for law enforcement officers.* Kansas City, Mo.: Ordin Press.

Taylor, C. 1982. *The complete book of combat handgunning.* Boulder, Colo.: Paladin Press.

———. 1984. *The fighting rifle.* Boulder, Colo.: Paladin Press.

———. 1985. *The combat shotgun and submachine gun as a special analysis.* Boulder, Colo.: Paladin Press.

U.S. Department of Justice. 1987. State and local law enforcement training needs in the United States, volume II technical report. Institutional Research and Development Unit, Training Division, Federal Bureau of Investigation, Quantico, Va.

Suggested Training Sources

Institute for Aerobic Research. Police fitness instructor training and certification. 12330 Preston Rd., Dallas, Tex. 75230.

Kaminsky and Associates. Police FTO training and instructor certification. 7616 Estate Circle Dr., Longmont, Colo. 80501.

Siddle, Bruce K. Defensive tactics training and instructor certification. PPCT Mgmt. Systems, Inc. P.O. Box 175, Waterloo, Ia. 62298.

Smith, Clint. Police firearms training and instructor certification. Intl. Training Consultants, Inc. P.O. Box 528, Huntertown, Ind. 46748.

Chapter 15

An Historical Perspective on the European System of Wildlife Management

(Originally published in: **Reneker, L. A. and R. J. Hudson, eds. 1991.** *Wildlife Production: Conservation and Sustainable Development.* AFES Misc. Pub. 91–6, Univ. of Alaska Fairbanks, Fairbanks, Alaska 601 pp.)

MICHAEL L. WOLFE,
Department of Fisheries and Wildlife,
Utah State University, Logan, Utah
84322–5210 USA.

Abstract: Hunting systems throughout most of Europe evolved in a tradition of hunting and game management as a prerogative of the privileged classes. The primal hunting freedom of the commoner was progressively expropriated, first by the ruling monarchs and later by feudal lords and the landed gentry. The social reforms of the eighteenth- and nineteenth-century revolutions largely abolished the excesses of feudalism and in some cases were followed by chaotic periods of virtually unrestricted freedom. Ultimately, however, modern systems developed along two or three basic lines. *De jure* at least, the concept of free-ranging wild animals as a common social possession shared by the citizens of a given country, constitutes a nearly universal element regardless of the system involved. The essential differences between the systems reside in the access to wildlife and its disposition once the animals have been reduced to possession. The countries of southern Europe are characterized by nearly absolute hunting freedom, limited only by minimal licensing and trespass restrictions. Throughout most of western Europe, the right to hunt and title to the animals once they have been killed are based on land tenure, although these rights may be leased to other persons. The central European variant of this system features the concept of hunting districts, with legally prescribed minimum holdings as a prerequisite to the exercise of a landowner's right to hunt. Other salient features include qualifying examinations for prospective hunters and mandatory shooting plans. In countries with centralized economies the state controls all hunting rights, but may lease these rights to hunting associations, with the management and harvest of economically important species frequently being conducted by state-operated collectives.

Many North Americans regard European wildlife management as a kind of monolith, and the title of this paper would appear to reinforce this oversimplification. While recognizing that several variants and exceptions to the general pattern exist, I believe that it is possible to distinguish certain characteristics that set European hunting systems apart from the one in place in North America, especially as summarized by Geist (1988). My notion of a more-or-less generalized European model embodies systems with:

a) land-tenure based hunting rights;

b) low participation rates;

c) intensive habitat manipulation for the benefit of certain game species; and

d) an open market in game meat and wildlife parts and/or products.

The purpose of this paper is to review the historical development of European system(s). I will focus on the central European system, because I consider it to be the most refined variant. Moreover, most of my examples will be drawn from the German situation, simply because of my greater familiarity with conditions there. Lund (1975) identified four major functions of wildlife law, namely:

a) facilitating a sustained yield;

b) weapons control;

c) class discrimination; and

d) protecting animal rights.

Undoubtedly, some of the elements in this typology may seem inappropriate or aversive in a modern, democratically oriented context, but they provide useful points of reference in the discussion that follows. A more enlightened typology might include the following elements:

a) conserving a diverse and viable wildlife resource;

b) protecting the interests of agriculture and forestry;

c) assuring public safety; and

d) protecting animal rights.

HISTORICAL PERSPECTIVE

Hunting systems throughout much of Europe evolved in a tradition of the chase as the prerogative of the privileged classes. Moreover, until relatively recently legal strictures relating to wildlife dealt primarily with the subject rather than the object of the chase. To examine the validity of these generalizations it is necessary to begin our historical overview with the collapse of the Roman Empire in the fourth century A.D. According to conventional legal doctrine, Roman law recognized free-ranging wild animals as *res nullius*. Literally translated, this term simply means that wildlife is an ownerless entity. However, some historical and contemporary interpretations have extended the concept to consider wild animals as a product of the soil. The distinction may seem academic, but has led to some semantic difficulties. I prefer the more parsimonious definition of the term.

At the time of Roman domination, the Celtic inhabitants of continental Europe had established a sedentary and prosperous agrarian culture, in which hunting had ceased to be of major significance in the procurement of food and had assumed a largely sporting character. Available evidence suggests that, although some class distinctions existed, virtually every free male member of the community enjoyed the right to hunt (Lindner 1937). The Teutonic tribes that invaded Europe from the north and east and replaced resident Celtic culture with their own, practiced by comparison a relatively crude level of agriculture, still relied on hunting for food and clothing, and had not yet developed strong concepts of private ownership. These circumstances point to the probable existence of a primal condition of near absolute hunting freedom for every free male member of a given community (Lindner 1940).

This condition persisted until the advent of inforestation by the Merovingian kings, beginning in the sixth century. Even the Germanic Codes (*leges barbarorum*), a compilation of existing tribal laws recorded in that era, imposed only nominal and largely methodological restrictions, which sought primarily to prevent human injury and crop damages due to hunting (Stahl 1979).

Beginning in the sixth century, the Frankish monarchs asserted their hegemony by claiming hunting rights for themselves or their designates on large tracts of land within their dominion, which came to be known as forest preserves. These districts were subject to protection by a legal code (forest law) that was distinct from common law. The institution of these imperial forest preserves marked the inception of a fundamental change in the legal status of wildlife, and shaped the course of hunting law in most of Europe until the revolutions of the eighteenth and nineteenth centuries. Available evidence suggests that the principal motivation for early inforestations was probably more political, for example, as a symbol of royal power—rather than venatic (Endres 1917). Initially, the forest preserves were established largely through expropriation of commons land. However, as their power grew in the eighth century, the Carolingian kings created forest preserves not only on their own lands, but also claimed unoccupied and conquered territory.

The centuries following the decline of the Carolingian dynasty witnessed a continuing decentralization of the Empire's executive power, and the establishment of a loose federation of minor principalities. Most of the sovereign rights and privileges previously reserved exclusively for the king were ceded to the provincial rulers. As the latter became increasingly independent of the king, their rule over their subjects became ever more absolute, a political trend exacerbated by the general intensification of feudalism (Wolfe 1970).

The prerogatives of forest law became vested in the individual provincial sovereigns, and in the following centuries these rulers progressively expropriated all the hunting rights in their kingdoms. Invoking self-serving legal interpretations of Roman law, they viewed themselves as sole proprietors of wild animals in their respective provinces, empowered to enact anything "necessary for the welfare of the state with respect to game and the chase." This power included the regulation of hunting seasons, destruction of ostensibly detrimental species (e.g., predators), and punishment of poachers.

Under the prevailing feudal system peasants and commoners—even those who owned land—were usually denied their traditional right to hunt or even to bear arms. In some cases, however, there existed a division of venatic privileges, with the nobility claiming the prerogative to hunt animals classed as "high game." This category commonly included the larger ungulates and terrestrial carnivores, as well as cranes, herons, eagles, and falcons. The right to hunt animals of the "lesser chase," which usually included roe deer (*Capreolus capreolus*) and a variety of smaller birds and mammals was sometimes conceded to members of the lower social classes.

The extensive husbandry measures for favored game animals practiced by the ruling class frequently resulted in densities which caused excessive crop damage. Violations of the "forest laws" were subject to draconic corporal or capital punitive measures. Moreover, the ruling class frequently extracted compulsory manual services and taxes from their subjects for the preparation and execution of large-scale medieval battues. As Stahl (1979) has noted, these intolerable conditions were in some cases sufficient to generate various peasant revolts, harbingers of the more sweeping revolutions of the eighteenth and nineteenth centuries. The succession of events described above is broadly synoptic of the evolution of hunting law throughout much of continental Europe with certain notable exceptions, namely Holland and parts of the Fennoscandian Peninsula, notably in Finland (Salo 1976). In these regions, feudalism did not develop to the extent that it did in central Europe and the hunting rights of the landowner were largely preserved.

William the Conquerer brought continental forest law to England in 1066 and with it the king's claim to wildlife over large areas of land. Subsequent events paralleled those on the continent, with the progressive elaboration of a system of "qualifications" based on social class or extensive land holdings as prerequisites to the right to hunt or even bear arms (Lund 1975). An act of 1671 extended the Crown's claim of wildlife ownership to animals outside the royal forests (Caughley 1983). The prerogative of the nobility to hunt persisted until 1831, when qualifications were finally abolished and hunting rights devolved to the landowner. The revolutions of the eighteenth and nineteenth centuries abolished many of the excesses of feudalism and wrested control of wildlife from the privileged classes throughout much of continental Europe. The resolutions that followed in the wake of these socio-political upheavals generally recognized the right to hunt as the prerogative of the landowner on his own property. These circumstances frequently brought with them elimination of all hunting restrictions and granted the owners of even the smallest parcels almost total hunting freedom. Given the small and fragmented land-tenure patterns prevailing in many parts of Europe, the dire consequences of these chaotic conditions soon manifested themselves in the near extermination of several species and drastic increases in the number of hunting accidents (Wolfe 1970).

Although the unrestricted right of the landowner to hunt on his own land persisted in some parts of Europe until the present, various countries of central Europe established laws that subject the prerogative to exercise this right to certain restrictions. For example, by 1850 Bavaria, Prussia, and Saxony all passed laws that linked the exercise of the landowner's right to hunt with the prerequisite of a specified minimum acreage of contiguous holdings (Ueckermann 1987). This separation of the right to hunt and the eligibility to exercise same is an important element of the central European hunting system (Bubenik 1989) often termed the district or Revier system. It should be noted, however, that the concept evolved more or less contemporaneously in Germany and parts of the Austro-Hungarian Empire.

The decades of the late nineteenth and early twentieth centuries witnessed the origins of a conservation ethic in various parts of central Europe, and with it the concept of wildlife as a common social possession shared by all the citizens of a given country. An important tenet of the legal philosophy underlying modern hunting law in these countries is the concept of wildlife as the hunter's charge and its preservation for the remainder of society as his moral and social obligation. This philosophy found expression in the improved hunting legislation of several central European countries. For example, the Reich's Hunting Law of 1934 centralized game administration throughout Germany and incorporated many of the positive features of its antecedents in the various German states. These included the provision for obligatory harvest plans and the requirement of hunter certification examinations for all prospective hunters.

The political restructuring of central Europe following World War II produced free enterprise and centralized analogs of the district system on the respective sides of the Iron Curtain. In West Germany for example, attempts by the occupying forces to impose a license system patterned after that in the United States were unsuccessful (Katzenmeier 1961). At the same time over exploitation of wildlife produced conditions reminiscent of those following the revolutions of a century earlier. A new law, passed in 1952, retained many of the positive features of the Reich's Hunting Law. In contrast to its precursor, however, the Federal Hunting Law decentralized wildlife administration, providing instead a legislative framework within which the individual German states were to enact their own hunting laws.

Following World War II, and continuing until very recently, the imposition of centralized economies in East Germany and several other countries of central and eastern Europe meant that hunting rights were divorced from land tenure and became controlled by the state. In many cases, this was accompanied by the expropriation of private land holdings into large collective enterprises. Hunting in these countries was based on a district system, with the major difference being that the districts were not operated as free enterprises but rather as collectives, with the proceeds from the sale of game going to the state. (cf. Isakovic 1970).

CONTEMPORARY SYSTEMS

Modern European hunting systems comprise two or three basic variants. The essential differences between the systems reside in the access to wildlife and its disposition once the animals have been reduced to possession. *De jure* at least, the concept of live, free-ranging wild animals as a common social possession shared by the citizens of a given country, constitutes an almost universal element regardless of the system involved.

This underlying philosophy is, however, subject to two differing legal conceptions regarding the precise juridical status of wildlife. One of these, characteristic of most western European countries, treats wild animals as essentially ownerless entities (*res nullius*). In those countries with centralized social-political systems, and since 1978 in Italy (Cassola 1981), wildlife belongs to the state (*res communitatis*).

It is important however to distinguish between the regulations imposed by a given country governing the eligibility to hunt and the "hunting rights holder" *per se*. Virtually all European countries prescribe that a prospective hunter acquire some form of hunting permit. However, the stringency of the conditions attached to obtaining a permit differ considerably between countries, ranging from minimal licensing requirements in Greece, Italy, Portugal, Spain, and Turkey to very rigorous in the countries of central Europe, and more recently France and the BENELUX countries. In the latter case, obtaining the initial hunting permit is subject to a certification, which generally entails extensive instruction and sometimes a formal apprenticeship, and finally passing a difficult examination.

The mere possession of a valid hunting license/permit does not of course guarantee the bearer an opportunity or place to hunt. To do so, he must either acquire the hunting rights for a given tract of land or be invited (either gratis or for a fee) by the person holding these rights.

Generally speaking in Europe, the "right to hunt" is considered as an inherent privilege associated with land tenure. The right to pursue and reduce free-ranging wildlife to possession resides with the landowner or the lessee of a given tract of land and that person acquires title to same, regardless of whether he actually killed it or not. Because the sale of game on the open market is legally sanctioned, the landowner/lessee has vested interests in the wildlife inhabiting his property and thus an economic incentive to manage same.

Hunting rights may be leased separately from other land uses' distinct legal prerogatives, and mere occupation of a tract of land by an agricultural or pastoral tenant does not convey a priori the right for him to hunt there (Wolfe 1979). Moreover, the tenant may not interfere with the landowner exercising this right or leasing same to a third party (Peterle 1958). In most cases, however, the agricultural tenant is legally entitled to compensation for verifiable damages to his crops by certain game animals or those resulting from hunting.

The most refined version of the land-tenure based system of hunting rights is the central European variant. The quintessential elements of this system are:

a) hunting districts with legally prescribed minimum size requirements;

b) the obligation of wildlife conservation and habitat management as an integral element of the right to hunt;

c) legally mandated limitations on game husbandry including compulsory harvest plans to protect other land-use interests; and

d) adherence to a strict code of venatic ethics.

The single most characteristic feature of this system is the hunting district or Revier. Strictly speaking, the right to hunt resides with the landowner, but the eligibility to exercise this prerogative is subject to various prerequisites. First, the person must possess a valid hunting permit. Secondly, the size of his contiguous holdings must meet some prescribed minimum. Should, however, the landowner's holdings fail to meet the minimum specifications, they are pooled with the parcels of other landowners from the same community to form a series of "communal" or cooperative hunting districts.

The proprietors of the lands in such a district comprise a legal corporation, and they may exercise the hunting rights in the Revier cooperatively or lease them to eligible persons. Revenues derived from leasing the district or from the economic returns from the sale of hunting concessions, meat, hides, etc., are apportioned among the members of the corporation according to fraction of the hunting district that their lands comprise. To be eligible to lease a communal hunting district, a person must have been a licensed hunter for at least three years.

The primary justification for the prescription of minimum acreages and lease terms is to promote consistent management on tracts that approach the spatial requirements of some of the more important game species. The system has been criticized because it excludes small landowners from hunting. Indeed, such stipulations could be construed as a throwback to the "qualifications" of early English wildlife law. While this criticism may be justified to some extent, the system is probably appropriate for many areas of western Europe, where land holdings are small and fragmented. In actuality, the minimum acreage requirements constitute a compromise. Nisslein (1964) noted that from a purely biological standpoint, they do not match the requirements of the larger and more mobile game species. On the other hand, specifying larger management units—especially for private hunting districts—would exclude all but a very small fraction of landowners from exercising the right to hunt on their own property.

The prescribed minimum sizes for hunting districts differ among countries and even to some degree between geopolitical regions of a given country. In general terms, however, hunting districts are larger in the northern and eastern parts of Europe. For example, the mandatory minimum size for hunting districts in Austria, Belgium, West Germany, and parts of Switzerland ranges from 50 to a few hundred hectares. Hunting districts in the former East Bloc countries are in the order of 1,000 to 5,000 hectares (Taber 1961; Isakovic 1970; Nagy and Bencze 1973; Bubenik 1989). These differences are due in part to the lower population density, and generally larger average size of agricultural units in the latter area. It should be pointed out, however, that under the centralized political systems prevailing until very recently in these countries, the establishment of hunting districts was not subject to the artificial constraints of private land ownership. Instead, the boundaries of hunting districts could be based on natural habitat units.

As noted earlier, an important feature of the central European system is the concept of the hunting-rights-holder as the steward of wildlife on his district and the mandate to manage the game animals within certain legally prescribed constraints. For example, in the Federal Republic of Germany, the objectives of wildlife husbandry are prescribed as:

a) maintaining a diverse and healthy (presumably viable populations) wildlife resource;

b) preserving the life requisites (habitat) for wildlife;

c) balancing wildlife population levels according to aesthetic considerations and other recognized land-use activities; and

d) avoiding game damages to agriculture, forestry, and aquaculture.

It is pertinent that the legal responsibility for husbandry extends only to species specifically defined by the law as "game animals" even if they have no hunting season. Roughly twenty-five percent of the birds and mammals in the Federal Republic of Germany fall within the ambit of hunting law (Hofer and Syrer 1990).

The procedural vehicle for realizing the objectives of wildlife husbandry are the provisions for harvest plans for selected wildlife species (usually big-game animals). These are based on annual population estimates, are reviewed by local wildlife boards, and can be enforced partially if not fulfilled by the party responsible for a given hunting district.

Perhaps the most striking characteristics of the central European system are the low participation rates, which average less than 0.05% of the population, approximately one twentieth of the comparable fraction in the United States. By contrast the countries of southern and northern Europe have generally higher participation rates, for example, Italy with 3.9% of its population hunting (Cassola 1981). Under the central European system the rigorous requirements for obtaining a hunting license, while ensuring that hunters are well versed in the biology of game species and technical aspects of hunting science, also serve to discourage casual participation. Using the Federal Republic of Germany as an example, the highly selective nature of hunting is demonstrated by the escalation in the prices for leasing hunting rights and the fact that a disproportionately large fraction (about thirty percent of the persons leasing hunting rights, derive from the ranks of the economically privileged, i.e., merchants and corporate businessmen). However, hunting opportunities for hunters from lower income strata do exist, primarily for small game and surplus deer. Indeed their participation may be important in fulfilling shooting plans. Thus, not unlike their historical antecedents, these hunting systems accomplish a sustained yield of game species principally by means of limiting participation rather than by restricting the harvest per individual participant (Matthews 1986). By contrast in North America, where hunting evolved in a tradition of virtually unlimited participation and access to large tracts of public land, the primary aim of regulations to ensure a sustained yield of game species has been that of restricting the take per individual, thereby promoting an equitable distribution of the harvest.

One testimonial to the efficiency of the Central European system is the fact that several countries of western and northern Europe have incorporated some of its features into their own systems of game administration. Variants of the system are presently in place in sixteen different European countries—or parts thereof—from Norway to northern Italy and Belgium to Bulgaria (Bubenik 1989). Even in France, where the landowner long enjoyed the right to hunt on his own property, however small and fragmented the tracts might be, recent decades have witnessed the implementation of various reforms. These include provisions for the establishment of cooperative hunting districts on a voluntary basis, harvest plans for big-game animals, and a qualifying examination for prospective hunters (Raffin and LeFeuvre 1982).

This widespread acceptance does not imply that the district system is not without its shortcomings. During the past fifteen years, the system has been subject to criticism from both within Europe as well as from other quarters (Geist 1985). Most criticisms are directed at its artificiality and emphasis on the production of certain favored species, especially as manifested by a preoccupation with trophy quality. Leopold (1936) commented on the artificial nature of the system and the impact of excessive deer husbandry on forest vegetation in Germany and Czechoslovakia over half a century ago, and Wolfe and Berg (1988) concluded that the current situation in the Federal Republic of Germany remained basically the same. Geist (1985) branded the German system of wildlife management as *defacto* game ranching, incorporating intensive predator control, artificial feeding, extensive habitat manipulation, introduction of non-native species, and genetic manipulation of wildlife. Although perhaps an overstatement, the assessment is not without some justification.

Traditionally, the German public has been largely acquiescent in its limited access to wildlife, seemingly content to accept hunters and foresters as the guardians of wildlife. However, increased awareness of environmental issues during the past two decades has led to the recognition that the interests of game (i.e., deer) management and wildlife conservation are not necessarily synonymous, which in turn has produced a polarization of the two interest groups. Concomitantly, public support of hunting and conventional wildlife management has eroded (Schroeder 1986).

I would caution against misinterpreting these criticisms as a condemnation of the central European system of wildlife management. As Wolfe and Berg (1988) have noted, its problems lie less in the law than in an overzealous practice by powerful hunting lobbies. Nor would it be just to infer that analogous problems do not exist under other systems, that is, the North American license system. I am by no means advocating elimination of the existing system, but rather specific reforms to rectify some of the problems just described. In my view, the most important measure is one of an attitude adjustment that must come from within the ranks of the hunters themselves. It is that of nurturing a greater adherence to the spirit rather than the letter of the law, with emphasis on the conservation of a diverse and viable wildlife resource rather than the production of a few selected species.

Literature Cited

Bubenik, A. B. 1989. Sport hunting in continental Europe. In *Wildlife Production Systems: Economic Utilisation of Wild Ungulates,* eds. R. J. Hudson, K. R. Drew, and L. M. Baskin, 115–132. Cambridge: Cambridge Univ. Press.

Cassola, F. 1981. *La caccia in Italia.* LaNuova Italia, Florence.

Caughley, G. C. 1983. *The deer wars.* Heinemann, Auckland.

Endres, M. 1917. Die Ableitung des Wortes ''Forst.'' *Forstwissenschaftliches Centralblatt* 39: 90–101.

Geist, V. 1985. Game ranching: threat to wildlife conservation in North America. *Wildlife Society Bull.* 13: 594–598.

———. 1988. How markets in wildlife meat and parts, and the sale of hunting privileges jeopardize conservation. *Conservation Biology* 2 (15–26).

Hofer, D. and E. Syrer. 1990. Wildtiere in Paragraphendschungel. *Allgeemeine Forst Zeitschrift* 12/13: 296–297.

Isakovic, I. 1970. Game management in Yugoslavia. *Journal Wildlife Mgmt.* 34: 800–812.

Katzenmeier, P. 1961. Die Entwicklung des Jagdrechts in Westdeutschland nach dem zweiten Weltkrieg. *Trans. Inter. Union Game Biologists* 5: 305–311.

Leopold, A. 1936. Deer and dauerwald in Germany. *Journal Forestry* 34: 366–375, 460–466.

Lindner, K. 1937. *Die Jagd der Vorzeit.* Berlin: Verlag Walter de Gruyter and Co.

———. 1940. *Die Jagd im fruehen Mittelalter.* Berlin: Verlag Walter de Gruyter and Co.

Lund, T. A. 1975. British wildlife law before the American Revolution: Lessons from the past. *Mich. Law Review* 74: 49–74.

Matthews, O. P. 1986. Who owns wildlife? *Journal Wildlife Mgmt.* 14: 459–465.

Nagy, J. G., and L. Bencze. 1973. Game management, administration, and harvest in Hungary. *Wildlife Society Bull.* 1: 121–127.

Nisslein, F. 1964. Jagdbetriebliche Folgerungen aus dem Verhaeltnis zwischen der Groesse der Jagdbezirke und den Wohnraum einiger jagdbarer Tiere. *Zeitschrift F. Jagdwissen* 10: 2–8.

Peterle, T. J. 1958. Game management in Scotland. *Journal Wildlife Mgmt.* 22: 221–231.

Raffin, J. P., and J. C. Lefeuvre. 1982. Chasse et conservation de la faune sauvage en France. *Biology Conservation* 23: 217–241.

Salo, L. J. 1976. History of wildlife management in Finland. *Wildlife Society Bull.* 4: 167–174.

Schroeder, W. 1986. Jagd 2000. *Die Pirsch* 38: 773–777.

Stahl, D. 1979. *Wild, Lebendige Umwelt.* Verlag K. Alber, Freiburg/Munich.

Taber, R. D. 1961. Wildlife administration and harvest in Poland. *Journal Wildlife Mgmt.* 25: 353–363.

Ueckermann, E. 1987. Managing German red deer (*Cervus elaphus* L) populations. In *Biology and Management of the Cervidae,* ed. C. M. Wemmer, 505–516. Washington, D.C.: Smithsonian Inst. Press.

Wolfe, M. L. 1970. The history of German game administration. *Forest History* 94: 6–16.

———. 1979. European game administration. In *Wildlife Law Enforcement,* ed. W. F. Sigler, 5–9. Dubuque: Wm. C. Brown Co.

Wolfe, M. L., and F. C. Berg. 1988. Deer and forestry in Germany 50 years after Aldo Leopold. *Journal Forestry* 86: 25–31.

Epilogue

Since this was written, major socio-political changes have occurred in Europe: the reunification of Germany; the dissolution of the former Soviet Union; the emergence of several independent nations; and a shaky confederation of Russia and some of the former Soviet republics in its place; as well as efforts to restructure a centralized economy throughout the entire East Bloc to one based on free enterprise. It remains to be seen whether this grand experiment will be successful or whether a conservative backlash will occur. The question may well be posed as to what will happen to wildlife and the infrastructure for managing it throughout this extensive region.

Times of socio-political change and economic hardship often bring with them the decimation of wildlife resources. The infrastructure, personnel, and resources to combat poaching and encroachment in nature preserves is often inadequate or lacking. There are indications that some species in parts of the new Commonwealth of Independent States (CIS) are suffering population declines due to over exploitation. Such is the case with the sturgeon in the Caspian Sea area, which is subject to extremely heavy illegal fishing pressure. The fall of the Iron Curtain has revealed that these countries had some of the worst pollution problems among the industrialized nations of the world (cf. Medvedev 1990).

On the positive side, most of the nature preserves of the former USSR and the East Bloc countries remain intact, and there exists considerable international interest in maintaining the ecological integrity of these areas (Cohn 1992). The dismantling of these extensive police military apparatus formerly used to maintain the communist system has also provided some unexpected benefits, notably a diversity of wildlife in such locations as in the extensive no man's land along the former border between the two Germanys and on military reserves in some of the Baltic countries.

For the long term, I view the outlook as probably optimistic. Russia, the Baltic nations, and the countries of central and eastern Europe had occasion to see the potential economic value of wildlife-based recreation (e.g., hunting and eco-tourism). I believe that this will compel the existing governments to protect their wildlife resources. Regarding the question as to what form the wildlife management and enforcement might take, I predict some free-enterprise based variant of the former system in place in Czechoslovakia and Hungary emerging, probably not unlike the Austrian-German system. Within this framework, the local constituencies of the national sportsmens' associations will likely continue to play a pivotal role in management and law enforcement efforts (Khokhlov 1992).

Some indication of the current state of affairs in the CIS was provided by Kohl (personal communication) who participated in a fact-finding trip to Russia and Kazakhstan in November 1992. Many of the former republics still function under the old laws of the former Soviet Union and will continue to do so until new laws have been implemented. The two largest republics, Russia and Kazakhstan, have taken the lead in this respect, creating new ministries of Ecology and Natural Resources in early 1992. In both republics draft legislation for the protection of animals and plants has been submitted and is expected to become effective in the near future. Presently, Russia acts as the designated successor state to the former USSR and serves as the representative for most of the CIS questions relating to the Convention on International Trade in Endangered Species (CITES). It is expected that Kazakhstan and the Ukraine will become signatories of the CITES Convention in the near future.

However, the wildlife problems facing the new nations of central and eastern Europe cannot be solved by protection alone. Environmental pollution may pose even more serious problems to wild animals and their habitat than legal and illegal exploitation. The mere substitution of a free-market economy for communism does not guarantee that conservation concerns will be better served.

Literature Cited

Cohn, J. P. 1992. Central and Eastern Europe aim to protect their ecological backbone. *BioScience* 42(11): 810–814.

Khokhlov, A. 1992. The present state of hunting and wildlife conservation in Russia. In *Proceedings International Conference on Improving Hunter Compliance with Wildlife Laws,* ed. K. H. Beattie, 20–22.

Medvedev, Z. A. 1990. The environmental destruction of the Soviet Union. *The Ecologist* 20: 24–30.

Chapter 16

Indian Hunting and Fishing Rights

by Daniel H. Israel[1]

INTRODUCTION

Cases 1–7

During recent years, no aspect of wildlife regulation has generated as many legal battles and as many newspaper headlines as the conflicts between the states and tribes over Indian hunting and fishing rights. Indian rights have surfaced in the Northwest, where Indian and non-Indian fishermen have struggled for the right to take valuable salmon for commercial purposes. Throughout the West, the tribes and the states are pitted in struggles to determine the extent to which the states can regulate hunting and fishing by Indians off their reservations on former Indian lands. And in the Southwest and East, tribes and the states are contesting who has the right to regulate on-reservation hunting and fishing by non-Indians. This chapter will consider the origins of Indian hunting and fishing rights, both on and off reservations. In addition, this chapter will summarize the current tribal and state disputes over wildlife regulation and enforcement, and will review the legal principles which relate to the competing authority of tribes, the states, and the United States to control Indian hunting and fishing.

ORIGINS OF ON-RESERVATION HUNTING AND FISHING RIGHTS

The United States has the exclusive authority to establish Indian reservations.[2] When the United States establishes a reservation by treaty, statute, or executive order, the reservation is created for the exclusive use of the Indians.[3] An important part of the bundle of rights which belong to Indians upon their reservations is the right to hunt and fish exclusive of state jurisdiction under tribal control.[4] Based on these principles, many tribes have exercised their powers of self-government recognized by the United States in the Indian Reorganization Act of 1934[5] to enact comprehensive hunting, fishing, and trapping codes for their reservations.[6] The management plans are an expression of the sovereign powers of the tribes, particularly their right to exercise general governmental authority over their reservations. In *Warren Trading Post v. Arizona Tax Commission*[7] the Supreme Court stated:

> Congress has, since the creation of the Navajo Reservation nearly a century ago, left the Indians on it largely free to run the reservation and its affairs without state control, a policy which has automatically relieved Arizona of all burdens for carrying on those same responsibilities.

Through the doctrine of ''reserved rights,'' the courts have acknowledged that when reservations were set aside, they constituted a reservation of rights by the Indians rather

Cases 8–16

than a grant of rights to them.[8] The right of Indians to control reservation wildlife exists regardless of whether the reservation was created by treaty, statute, or executive order.[9] Further, the right to hunt and fish free of state regulation is reserved to Indians on their reservations regardless of whether the treaty, statute, or executive order expressly mentions hunting and fishing.[10]

Indian hunting and fishing rights implicit in the establishment of a reservation have been found to extend to waters bordering a reservation. In such situations, the courts look to the circumstances in which the reservation was created to determine whether the purpose of making the reservation was to include rights to utilize adjacent waters. In *Alaska Pacific Fisheries v. United States,*[11] the Supreme Court found that the reservation of ''the body of land known as Annette Islands'' included the adjacent fishing ground as well as the upland because:

> . . . the Indians could not sustain themselves from use of the upland alone. The use of the adjacent fishing grounds was equally essential. . . . The Indians naturally looked on the fishing grounds as part of the islands and proceeded on that theory in soliciting the reservation.

As with rights to the land itself and to water, timber, and minerals, Indian hunting and fishing rights are property rights of the particular tribe. Any destruction or diminishment of those rights would constitute a taking within the meaning of the Fifth Amendment to the U.S. Constitution and would entitle the tribe to compensation.[12] Not only are hunting and fishing rights protected by the Fifth Amendment, but the United States, through the course of its dealings with the Indians, has an obligation to protect property rights, including hunting and fishing rights, secured to the tribes. That relationship is one of a trusteeship or guardianship. It binds the United States to deal fairly and protectively with all Indian rights and to protect those rights, including hunting and fishing rights, from unlawful intrusion from state or private interests.

TRIBAL REGULATION OF ON-RESERVATION HUNTING AND FISHING

By reason of the sovereignty of Indian tribes,[13] and through congressional legislation,[14] Indian tribes retain the exclusive authority to regulate tribal property, including reservation wildlife.[15]

In the *United States v. Wheeler,*[16] the U.S. Supreme Court observed:

> The powers of Indian tribes are, in general *''inherent powers of a limited sovereignty which has never been extinguished.''* F. Cohen, Handbook of Federal Indian Law 122 (1941) (emphasis in original). Before the coming of the Europeans, the tribes were self-governing sovereign political communities. . . . Like all sovereign bodies, they then had the inherent power to prescribe the laws for their members and to punish infractions of those laws.
>
> Indian tribes are, of course, no longer ''possessed of the full attributes of sovereignty.'' . . . Their incorporation within the territory of the United States, and their acceptance of its protection, necessarily divested them of some aspects of the sovereignty which they had previously exercised. By specific treaty provision they yielded up other sovereign powers; by statute, in the exercise of its plenary control, Congress has removed still others.
>
> But our cases recognize that the Indian tribes have not given up their full sovereignty. We have recently said: ''Indian tribes are unique aggregations possessing attributes of sovereignty over both their members and their territory. . . . [They] . . . are a good deal more than 'private, voluntary organizations.' '' . . . The sovereignty that the Indian tribes retain is of a unique and limited character. It exists only at the sufferance of Congress and is subject to complete defeasance. But until Congress acts, the tribes retain their existing sovereign powers. *In sum, Indian tribes still possess those aspects of sovereignty not withdrawn by treaty or statute, or by implication as a necessary result of their dependent status.*

Cases 17–27

Several courts have held that the tribes' sovereign powers include the right to protect reservation wildlife and to control the exercise of hunting and fishing by both members[17] and nonmembers.[18] In most cases, the tribal power to regulate hunting and fishing takes the form of a tribal ordinance and tribal management plan. The management plans develop conservation measures, including the fixing of seasons and bag limits, and the manner of taking. Normally, the management plans impose license fees. Further, the management plans authorize tribal game wardens to enforce rules and regulations on hunting and fishing by both members and nonmembers. Members are prosecuted in the tribal criminal courts and nonmembers are either dealt with through a civil fine or prosecuted by the U.S. Attorney in the federal courts pursuant to 18 U.S.C. 1165.[19]

STATE REGULATION OF ON-RESERVATION HUNTING AND FISHING

Typically, the treaty, statute, or executive order which created an Indian reservation did not deal specifically with the question of state jurisdiction. Nevertheless, the courts have construed the creation of Indian reservations to preclude by implication extensions of state law to Indians on the reservation.[20] Courts have fashioned certain axioms of treaty construction which preclude the implication that the United States intended the state to control reservation affairs. Treaties must be interpreted as the Indians would have understood them,[21] doubtful expressions must be resolved in favor of Indians,[22] and the treaties must be construed liberally in favor of the Indians.[23] The general rule then is that the exercise of state power is preempted by the creation of a reservation pursuant to federal law for the use and occupation of Indians.[24]

Indian hunting, fishing, and trapping rights are not protected from state control or regulation by the creation of the reservation in the first instance, but when the right is embodied in a treaty, act, or agreement (either expressly or by implication), state jurisdiction is again prohibited on the theory that the state interference would constitute an interference with and taking of private property. The Supreme Court, in rejecting the notion that Congress impliedly authorized the State of Wisconsin to regulate Menominee hunting and fishing after the severance of the federal relationship between the Menominee Tribe and the United States stated:

> We find it difficult to believe that Congress, without explicit statement, would subject the United States to claim for compensation by destroying property rights conferred by treaty.[25]

Under these modern principles whenever an Indian reservation is created, hunting and fishing rights attach within the reservation boundaries, and unless specifically limited by the treaty, they belong exclusively to the tribe and they may be exercised free of state law. Further, the tribal hunting and fishing rights are not lost because Indian land within the reservation has passed out of Indian title and into non-Indian ownership. To the contrary, a change in land tenure merely converts the Indian rights from exclusive to nonexclusive.[26] For example, in *Leech Lake Band of Chippewa Indians v. Herbst,* the Court upheld the retention of Indian hunting and fishing rights even though Congress effected "a complete extinguishment of the Indian title" based upon an agreement between the United States and the Indians in which the Indians agreed to "grant, cede and relinquish, and convey . . . all our rights, title, and interest in and to the land."[27]

In some states Congress has authorized the presence of state civil and criminal jurisdiction over the reservations pursuant to Public Law 280. This statute, enacted in 1953, provides that in certain mandatory states and other optional states, state civil and

Cases 28–36

criminal laws shall apply as against Indians on the reservations. But Public Law 280 contains an express proviso for Indian hunting, fishing, and trapping rights and for the regulation of Indian hunting, fishing, and trapping. Public Law 280 provides:

> Nothing in this section . . . shall deprive any Indian or any Indian tribe, band, or community of any right, privilege, or immunity afforded under Federal treaty, agreement, or statute with respect to hunting, trapping, or fishing or the control, licensing, or regulation thereof.[28]

The courts that have considered the effect of Public Law 280 have consistently ruled that state jurisdiction in the area of hunting and fishing is not permitted, and that Public Law 280 Indian reservations have the same status as non-Public Law 280 reservations with respect to the regulation and control of hunting and fishing.[29]

On-reservation hunting and fishing, therefore, is within the exclusive authority of the Indian tribe or band governing the reservation. This exclusive authority exists regardless of whether the reservation was created by treaty, statute, or executive order. As a part of the authority of tribes to run their reservations, tribal governments are empowered to enact ordinances and fish and game management plans for the purpose of creating a comprehensive management and enforcement scheme over reservation wildlife. Tribal regulation is exclusive and state authority over reservation wildlife is not permitted in the absence of an express act of Congress.[30]

FEDERAL REGULATION OF ON-RESERVATION HUNTING AND FISHING

Consistent with the notion that tribes retain the exclusive authority to regulate on-reservation hunting and fishing, courts have ruled that the federal government will not be permitted to regulate hunting and fishing on reservations. A court has held that regulations promulgated by the Commissioner of Indian Affairs and the Secretary of the Interior concerning on-reservation fishing were beyond their authority in that they were not authorized under the treaty establishing the reservation.[31] Similarly, a federal tax on the exercise of treaty fishing rights within the waters of the reservation was found by one court to be unlawful.[32] Further, it has been held that the federal Migratory Bird Treaty entered into subsequent to the treaties creating the Indian reservations does not modify or extinguish the Indians' right to hunt.[33] Finally, a court has held that the hunting and fishing rights of Indians on-reservation are not restricted by the Bald Eagle Protection Act because the Bald Eagle Protection Act did not expressly abrogate the previously existing Indian rights.[34]

Generally, then, courts will not restrict or limit the exercise of reservation hunting and fishing in the absence of express congressional authority to do so. And while it is clear that Congress has the power to abrogate or modify **Indian** hunting and fishing rights in favor of federal or state regulation,[35] the courts have not permitted an abrogation of hunting and fishing rights unless there is clear and express congressional language or legislative history indicating an intent to extinguish.[36]

ON-RESERVATION HUNTING AND FISHING BY NON-INDIANS

The extent to which Indian tribes can extend their authority over reservation wildlife to exclusively regulate and tax non-Indian hunters and fishermen is unresolved. Ultimately, the issue may be resolved by the U.S. Supreme Court. In all likelihood, the final resolution of this conflict will depend upon the weighing of tribal versus state interests with respect to reservation wildlife. The tribes contend that by law and by practice they have undertaken, particularly in recent years, comprehensive management

Cases 37–44

and regulatory responsibility for the reservations and that as a result it is impractical and against sound conservation practices to distinguish between non-Indian and Indian sportsmen on the reservations. In contrast, the states contend that because the wildlife know no boundaries and because the reservations are located within the territorial limits of the states, the states should extend their general authority over non-Indians even when the non-Indians hunt and fish on the reservations.

The courts have consistently recognized that the tribes can bar non-Indians from the reservation.[37] Further, the courts have consistently recognized that the tribes can duplicate state law by imposing fees on non-Indian hunting and fishing and can create non-Indian seasons and bag limits.[38] Also, the courts have recognized that Congress, in enacting 18 U.S.C. 1165, intended to use a federal trespass statute enforceable by the U.S. Attorney in the federal courts as the exclusive vehicle for punishing non-Indian violators of tribal hunting and fishing codes.[39]

Currently, there are four cases before the courts presenting the issue of whether the tribes have the *exclusive* right to regulate all aspects of reservation hunting and fishing by nonmembers. In *Eastern Band of Cherokee Indians v. North Carolina*,[40] the U.S. Court of Appeals for the Fourth Circuit ruled in the context of a put-and-take fishing recreation program on nonnavigable streams located within the Eastern Cherokee Reservation that the state of North Carolina did not have a state interest sufficient to support its fees of non-Indian fishermen. The court gave considerable emphasis to the fact that the Eastern Band of Cherokee Indians, together with the U.S. Department of the Interior, had developed a comprehensive stocking and management program. The court concluded there was simply no state interest with respect to reservation fishing. A similar decision was reached with respect to hunting in *Mescalero Apache Tribe v. State of New Mexico*.[41]

Two U.S. District Court cases reached a contrary result. In *White Mountain Apache Tribe v. State of Arizona*,[42] the Arizona federal court ruled that Arizona, and not the White Mountain Apache Tribe, had the right to regulate and enforce non-Indian hunting and fishing on the Fort Apache Reservation. The court relied on the principle that states retain general authority over non-Indians throughout their territory, and on a finding that the imposition of state fees and regulatory authority on the Fort Apache Reservation caused no governmental or economic injury to the tribe. A similar decision was reached by the U.S. District Court for the Southern District of California.[43]

Central to this dispute is the weighing of tribal versus state interests over wildlife management on the reservation. Hence, the extent to which game and fish migrate on and off the reservation is critical. Also, the extent to which the state provides ongoing management and enforcement personnel and financing on the reservation is relevant to sustaining a state presence. Finally, the extent to which a given tribe has assumed on its own, or in cooperation with the United States, a comprehensive program for stocking and management of the reservation is significant to determining the degree of tribal involvement in the area. For example, recently, the U.S. Court of Appeals for the Ninth Circuit refused to prevent the state of Washington from enforcing its laws on Colville Indian Reservation fishing by non-Indians because the court found that the Colville Tribe had not evidenced an intent to remove all state interests from the reservation. To the contrary, the court found that while the Colvilles had developed a comprehensive management plan, the tribe had expressed an intent to allow the state of Washington to continue to impose its fees and to continue to enforce state laws regarding hunting and fishing as against non-Indians.[44]

These disputes have increased tensions between the various state wildlife agencies and the tribes. It is hoped that in the near future the courts will establish a guiding set of principles. Once that occurs, the tribes and the states can develop, hopefully on a

Cases 45–54

cooperative basis, a sound management program for the reservations. Many tribes may desire to retain exclusive financial and regulatory responsibility, while others may choose to share that responsibility with the states. In the end, hopefully, the common concern of the tribes and the states for the welfare of the reservation wildlife will lead to an improved situation.

OFF-RESERVATION HUNTING AND FISHING

Throughout the last half of the nineteenth century, most treaties entered into between the United States and the Indian tribes involved a substantial relinquishment of land by the Indians. And in many of those cases, the Indians retained, through express language in their treaties, the right to hunt and fish on the lands and waters of their former homelands while agreeing to live permanently on small reserves. The original hunting and fishing rights thus reserved have been characterized as off-reservation hunting and fishing rights. In recent years, these rights have generated a number of legal disputes and on occasion intense Indian and non-Indian confrontations.

States retain the inherent authority to regulate the taking of fish and game within their boundaries.[45] Normally, state law can be applied to Indians who are outside the reservation, but state law cannot be applied if it would ''impair a right granted or reserved by federal law.''[46] As a result, therefore, treaties entered into between the United States and the Indians providing for off-reservation fishing or hunting rights can override the state power to regulate the taking of game within its borders.[47]

It is necessary to examine carefully each treaty which provides for an off-reservation hunting or fishing right to determine how that treaty right limits state authority. Some of the most common reserved off-reservation rights are found in treaties with the Indians of the Northwest. Those treaties often reserved a right to fish ''at usual and accustomed places'' which is ''in common with the citizens of the territory.''[48] These treaties also reserved, on occasion, hunting rights, stating ''the privilege of hunting . . . on open and unclaimed lands'' is hereby reserved.[49] At other times, the United States and the Indians reserved the right of the Indians ''to hunt on the unoccupied lands of the United States so long as the game may be found thereon and so long as peace subsists among the whites and the Indians on the borders of the hunting districts.''[50] Similar treaty language can be found in other treaties of this era involving the Navajo, the Crow, and the Southern Cheyenne. Other off-reservation rights have been secured by agreement rather than by treaty.[51]

Off-reservation fishing rights have also been secured on the Great Lakes. In a series of treaties in 1837, 1842, and 1854, the Chippewa Indians in Wisconsin preserved on a temporary basis off-reservation rights in the interior lands and lakes, and permanent off-reservation rights in Lake Superior.[52] Because of the great importance to the Chippewa Indians of commercial fishing in the Great Lakes, the courts have implied commercial fishing rights from a treaty which relinquished territory,[53] and a treaty which set aside lands ''for the use of the Chippewas of Lake Superior.''[54]

Off-reservation hunting and fishing rights, like on-reservation hunting and fishing rights, raise questions about the competing authority of the tribes, the United States, and the states to regulate treaty hunting and fishing. The courts, in struggling with this problem, have not reached a final solution. Nevertheless, a principle is evolving which suggests that where off-reservation hunting and fishing rights exist, a dual system of regulation will develop whereby the tribes regulate their members and the states regulate the non-Indians. In such situations, both the tribes and the states must regulate the resources in question so as to protect the right of the other to exploit the wildlife.

TRIBAL REGULATION OF OFF-RESERVATION HUNTING AND FISHING

Cases 55–57

Consistent with the notion that tribes retain the exclusive right to regulate hunting and fishing by their members on-reservation is the concept that in reserving their historical off-reservation hunting and fishing rights at the time when significant amounts of reservation land were ceded, the tribes as a part of their off-reservation treaty right retained the power to regulate their members in the exercise of those rights.[55] In *Settler v. Lameer,* the Court of Appeals held that a tribe with an off-reservation right ''in common with citizens of the territory'' has the authority to arrest and prosecute tribal members outside the reservation for violation of tribal fishing regulations. The tribe's right to regulate its members as a part of the off-reservation treaty right itself is protected from state law under the Supremacy Clause of the U.S. Constitution.

A significant part of the tribes' right to regulate off-reservation hunting and fishing is the implementation of a comprehensive management plan for off-reservation hunting and fishing similar to the comprehensive management plan for on-reservation hunting and fishing. Tribes must issue identification permits for their members, must identify seasons and bag limits, usually in cooperation with the states, and must adopt adequate law enforcement and judicial machinery to arrest and punish members who violate their codes. At the same time, since the hunting and fishing activities are taking place off the reservation in an area under the general police power of the state, it is essential that tribes develop cooperative arrangements with the states for management control. Cross-deputization of tribal and state law enforcement officials is desirable as well as cooperative wildlife management agreements. An essential component to the notion of the tribe regulating the Indian activity and the state regulating the non-Indian activity is that each government has a responsibility to protect the wildlife resources for utilization by the other.

In *Ute Mountain Tribe of Indians v. State of Colorado,*[56] the Ute Mountain Tribe and the state of Colorado entered into a comprehensive consent decree dealing with the exercise of Ute Mountain hunting and trapping rights in a four-million-acre area of southwest Colorado preserved by the Brunot Agreement of 1874. The consent decree authorizes a joint tribal and state management plan for the area and requires the Ute Mountain Tribe to enact a comprehensive ordinance controlling all aspects of off-reservation hunting and trapping. The consent decree further describes in detail the particular small and large game covered by the off-reservation hunting rights and attempts to provide the Indians with the game which they require for their subsistence, ceremonial, and religious purposes, while limiting the Indians' right to hunt game not generally available or game not existing at the time of the establishment of the treaty right. Finally, the consent decree provides that the tribe must enact penalties and undertake active enforcement and prosecution of members who violate the tribal code so as to guarantee that the system of criminal enforcement undertaken by the tribe is comparable in all respects to that undertaken by the state of Colorado.

STATE REGULATION OF OFF-RESERVATION HUNTING AND FISHING

The most controversial aspect of off-reservation hunting and fishing is the extent to which the states may limit Indian fishing in order to protect important non-Indian commercial interests in the Northwest. The U.S. Supreme Court has dealt with the meaning of Northwest Indians' fishing rights ''at usual and accustomed places'' ''in common with the citizens of the territory'' on numerous occasions.[57]

Cases 58–65

On July 2, 1979, the U.S. Supreme Court upheld the right of Northwest Indians to take up to fifty percent of the annual harvest of salmon and steelhead in the waters of the state of Washington in *Washington v. Washington State Commercial Passenger Fishing Vessel Association.*[58] The 1979 decision of the Supreme Court climaxed a decade of litigation designed to determine the nature and scope of the ''in common with'' commercial fishing treaty rights of Washington State Indians. The opinion upheld virtually all of the treaty claims of the Indians and secured for the Northwest tribes their right to take up to fifty percent of the fish which pass annually through their usual and accustomed fishing stations. Hopefully, this decision by the U.S. Supreme Court will bring a greater degree of certainty and predictability to off-reservation fishing rights than has existed in the one hundred and twenty-four years since the treaties were negotiated in 1855.

While *Washington v. Washington State Commercial Passenger Fishing Vessel Association* upheld the off-reservation treaty rights of the Indians, many of the parameters of the fishing rights will still be determined by reference to earlier decisions. Let us now turn to those earlier decisions. The Supreme Court in analyzing the Northwest treaties has held that the Indians may be regulated by the state where such regulation is reasonable, necessary for conservation, and does not discriminate against the Indians. *Puyallup Tribe, Inc. v. Dept. of Game.*[59] In a subsequent decision in the same case, the Supreme Court made it clear that only state regulation which had been shown to be necessary to prevent destruction of the fish resources fit the ''necessary for conservation'' standard.[60]

The Puyallup cases are consistent with an earlier holding of the Supreme Court based on the same treaties.[61] In *United States v. Winans,* the court held that Indian rights were more extensive than those of the average citizen of the state and that any attempt by the state to treat the Indians on an equal basis would constitute ''an impotent outcome to negotiations and the convention, which seems to promise and give the word of the nation for more.''[62] The Supreme Court, also prior to the Puyallup cases, had recognized that the right of Indians to fish could not be conditioned upon the purchase of a state license.[63] While allowing state regulation of ''the manner of fishing, the size of the take, the restriction of commercial fishing and the like,'' the Supreme Court has restricted the type of regulations to which the Indians may be subjected to those which are required to conserve the resource. Thus, regulations applicable to Indians are not judged by the normal standards which govern the applicability of state law to its citizens generally. To the contrary, the states may regulate Indian treaty rights only when ''necessary for conservation.''

Decisions of the U.S. Court of Appeals for the Ninth Circuit have refined the limits under which states may regulate the off-reservation fishing rights in the Northwest. In *United States v. Washington* and *Sohappy v. Smith,*[64] the court stated that regulations necessary for the conservation of the resource meant that before a state can limit or restrict off-reservation fishing, it must show: (1) that the specific state regulation is required to prevent demonstrable harm to the actual conservation of the fish; (2) that the measure is appropriate to its purpose; (3) that tribal regulation or enforcement is inadequate to prevent demonstrable harm to the fish; and (4) that the conservation goals of the state cannot be achieved to the full extent necessary by restricting nontreaty sportsmen.

The limitations on state regulation developed in the context of the ''in common with'' fishing treaty rights of the Northwest have begun to find their way into off-reservation rights elsewhere. These standards have been applied to hunting in Oregon[65]

Cases 66–75

and in Colorado.[66] The courts have correctly understood that the concept of off-reservation rights requires a shared tribal and state responsibility regardless of what treaty language creates the off-reservation right. The Supreme Court itself has adopted a similar rule respecting Indian hunting on a portion of land which used to be a part of the Colville Indian Reservation.[67]

In *State v. Tinno*,[68] the Supreme Court of Idaho construed a treaty between the United States and the Shoshone-Bannock Tribe which gave the Indians "the right to hunt on the unoccupied lands of the United States so long as game may be found thereon and so long as peace subsists among the whites and Indians on the borders of the hunting districts" to provide an unqualified off-reservation fishing right. The court construed the words "right to hunt" as including fishing because of the Indians' understanding at the time of the treaty. While the court suggested that the right was without limitation, restriction, or burden, the court subsequently held that the treaty right to fish could be regulated by the state when necessary to preserve the fishery.[69] Similarly, the Michigan Supreme Court in *People v. LeBlanc*[70] held that Chippewas retained commercial fishing rights on Lake Superior and Lake Michigan under an 1835 Chippewa and Ottawa treaty which provided them with the "right of hunting on the lands ceded." The court, however, ruled that Michigan could regulate this fishing when necessary to prevent a substantial depletion of the fishery and in so doing adopted the limitations on state authority found in the Northwest cases.[71]

Off-reservation hunting and fishing rights may also be aboriginal in nature rather than secured by treaty. Aboriginal rights are rights of occupancy which have never been extinguished by an Act of Congress.[72] The Supreme Court of the State of Idaho in *State v. Coffee*[73] ruled that the Kootenai Indians retained an aboriginal right to hunt on open and unclaimed lands of their former aboriginal domain. In so holding, the Idaho Supreme Court adopted the off-reservation limitations on state jurisdiction over treaty rights and applied them to aboriginal rights.

FEDERAL REGULATION OF OFF-RESERVATION HUNTING AND FISHING

The United States has promulgated regulations for off-reservation treaty fishing.[74] The Department of Interior regulations provide for identification cards for Indians, identification of fishing equipment, and the framework for the issuance of substantive regulations to govern the exercise of treaty fishing rights. To date these regulations have never been fully implemented. Further, it is not altogether clear that given the authority of the tribes to regulate Indian hunting and fishing on or off the reservation that the United States has any inherent authority to regulate treaty rights.[75]

CONCLUSION

Indian hunting and fishing rights have been the subject of great turmoil over the past three decades. Numerous decisions have reached the U.S. Supreme Court. It is now apparent that tribes exercise broad powers over on-reservation hunting and fishing by their members. Further, it is also settled that tribes can exercise tribal authority over off-reservation hunting and fishing subject to state conservation restrictions when necessary to preserve a species of fish or game. However, the extent to which tribes can comprehensively regulate reservation wildlife is still unclear, for the courts have not reached a consensus on whether tribes can oust state jurisdiction and regulate non-Indian visitors who come to the reservation to hunt and fish.

Notes

1. Daniel H. Israel, Attorney, Native American Rights Fund, Boulder, Colo., 1980. Updated by William F. Sigler, 1994.

2. *McClanahan v. Arizona Tax Comm.*, 411 U.S. 164 (1973).

3. *United States v. John*, 437 U.S. 634 (1978); *United States v. McGowan*, 302 U.S. 535 (1938).

4. *Menominee Tribe v. United States*, 391 U.S. 404 (1968); *Alaska Pacific Fisheries v. United States*, 248 U.S. 78 (1918); *Spalding v. Chandler*, 160 U.S. 394 (1896).

5. *See* 25 U.S.C. 476, *et seq.; and see* the contemporaneous Solicitor's Opinion which describes the powers of Indian tribes, 55 I. D. 14 (1934).

6. Examples of tribes which have developed comprehensive hunting and fishing codes are the White Mountain Apache Tribe of the Fort Apache Reservation in Arizona, the Mescalero Apache Tribe of the Mescalero Reservation in New Mexico, the Eastern Band of Cherokee Indians of the Eastern Cherokee Reservation in North Carolina, and the Confederated Tribes of the Colville Reservation in the State of Washington.

7. 380 U.S. 685 (1965).

8. *See United States v. Winans*, 198 U.S. 371 (1905).

9. *See* n.3 *supra*.

10. *See Menominee Tribe v. United States*, 391 U.S. 404 (1968) and *Kimball v. Callahan*, 493 F. 2d 564 (9th Cir. 1974), *cert. denied*, 419 U.S. 1019 (1974).

11. 248 U.S. 78, 89 (1918).

12. *Menominee Tribe v. United States*, 391 U.S. 404 (1968); *Hynes v. Grimes Packing Co.*, 377 U.S. 86, 105 (1949); *Whitefoot v. United States*, 293 F. 2d 658 (Ct. Cl. 1961), *cert. denied*, 369 U.S. 818 (1962).

13. Compare *Worcester v. Georgia*, 31 U.S. (6 Pet.) 515 (1832), which first recognized the sovereign status of Indian tribes, with *McClanahan v. Arizona Tax Comm.*, 411 U.S. 164, 172 (1973), which modified the notion of sovereign power and focused on the source of tribal authority in specific acts of Congress.

14. *See*, e.g., Indian Reorganization Act of 1934, 25 U.S.C. 476, *et seq.*, and 18 U.S.C. 1165.

15. *See McClanahan v. Arizona Tax Comm.*, 411 U.S. 164, 181 (1973); *United States v. Washington*, 520 F. 2d 676, 688 (9th Cir. 1975); *Settler v. Lameer*, 507 F. 2d 231, 236–37 (9th Cir. 1974); *Quechan Tribe of Indians v. Rowe*, 531 F. 2d 408, 411 (9th Cir. 1976).

16. 435 U.S. 313 (1978).

17. *Settler v. Lameer*, 507 F. 2d 231 (9th Cir. 1974) and *United States v. Washington*, 520 F. 2d 676 (9th Cir. 1975); *see also Menominee Tribe v. United States*, 391 U.S. 404 (1968).

18. *Eastern Band of Cherokee Indians v. North Carolina*, 588 F. 2d 75 (4th Cir. 1978) *affirmed*. 9 ELR 20106. and *Quechan Tribe of Indians v. Rowe*, 531 F. 2d 408 (9th Cir. 1976).

19. Recently the U.S. Supreme Court in *Oliphant v. Suquamish Indian Tribe*, 435 U.S. 191 (1978), held that tribes do not retain the sovereign power to prosecute non-Indians in tribal courts for criminal violations of tribal law. Because the Supreme Court decision forbids only tribal criminal power, certain tribes have developed civil forfeiture and civil enforcement schemes utilizing citations and the taking of property as collateral for the payment of fines. Whether or not this method of enforcing tribal policy as it relates to reservation wildlife will be upheld by the courts remains to be seen.

20. *See*, e.g., *McClanahan v. Arizona Tax Comm.*, 411 U.S. 164 at 174–75 (1973). Many of the ideas discussed in this chapter on state and federal jurisdiction over Indian hunting and fishing were derived from an excellent presentation on Indian hunting and fishing rights prepared for the American Indian Policy Review Commission by David H. Getches, Esq. in 1976.

21. *United States v. Winans*, 198 U.S. 371 (1905).

22. *Alaska Pacific Fisheries v. United States*, 248 U.S. 78, 89 (1918).

23. *Tulee v. Washington*, 315 U.S. 681, 684–85 (1942).

24. This rule is in contrast to the rule that where a non-Indian federal enclave is established, state laws continue in effect. *Kleppe v. New Mexico*, 426 U.S. 529, 543 (1976); *Surplus Trading Co. v. Cook*, 281 U.S. 647, 650–51 (1930).

25. *See Menominee Tribe v. United States*, 391 U.S. 404, 413 (1968); *Kimball v. Callahan*, 493 F. 2d 564 (9th Cir. 1974), *cert. denied*, 419 U.S. 1019 (1974); *Kimball v. Callahan* (II), 590 F. 2d 768 (9th Cir. 1979).

26. *See Kimball v. Callahan* (II), 590 F. 2d 768 (9th Cir. 1979); *Puyallup Tribe, Inc. v. Washington Game Dept.* (Puyallup III), 433 U.S. 165 (1977); *Leech Lake Band of Chippewa Indians v. Herbst*, 334 F. Supp. 1001 (D. Minn. 1971).

27. This holding is consistent with the definition of ''Indian country'' for jurisdiction purposes found in the federal criminal statutes which extends Indian country to all lands within a reservation and allotments notwithstanding the issuance of any patent and including rights-of-way. 18 U.S.C. 1151.

28. *See* U.S.C. 1162 (b).

29. *See Quechan Tribe of Indians v. Rowe*, 5431 F. 2d 408 (9th Cir. 1976) and *Klamath & Modoc Tribes v. Maison*, 139 F. Supp. 634 (D. Ore. 1956).

30. In *Puyallup Tribe, Inc. v. Washington Game Dept.* (Puyallup III), 433 U.S. 165 (1977), the U.S. Supreme Court carved out a narrow exception to this rule. In Puyallup III, the Supreme Court allowed the State of Washington to regulate on-reservation Indian and non-Indian fishing because the Supreme Court found that all but 22 acres of the Puyallup Reservation had been transferred out of Indian ownership, leaving no exclusively Indian land or waters behind. A second reason for authorizing on-reservation state authority was the important finding by the Supreme Court that an off-reservation treaty right for the benefit of both Indians and non-Indians could be destroyed by unregulated on-reservation fishing. To date, this case represents the only occasion where the Supreme Court has authorized state regulation of on-reservation hunting and fishing.

31. *Mason v. Sams,* 5 F. 2d 255 (W.D. Wash. 1925).

32. *Strom v. Commissioner,* 6 T.C. 621 (1946).

33. *United States v. Cutler,* 37 F. Supp. 724 (D. Idaho 1941).

34. *United States v. White,* 508 F. 2d 453 (8th Cir. 197+).

35. *Lone Wolf v. Hitchcock,* 187 U.S. 553 (1903).

36. *Menominee Tribe v. United States,* 391 U.S. 404 (1968). For an excellent discussion of the creation of treaty rights and their subsequent modification or extinguishment, *see* Wilkinson and Volkmann, ''Judicial Review of Indian Treaty Abrogation: 'As Long as Water Flows, or Grass Grows Upon the Earth'—How Long a Time is That?,'' 63 Calif. L. Rev. 601.

37. *Quechan Tribe of Indians v. Rowe,* 531 F. 2d 408 (9th Cir. 1976); *State of California v. Quechan Tribe,* 424 F. Supp. 969 (D.S. Calif. 1977), *vacated,* 595 F. 2d 1153.

38. *Quechan Tribe of Indians v. Rowe,* 531 F. 2d 408 (9th Cir. 1976); *Eastern Band of Cherokee Indians v. North Carolina,* 588 F. 2d 75 (4th Cir. 1978) *affirmed.* 9 ELR 20106. *Confederated Tribes of the Colville Indian Reservation v. State of Washington,* 591 F. 2d 89 (9th Cir. 1979).

39. *See United States v. Finch,* 548 F. 2d 822 (9th Cir. 1976), *remanded with order to dismiss,* 433 U.S. 676 (1977); *cf., United States v. Sanford,* 547 F. 2d 1085 (9th Cir. 1976). Prior to the decision in *Oliphant v. Suquamish Indian Tribe,* 435 U.S. 191 (1978), which ruled that tribal courts do not have the authority to impose criminal penalties on non-Indians, it was widely assumed that non-Indian violators of tribal hunting and fishing codes could be prosecuted either in tribal court or by the U.S. Attorney in federal court pursuant to 18 U.S.C. 1165. Subsequent to *Oliphant,* it is now clear that offenders can be prosecuted only by the U.S. Attorney.

40. 588 F. 2d 75 (4th Cir. 1978). *affirmed.* 9 ELR 20106.

41. U.S.D.C. N.M. No. 77–395–M (Civil), decided Aug. 7, 1978. (10th Cir. 1982). *affirmed, reinstated.* 677 F. 2d 55.

42. U.S.D.C. Ariz. No. 77–867 Civ., decided June 13, 1978. (9th Cir. 1981). *vacated and remanded.* 649 F. 2d 1274.

43. *State of California v. Quechan Tribe,* 424 F. Supp. 969 (D.S. Calif. 1977), *vacated,* 595 F. 2d 1153. *See also State v. Danielson,* 427 P. 2d 689 (Mont. 1967); cf., *Donahue v. California Justice Court,* 15 C.A. 3d 557 (Ct. App. 1971), 93 Calif. Rptr. 310, *cert. denied,* 404 U.S. 990.

44. *Confederated Tribes of the Colville Indian Reservation v. State of Washington,* 591 F. 2d 89 (9th Cir. 1979).

45. *Hughes v. Oklahoma* ____ U.S. ____ , 47 U.S.L.W. 4447 (Decided Apr. 24, 1979).

46. *Mescalero Apache Tribe v. Jones,* 411 U.S. 145, 148 (1973).

47. *Missouri v. Holland,* 252 U.S. 416 (1920).

48. *See* Treaty with the Yakimas, 12 Stat. 951.

49. *See* Treaty of Medicine Creek, 10 Stat. 1132.

50. *See* Treaty with Shoshone and Bannock, 15 Stat. 673.

51. *See* e.g., the agreement between the Blackfeet Indians and the United States in 1885 and the agreement between the United States and the Ute Indians in 1874. Agreements rather than treaties were utilized by the United States and the Indian tribes to cede lands after 1871. There is no legal differentiation as between rights secured by agreement as opposed to those secured by treaty, because all agreements had to be ratified by Congress.

52. *See,* e.g., Chippewa Treaty of 1854, 10 Stat. 1109.

53. For example, in Michigan the Chippewa Indians retain wide commercial fishing rights in Lake Superior under a treaty which provided: The Indians stipulate for the right of hunting on the lands ceded, with the other usual privileges of occupancy, until the land is required for settlement. *See People v. Leblanc,* 399 Mich. 31, 248 N.W. 2d 199 (1976).

54. *See State v. Gurnoe,* 53 Wis. 2d 390, 192 N.W. 2d 892 (1972) and *People v. Jondreau,* 384 Mich. 539, 185 N.W. 2d 375 (1971).

55. *Settler v. Lameer,* 507 F. 2d 231 (9th Cir. 1974); *United States v. Washington,* 520 F. 2d 676 (9th Cir. 1975); *Kimball v. Callahan* (II), 590 F. 2d 768 (9th Cir. 1979).

56. D. Colo. 1978, Civil Action No. 78–C–0220.

57. *See* e.g., *United States v. Winans,* 198 U.S. 371 (1905); *Seufert Bros. Co. v. United States,* 249 U.S. 194 (1919); *Tulee v. Washington,* 315 U.S. 681 (1942); *Puyallup Tribe, Inc. v. Dept. of Game* (Puyallup I), 391 U.S. 392 (1968); *Department of Game v. Puyallup Tribe* (Puyallup II), 414 U.S.

44 (1973); *Puyallup Tribe, Inc. v. Washington Game Dept.* (Puyallup III), 433 U.S. 165 (1977).

58. ____ U.S. ____ , 47 U.S.L.W. 4978 (Decided July 2, 1979).

59. (Puyallup I), 391 U.S. 392 (1968).

60. *Department of Game v. Puyallup Tribe* (Puyallup II), 414 U.S. 44 (1973).

61. *United States v. Winans,* 198 U.S. 371 (1905).

62. *Supra* at 198 U.S. 380.

63. *Tulee v. State,* 315 U.S. 681 (1942).

64. 520 F. 2d 676 (9th Cir. 1975); 529 F. 2d 570 (9th Cir. 1976).

65. *See Kimball v. Callahan* (II), 590 F. 2d 768 (9th Cir. 1979).

66. *See Ute Mountain Tribe of Indians v. State of Colorado,* D. Colo. 1978, Civil Action No. 78–C–0220, Consent Decree.

67. *Antoine v. Washington,* 420 U.S. 194 (1975). *See also Holcomb v. Confederated Tribes of the Umatilla Indian Reservation,* 382 F. 2d 1019 (9th Cir. 1967).

68. 94 Id. 759, 497 P. 2d 1386 (1972).

69. 497 P. 2d at 1393.

70. 399 Mich. 31, 248 N.W. 2d 199 (1976).

71. In reaching this decision, the Michigan Supreme Court appears to have modified its earlier ruling in *People v. Jondreau,* 384 Mich. 539, 185 N.W. 2d 375 (1971), dealing with a different treaty on Lake Superior where the court refrained from acknowledging any state right to limit Indian fishing.

72. *See Oneida Indian Nation v. County of Oneida,* 414 U.S. 661 (1974) and *Johnson v. M'Intosh,* 21 U.S. (8 Wheat.) 543 (1823).

73. 97 Id. 905, 556 P. 2d 1185 (1976).

74. 25 C.F.R. part 256.

75. *See* Hobbs, ''Indian Hunting and Fishing Rights II,'' 37 G. Wash. L. Rev. 1251, 1266. *See also Kake v. Egan,* 369 U.S. 60 (1962), where the Supreme Court held that fishing rights of Alaska Natives were not exclusive and that certain federal regulations could not exempt the Alaska Natives from state trapping regulations.

Chapter 17

Aircraft in Wildlife Law Enforcement

TYPE OF AIRCRAFT USED

The North Carolina Wildlife Resources Commission owns and operates four PA–18–150:s IFR certified—often called the ''supercub.'' It works well in law enforcement for several reasons: the observer sits behind the pilot, allowing an unobstructed view of both sides; it can be flown slowly, allowing both pilot and observer time to scan; it is quieter than other planes and helicopters, and less likely to attract attention to itself; takeoff and landing distances are short; and the operating costs and upkeep are less than larger planes.

The August Aviation Corporation offers an A109 SP helicopter for special performance. It can be equipped for police work and is convertible to an EMS configuration in several minutes. It can also be equipped for single-pilot IFR operation. Where the aircraft is to cover large areas such as the Gulf of Mexico, or is to fly at night over large land areas, a twin-engine aircraft is preferred to the single-engine. Twin-engine and some single-engine planes need:

(1) a digital distance-measuring device;

(2) Omni-directional radio equipment;

(3) airborne radar with a mapping capability;

(4) large fuel tanks;

(5) a radio on frequency with patrol craft or other ground vehicles; and

(6) drop buoys and other location aids, including 7 $\times$ 50 high-quality binoculars (Goodson et al. 1975).

PILOT QUALIFICATIONS

Pilots are generally hired from the ranks of the wildlife enforcement division. They should have an exceptional knowledge of fish and wildlife, boat laws, and the terrain over which they are flying. The pilot must know how to operate a plane by reading instruments and how to relay exact locations to ground crews (Rich and Shankle 1988).

The pilot must be able to fly designated routes and make decisions as when to conclude the mission due to weather, aircraft, or personnel problems. The pilot must know the position of the plane once a possible violator has been spotted. He takes Omni radio station and DME readings from predesignated stations as directed by the observer and concludes the surveillance when so directed. The observer, who is a co-partner, keeps a continual lookout for signs of violation; directs the pilot and takes Omni radio station and DME readings; records the readings as given by the pilot;

White-tailed deer.
Courtesy U.S. Fish and Wildlife Service, Luther Goldman.

maintains contact with the base radio station giving information on when the aircraft is airborne, where it is located at each precise time, possible violations stated by code name, and direction subject is moving; directs the pilot so that the subject will be in view; and takes charge of pursuit communications when both the mobile unit and subject are in view (Goodson et al. 1975).

ADVANTAGES OF AIRCRAFT IN LAW ENFORCEMENT

The obvious advantage of an aircraft is that it can cover large areas—often while undetected—at night and sometimes in daylight. Activities looked for from the air include illegal hunting of small and big game, trapping, and fishing. Plane personnel also look for waterfowl or dove hunters in baited areas. At night observers are mainly looking for lights, but during the day they look for vehicles, boats, and people (Rich and Shankle 1988).

Uses of an aircraft for other than law enforcement include tracking bear, deer, raccoon, otters; observation of eagles and their movement; pattern and nesting of birds; and waterfowl counts. Aircraft can also be used to determine extent of fish kills and occasionally find the source of the killing agent. Search and rescue is and always will be an important use of aircraft (Rich and Shankle 1988).

The use of aircraft in enforcement and other wildlife activities also has a side benefit. Very often violators who are aware that the division uses planes in enforcement believe that civilian planes flying the area are enforcement planes and, therefore, act in a more circumspect manner.

CITED CASES

The use of aircraft as an observation platform in law enforcement has become increasingly valuable in the last few years. Since it is generally a warrantless search, it may come under the umbrella of the Fourth Amendment. In recent U.S. Supreme Court

Cases 1–4

cases a warrantless aerial surveillance was upheld because it was ruled the government's conduct did not invade into a reasonable expectation of privacy. Therefore, it did not constitute a search or seizure under the Fourth Amendment to the Constitution (DiPietro 1989).

Most of the following cases are drug related rather than wildlife related. However, the Fourth Amendment principle remains the same. In *Ciraolo v. California,*[1] Santa Clara, California, police received an anonymous phone tip that marijuana was growing in Ciraolo's backyard. The police were unable to observe the yard from ground level because of a six-foot outer fence and a ten-foot inner fence completely enclosing the yard. The officers obtained a plane and flew over the house at an altitude of one thousand feet and readily identified marijuana plants. They also used a 35-mm camera to take photos. A warrant search based on these naked-eye observations was executed, and the marijuana plants were seized. In ruling that the warrantless aerial observation in Ciraolo's backyard did not violate the Fourth Amendment, the U.S. Supreme Court applied the *Katz v. United States*[2] test. The two-pronged test in *Katz* requires the Court to determine whether or not Ciraolo by his conduct exhibited a subjective expectation of privacy in his backyard. The Court began its analysis by concluding that Ciraolo met the subjective prong of the *Katz* test and clearly manifest his subjective intent and desire to maintain privacy as demonstrated by the two fences. The Court then addressed the second prong of the *Katz* principle which is whether or not society as a whole would recognize Ciraolo's expectation as objectively reasonable. Ciraolo argued that the yard was within the curtilage of his home (curtilage is generally considered to be the area immediately adjacent to a residence and to which the intimate activity related to the home and to domestic life are associated). He contended that he had done all they could reasonably expect of him to maintain privacy without actually covering the yard, which would defeat its purpose as an outside living area. The Court stated that even though the marijuana fell within the curtilage, the fact that the police could observe it meant that a person was knowingly exposing himself to the public even in his own home or office, and this is not subject to Fourth Amendment protection. The Court observed that a passing airplane or a power company repair mechanic on a pole could see into the yard. The Court continued that the security measures taken by Ciraolo were inadequate to prevent observations of his backyard. The mere fact that he had taken some measures to restrict view did not preclude an officer's observation from a public vantage point. The Court pointed out that the Fourth Amendment has never required law enforcement officers to shield their eyes when passing by a home on a public thoroughfare, and any member of the flying public in this airspace could have seen everything that the officers saw. In summary, the Court concluded that Ciraolo's expectation that his garden was constitutionally protected from naked-eye observation from a one thousand-foot altitude was unreasonable and not one society will honor. The Fourth Amendment does not require police who travel in public airways at an altitude of one thousand feet to obtain a warrant in order to observe what is visible to the naked eye (DiPietro 1989).

In *Florida v. Riley,*[3] the Supreme Court approved a warrantless surveillance of a partially covered greenhouse within resident curtilage from a helicopter at four hundred feet. This greenhouse was located ten to twenty feet behind Riley's mobile home on five rural acres. Two sides of the greenhouse were not enclosed, but the contents were obscured from view by surrounding property. The greenhouse was covered by corrugated roofing but had two panels missing. Officers, on an anonymous tip, determined that they could not discern the contents of the greenhouse and, therefore, used a helicopter flying at a height of four hundred feet. From this deck they observed with their naked eye[4] that marijuana was growing inside the greenhouse as viewed through the

Turkeys are among the largest and most difficult of all birds to hunt.
Courtesy U.S. Fish and Wildlife Service, Luther Goldman.

two missing panels. A search warrant based on these observations was executed and marijuana seized. The Court ruled that Riley could not reasonably claim the contents of his greenhouse were protected from aerial surveillance since he had not replaced the two panels. Although the inspection in *Riley* was made from a helicopter, the Court said that was constitutionally irrelevant since private and commercial helicopter flights in public airways are routine. The Court further noted that the FAA regulations permit helicopters to fly below limits established for fixed-wing aircraft if the operation is conducted without hazard to persons or property at ground level. The Court questioned, however, against assuming that compliance with FAA regulations automatically satisfies Fourth Amendment requirements. This is not necessarily the case since those regulations are intended to promote air safety rather than to protect the right against unreasonable search and seizure. The fact that a helicopter can fly over someone's home at virtually any altitude or angle without violating FAA regulations does not automatically defeat an individual's reasonable expectation of privacy under the Fourth Amendment. The question was whether or not the helicopter was in a public airway at an altitude at which members of the public regularly travel. The Court found that Riley failed to produce evidence that helicopters flying at four hundred feet are so rare that he could not reasonably have anticipated their observing his greenhouse. Officers using aerial observations below altitudes specified for fixed-wing aircraft should seek evidence of frequency and/or routine nature of such overflights from other sources such as local airport managers and flying schools. The Court stated that the intrusiveness of an aerial surveillance and the degree of disruption caused by it are relevant in assessing Fourth Amendment rights. In that regard the Court found there was no interference with Riley's use of his greenhouse or other parts of his curtilage. No intimate details connected with house or curtilage were observed nor was there any undue noise, wind, dust, and so forth (DiPietro 1989).

Case 5

In *Dow Chemical v. United States,*[5] the Court noted that one "may not legitimately demand privacy for activities out-of-doors in the fields, except in the area immediately surrounding the house." Dow had denied a request by the Environmental Protection Agency for an onsite inspection of its two thousand-acre facilities for manufacturing chemicals, which consisted of numerous covered buildings. The EPA employed a commercial aerial photographer using a standard floor-mounted aerial mapping camera and took photographs of the facility from altitudes of twelve thousand, three thousand, and twelve hundred feet. At all times the aircraft was lawfully within navigable airspace. Dow alleged that the EPA warrantless aerial photography violated the Fourth Amendment. In rejecting Dow's claims, the Court observed that the government has greater latitude to conduct warrantless aerial inspections of commercial property because expectations of privacy of an owner of a commercial property is significantly less than that accorded a private individual. While acknowledging Dow's reasonable and legitimate expectations of privacy in its covered buildings, the Court held that the warrantless taking of aerial photographs of the open area of Dow's plant complex from an aircraft lawfully in public navigable airspace was not a search. The Court also noted that this was without physical entry. The Court said admittedly the photographs gave EPA more detailed information than could be obtained with the naked eye. The photographs were limited to an outline of the buildings and equipment and were not of a revealing and intimate nature as protected by the Constitution. The Court noted that the camera, while expensive, was available to the general public and further suggested that a search per se would have occurred had the government used more highly sophisticated surveillance equipment not generally available to the public (DiPietro 1989).

Case 6

In *United States v. Oliver*,[6] the Supreme Court ruled that the Fourth Amendment does not recognize an expectation of privacy in open fields and it is unlikely that aerial observation of property in these fields would intrude upon a reasonable expectation of privacy. However, factors such as altitude, sensory enhancement, and intrusiveness take on greater significance with respect to observations of a curtilage. In open fields the courts look to factors such as public access to the area and the historical protection traditionally given a particular area (DiPietro 1989).

Literature Cited

DiPietro, L. A. 1989. Aerial surveillance: Fourth Amendment considerations. *FBI Law Enforcement Bull.* (Federal Bureau of Investigation, U.S. Dept. of Justice, Washington, D.C.) 58 (12): 18–24.

Goodson, B., M. Hogan, D. E. Curtis, D. C. Harris, and T. C. Lewis. 1975. In *Proceedings of the Annual Conference, Southeast Assn. of Game and Fish Comms.*

Rich, J. R., and T. E. Shankle. 1988. Aircraft in wildlife law enforcement. In *Proceedings of the Annual Conference, Southern Assn. of Fish and Wildlife Agencies.*

Notes

1. 476 U.S. 207 (1986).
2. 389 U.S. 347 (1967).
3. 109 S. Ct. 693 (1989).

4. 109 S. Ct. 693 (1989).
5. 476 U.S. 227 (1986).
6. 466 U.S. 170 (1984).

Recommended Reading

Haines, H. V. 1964. Use of aircraft in wildlife law enforcement. In *Proceedings Southeast Assn. of Game and Fish Comms.*, Vol. 18.

McQuerry, J. A., Sr. 1957. The most effective way to fight or prevent headlighting. In *Proceedings Southeast Assn. of Game and Fish Comms.*, Vol. 11.

Milstead, R. E. 1964. Aircraft employment in wildlife law enforcement. In *Proceedings Southeast Assn. of Game and Fish Comms.*, Vol. 18.

Swindell, T. E. 1957. Factors affecting and methods used in combatting the night hunting of deer in Florida. In *Proceedings Southeast Assn. of Game and Fish Comms.*, Vol. 11.

Multiple-Choice Questions

Mark only one answer.

1. Which statement is correct?
   a. advantages of aircraft use in wildlife law enforcement include ability to cover large areas quickly
   b. observed illegal activity may extend from trapping to baiting migratory birds
   c. nonenforcement uses of aircraft include census and life history studies
   d. wildlife agencies use aircraft in search and rescue operations
   e. all of the above

2. Which statement is true in regard to the Katz test (*Katz v. United States*)?
   a. enforcement officers may use aircraft to search private property if it does not violate the Fourth Amendment
   b. the Katz test is a three-pronged test
   c. the Katz test is a single test
   d. the Katz test is a two-pronged test
   e. it may not be used inside curtilage

3. In *Florida v. Riley* the defendant was convicted because:
   a. he carelessly did not replace two roof panels on his greenhouse where he was growing marijuana
   b. the area was not resident curtilage
   c. the government used a helicopter
   d. the search warrant was not tainted
   e. the substance was marijuana

4. Which statement is true regarding aircraft in wildlife law enforcement?
   a. the pilot does nothing but fly the plane
   b. the pilot may rarely take position readings
   c. the pilot needs little or no knowledge about wildlife and wildlife law
   d. the pilot must be a civilian
   e. the pilot should be a knowledgeable wildlife officer who can make personal observations and take point readings

5. Law enforcement personnel use what type of aircraft in enforcement?
   a. single-engine planes only
   b. twin-engines only
   c. rarely anything but helicopters
   d. four-engine planes usually
   e. none of the above is correct

Chapter 18

Recreational Law Enforcement

OVERVIEW

Recreational law enforcement is arbitrarily defined as dealing with any ''out-of-doors'' laws other than those that are narrowly defined as applying to fish and wildlife. Damage to the ecosystem can take many forms. The mechanics are varied and the reasons are generally ignorance, carelessness, maliciousness, or greed.

Ruts in fragile soil such as that in deserts, at high altitudes, or in the far north may last for decades or even centuries. Downhill ruts may start as shallow ditches and eventually become gulches. Ruts are commonly caused by snowmobiles, ATVs, snow cats and 4x4s, fossil-fueled vehicles, and, rarely, by skiers. Fast boats operated by anglers, hunters, bird watchers, or recreationists may at times harass waterfowl and shorebirds and increase erosion of shores or dikes. Cutting timber for lumber, firewood, or Christmas trees may be by professional thieves or by families on an outing. There is also a continuing problem with thieves who steal artifacts, cacti, and other plants that can be used in ornamental gardens.

Man-made fires can be very destructive to a watershed, especially if they are in a steep canyon. Both wind and water erosion are accelerated by fires and other factors adverse to the watershed. Riparian damage by livestock and, to a lesser extent, by people is quite destructive on some western lands.

Soil, water, and air quality can be discussed together since air pollutants may be, and often are, deposited on the soil, part of which will be carried during rains into nearby bodies of water. Pollutants originate from many sources: fossil-fueled vehicles; industry that exhausts sulfates and nitrates; manufacturers that produce toxic byproducts; radioactive products of uranium mines and acid effluents from inactive mines; and from nonpoint pollution agricultural activities. One of the most insidious problems today is the loss of wetlands, in part because some policymakers and politicians do not know, or refuse to believe, how much swamps and marshes contribute to the overall health of the ecosystem and, therefore, to humanity.

CRIMES ON FEDERAL LANDS

In 1950, the U.S. Forest Service hired its first criminal investigators to track down timber thieves, arsonists, antiquities poachers, drug manufacturers, and other woodland lawbreakers. During the following decade, when marijuana growers began sowing crops in remote forests, the enforcement branch of the Forest Service grew dramatically. The crimes the agents faced varied considerably. In the Southwest, agents found themselves matching wits with pot hunters, that is, thieves who steal artifacts from ancient Indian sites. One person was caught attempting to peddle the mummy of an

Indian infant. The asking price for this 1,350-year-old mummy was thirty-five thousand dollars. Thieves also steal dinosaur bones that command a high price in the United States and abroad. In some cases, people use bulldozers to rip apart ancient sites—destroying everything as they look for pots which can be grabbed readily—and then move on before being caught. Illegal amphetamine laboratories, which emit distinctive odors, sometimes locate in national forests because they are isolated from the general public. In 1987, one-third of the arrests were for the crime of growing marijuana. The Forest Service believes that 835,000 acres of Forest Service land are dangerous to the general public because the growers have installed booby traps, fired guns at people, and generally intimidate campers, hunters, bird watchers, and other law-abiding people attempting to use the recreational facilities of the national forests. Humans aren't the only victims of marijuana growers. The growers shoot deer and poison squirrels and rabbits which damage their plants. In spite of all this, the Forest Service believes that arson is the most destructive woodland crime. Unfortunately, these crimes are growing rather than receding (Wolkomir and Wolkomir 1988).

Drug runners are most frequently caught on county, state, or federal highways. In many cases these people are dangerous.

ILLEGAL USE OR DISPOSAL OF HAZARDOUS AND TOXIC MATERIALS

Under cover of darkness or behind the protective barrier of fences, guard stations, and corporate structures, some of the most dangerous and far-reaching crimes that law enforcement agencies have faced are occurring. These crimes affect not only the present—but future—generations, down to the basic necessities of life. They involve the illegal use and disposal of hazardous and toxic materials. The cost to taxpayers is billions of dollars and growing. For a long time most citizens viewed these materials simply as waste to be disposed of in the cheapest possible way. Many did not realize the negative, long-term public health effects. Early on it was realized that the environmental enforcement program in many states, run by technicians and scientists, was not adequate from a legal and investigative standpoint. New enforcement techniques had to be developed to handle the complexity of environmental laws. Doubting prosecutors had to be convinced that the program was worth their time and that the evidence was not flawed. The crucial part of developing a case against a violator is ensuring that all personnel adhere to the legalities so that the evidence can be admitted into court. Corporations, regardless of size, can make it difficult to interview people or obtain documents, especially if the officer is trying to compile a criminal case against them. Large corporate structures with batteries of attorneys and mind-boggling record keeping make environmental investigations a real challenge. Corporate legal staffs make it difficult for the investigators to obtain a search warrant. Environmental investigators also encounter problems in the area of jobs and public opinion versus the environment. In the economy today, this is a very real obstacle to a prosecuting attorney. The philosophy is still prevalent in some places in the criminal justice system that environmental laws should not be enforced in the same manner or to the same degree as regular criminal law classifications. Unfortunately, this attitude has a deadly price tag (Murphy 1983).

AGRICULTURAL CRIMES

Conventional belief is that crime is a city phenomenon. Unfortunately for farmers and ranchers, this is not true. Rural crime has increased drastically in the last twenty years. Crimes such as theft of farm equipment, grain, gasoline, livestock, lumber, and pesticides are more commonplace than ever before. In Dade County, Florida, the value of avocado, lime, and mango fruit stolen annually is one million dollars. Organized crime is heavily into agricultural theft and fraud. One "family" defaulted on payments to a group of Wisconsin dairy farmers, and in Indiana this same group defaulted on payments to a farmer's cooperative (Swanson and Territo 1980). In South Dakota, the attorney general obtained a grant to investigate organized criminal activities involving multistate grain frauds. Timber losses in western Washington have been reported to run a million dollars annually and it is estimated that seventy percent of the vandalism and theft of timber go unreported. Pesticide thefts from ranches and farms reportedly exceed two million dollars a year. Stolen are such things as cars, trucks, earthmovers, farm tractors, poultry, and beehives.

What can be done about this increased crime? Suggestions for protecting livestock include permanent brands, tatoos, or other identification markings, and signs posted conspicuously around the property indicating that all animals are permanently branded. When possible, livestock should be counted on a daily basis, and when the owner is absent neighbors should be asked to check the property. Fences and gates should be checked regularly, and farmers and ranchers should be encouraged to report thefts immediately. One company is marketing a uniquely numbered confetti, which is mixed with grain to reduce the desirability of stealing (Swanson and Territo 1980).

THEFT OF HEAVY EQUIPMENT

Theft of construction and farm equipment is a growing problem among manufacturers, distributors, owners, and law enforcement personnel. Construction and farm equipment thefts are often by professional thieves stealing on order, for stripping or for export. These pieces of equipment, unlike conventional motor vehicles, have no standard, permanently affixed identification numbers. Each manufacturer has its own numbering system which can vary from four to fifteen characters. Identification numbers vary in height, composition, and location, and they are easily removed. Heavy equipment may also have several identification plates for component parts. These can easily be confused with plates describing the overall equipment. Professional thieves can merely remove these identification plates and substitute counterfeit ones. Confidential numbers are welded over. Anti-theft devices that would help solve problems are costly options. Identification of individual pieces of equipment is further complicated by the fact that there is generally no registration or title required for off-road equipment (Lyford 1981).

Further, owners have the problem of inventory control since construction and farm equipment may be spread over several sites or fields and left for days or weeks at a time in isolated areas. The National Crime Information Center (NCIC) maintains computerized files of wanted and missing articles such as automobiles and heavy equipment. The purpose of the NCIC files is to provide law enforcement officers with timely and accurate information which enables them to determine whether or not an article or vehicle is stolen. However, if improper information is given, it may lead to inaccurate and, therefore, worthless data. Another problem that officers have is that heavy equipment may be known by different names in different parts of the country. The best thing a police officer can do is ask the equipment dealer to assist in locating and recording identification numbers, including the confidential ones. Owners can install fencing and adequate lighting near equipment, inventory frequently, and report losses as early as possible (Lyford 1981).

MARIJUANA IN NATIONAL FORESTS

Growing marijuana in the one hundred fifty-six national forests is a billion-dollar criminal industry. Forest Service efforts to eradicate marijuana are, in themselves, insufficient to control the problem. What is needed as a deterrent is to arrest and *convict* the growers and sellers of the plant. Seventy percent of the arrests in California in 1986 produced no convictions, in part because of inadequate investigations. Over the past two years the Forest Service has declared nearly one million acres of national forest off limits to the public because of a threat to their safety from hand grenades and booby traps. The growers may also be armed with machine guns and other automatic weapons and protected by Dobermans and pitbulls. Eradication helicopters have been fired on. Booby traps have seriously wounded hunters, and poisons are set out by the growers to kill wildlife to keep them from eating their plants. Forest Service officials believe that criminal cartels produce the marijuana, but they lack specific information. This relatively new and bizarre activity has become a hazard for professional foresters as well as hunters, anglers, and outdoor recreationists (*The Herald Journal,* Logan, Utah, Friday, May 27, 1988).

OVERUSE OF THE NATION'S RECREATION AREAS

Out-of-doors people are literally loving to death some of the great wonders of nature. National parks and national monuments, national forests, and Bureau of Land Management areas are threatened by the multitude of visitors. Sometimes it appears that the best way to cause an area to be overused is to declare it a national park. Overcrowding appears from the climber-crowded Colorado Rockies to the diver-packed Florida reefs. At America's second most visited national park, Acadia National Park on the coast of Maine, four million tourists jam twenty-seven miles of road each year. Engine emissions join with industrial contaminants from neighboring states to give Acadia the distinction of being the only national park with air pollution problems that require health warnings such as those issued in Denver and Los Angeles. At Canyon Lands National Park in Southern Utah crowds of four-wheel-drive enthusiasts each day spend three nerve-wracking hours crossing the nightmarish, rubble-laden, eight-mile road over Elephant Hill to the once-isolated Anasazi Indian ruins in the back country. When they arrive it is only to find that all campsites are taken.

This trampling of the public's "backyard" was never envisioned when the United States became the first country to realize that civilizing its frontier would quickly obliterate natural wonders and resources unless steps were taken to preserve them. Unlike other nations, the United States formally set aside hundreds of millions of acres to be protected. Today the national park system alone includes 24.6 million acres in fifty parks.

The mission of the Park Service has always been to preserve the park "unimpaired for future generations." Two other federal agencies, the Forest Service and the Bureau of Land Management, operate on the principle of multiple use, which means recreation must compete with activities such as logging, grazing, and mining. If the present trend in recreational activity continues it will probably take precedence over all other multiple uses. In 1990, the Bureau of Land Management recorded seventy-one million visitors, five times as many as there were ten years before.

The Park Service believes that by the year 2010 an estimated ninety million people will visit their already heavily stressed sanctuaries. The Bureau of Land Management states that by the year 2000 outdoor vacationer numbers will jump from forty-two to

Ice fishermen appear to enjoy freezing.
Courtesy Utah Division of Wildlife Resources.

sixty-four percent. Two groups of people, the baby boomers and the retired boomers, are hitting recreational areas hardest where they demand lodging, fast food outlets, and other amenities that are especially intrusive in natural settings. The birth explosion accompanied by the returning of World War II troops continued until the advent of the birth control pill in the early 1960s. During this period seventy-seven million Americans were born, thirty-one percent of the current U.S. population. The Census Bureau predicts that the number of Americans over the age of fifty-five, now at fifty million, will have grown to seventy-five million by the year 2010. These older Americans will account for one out of every four citizens (J. Coates, Droves of nature lovers trampling Nation's backyard. *The Salt Lake Tribune,* April 28, 1991).

Thousands of motorcyclists are being blamed for damaging the scenic attractions around Moab, Utah, which made it so appealing in the past. Federal land managers, law enforcement officials, and some town folks say the motorbike recreation activities are becoming too popular. Others like the bikers, who just spend their money and go home. In 1989, the first year counts were taken, 20,000 bikers went through the turnstile gate at trailhead. In 1991, 50,000 bikers used the trail, and by 31 May 1992, 46,500 bikers had already traveled the slick rock trail. This led Bureau of Land Management officials to predict a near doubling of use in 1992. The impact on the trail is visible. The smooth mounds of pink rock are now stained with a black path, the residue of thousands of knobby rubber cycle tires.

The Bureau of Land Management is more concerned about the recreational hazards of present camping practices than the increase in motorbike traffic. They cite such things as hacking down a two hundred-year-old juniper tree, cutting branches in undeveloped areas for firewood, and even cutting down trees so that camp trailers can be moved back into the area. While not criticizing any particular group of users of the public land, there must be protection for the very things that the recreationists travel from near and far to see (C. Smith, Bikes ride roughshod over land, says critics. *The Salt Lake Tribune,* Friday, July 10, 1992).

Desert bighorn sheep.
Courtesy U.S. Fish and
Wildlife Service, Charles
G. Hansen.

Literature Cited

Lyford, G. J. 1981. Heavy equipment theft. *FBI Law Enforcement Bull.* (Federal Bureau of Investigation, U.S. Dept. of Justice, Washington, D.C.) 50 (3).

Murphy, W. M. 1983. Enforcing environmental laws—a modern day challenge. *FBI Law Enforcement Bull.* (Federal Bureau of Investigation, U.S. Dept. of Justice, Washington, D.C.) 52 (11): 15–19.

Swanson, C. R., and L. Territo. 1980. Agricultural crime: Its extent, prevention and control. *FBI Law Enforcement Bull.* (Federal Bureau of Investigation, U.S. Dept. of Justice, Washington, D.C.) 49 (5): 8–12.

Wolkomir, J., and R. Wolkomir. 1988. Tree cops: Putting the pieces together. *Natl. Wildlife* 26 (4): 12–13.

Case Notes

Utah v. Bartley, 784 P. 2d 1231 (1989).

Defendant Bartley was convicted of theft, a third-degree felony in violation of Utah Code Ann. (76–6–404,–412(b)(i)(1978) by the Seventh District Utah Court, San Juan County. Bartley appealed, but the appeals court concurred with the decision of the district court.

The southern portion of San Juan County, Utah, is sparsely populated, consisting mainly of farms, ranches, and oil and gas wells. The sheriff's office had received reports for several months in late 1986 that "drip gas" thefts were occurring in the area. Drip gas is also known as "gas condensate." The sheriff received a specific report that three cars with liquid storage tanks were travelling near an oil-producing area. The sheriff proceeded to the area and observed that the three vehicles appeared to be lugged down and heavily laden. There were signs that the drip gas storage had been drained. The sheriff also noted that there was a rank odor of drip gas near the heavily laden cars. The defendant claimed the roadblock was illegal and that the sheriff did not have probable cause to search the vehicle because he did not know that a crime had been committed. The court ruled that evidence known to the sheriff, plus the distinct odor of drip gasoline, gave the sheriff probable cause to search the vehicles and the arrest was effected.

Wyoming Wildlife Federation v. United States, 792 F. 2D 981 (1986).

Wildlife federations brought action against the government and public utilities board with alleged violation of the federal environmental laws arising out of an easement granted to construct a water project. The United States District Court of Wyoming entered judgment on settlement agreement and awarded attorney fees to the wildlife federations. The government (U.S. Forest Service) appealed. The court of appeals held that: (1) wildlife federations achieved most of the objectives of the complaint and were "prevailing parties" entitled to attorney fees and costs, and (2) the government's litigation positions were not substantially justified and were insufficient to overcome the wildlife federations' entitlement to attorney fees and costs.

The litigation underlying the award of attorneys' fees, which is the subject of this appeal by the U.S. Forest Service, by and through defendant, the regional forester, who granted an easement to the defendant, City of Cheyenne Board of Public Utilities, to use land in the Medicine Bow National Forest for construction of stage II of the Cheyenne Water Diversion Project. The plaintiffs filed suit against the United States in the United States District Court for the District of Colorado, alleging violations of the Federal Land Policy and Management Act. The plaintiffs filed an application for costs and attorneys' fees under the Equal Access to Justice Act. In the settlement agreement, the Public Utilities Board promised to replace wetland values and the government agreed to enforce the board's promise to achieve some benefits of complaint which alleged that easement to construct water projects would violate executive order by failing to adequately protect and preserve wetlands and, therefore, the wildlife federations were entitled to fees and costs and "prevailing parties" even though the board made concession. The government also agreed to close newly constructed roads by the utility company to achieve objective of complaint which sought to prevent construction of the road in the area. The government also promised to enforce all mitigation measures including minimum stream flow goal of the wildlife federations. Engineers had omitted minimum required acreage of wetland mitigation which was deemed as not substantially justified. The circuit court stated that since the plaintiffs had won the original case they were entitled to payment of attorneys' fees and costs.

United States v. Coker, No. 78–2032, United States Court of Appeals, Tenth Circuit, June 15, 1979.

The United States District Court for the Eastern District of Oklahoma granted the defendant a motion to suppress certain evidence seized after a warrantless arrest and a warrantless search incident to arrest. The court of appeals held that there was no probable cause for arrest of the defendant, and thus the evidence seized incidental to the search was properly suppressed. The court of appeals noted that the defendant had been seen in the general area of a marijuana patch within a federal wildlife refuge, rumor had linked him to the patch and he was stopped because he ducked down in his car, but there was no reasonable basis for believing that he was responsible for cultivating the marijuana patch. The court further stated that the government bears a heavy responsibility when it seeks to justify warrantless arrests and searches. Further, in evaluating correctness of trial court's conclusion that there was no probable cause for arrest, the court of appeals is bound by the trial court's factual and credibility determinations unless they are clearly erroneous. The court of appeals said the only issue presented by this appeal is whether or not the trial court erred in granting the defendant's motion to suppress certain evidence seized by a warrantless search and warrantless search incident to arrest. In summary, the court of appeals said that at the time of the defendant's arrest there were lacking facts and circumstances sufficient to warrant a prudent man in believing that the defendant had committed or was committing an offense (*Gerstein v. Pugh*[1]). The court further noted that probable cause to arrest the defendant was lacking inasmuch as the search was incidental to the defendant's arrest. Evidence in the search was properly suppressed by the district court.

[1]*Gerstein v. Pugh*, 420 U.S. 103, 111, 95 S. Ct. 854, 862, 43 L. Ed. 2d 54 (1975).

State v. Baird, 763 P. 2d 1214 (Utah App. 1988)

The defendant was convicted of unlawful possession of a controlled substance before the Fourth District Court, Juab County, Utah. The defendant appealed and the court held that: (1) the officer lacked reasonable and articulate suspicion for the initial stop of the vehicle, and (2) the evidence discovered after the stop could not be used to justify it. The highway patrol officer was unaware of foreign state's color scheme for determining license plate sticker validity and, therefore, lacked reasonable and articulate suspicion to make the investigative stop of the automobile. The officer later testified that he stopped the automobile because something just struck him as funny about the sticker. After the stop, the officer noted the car had new tires and shocks, as well as a twisted-off gas cap and jack in the back seat of the automobile. The driver seemed to be confused about ownership of the car and there was a smell of marijuana. The officer, without the consent of the defendant, took the car keys and with the county sheriff and the county attorney conducted an inventory search of the car. The locked trunk was opened and one hundred sixty-five pounds of marijuana was found.

The appeals court noted that there are three levels of police-citizen encounters requiring different degrees of justification to be constitutionally permissible. The Utah Supreme Court lists these as: (1) a police officer may approach a citizen at any time and pose questions so long as a citizen is not detained against his will; (2) an officer may seize a person if the officer has an articulable suspicion that the person has committed or is about to commit a crime; however, the detention must be temporary and last no longer than necessary to effect the purpose of the stop; and (3) an officer may arrest a suspect if the officer has probable cause to believe an offense has been committed or is being committed. The ''reasonable suspicion'' standard has been applied in Utah courts in *Mendoza*[1] as follows: Mexican appearance, California plates, route, time, erratic driving with police car tailing two to six feet behind, and nervous behavior after the stop.

[1]*State v. Mendoza*, 748 P. 2d 181 (Utah 1987).

State v. Aquilar, 758 P. 2d 457 (Utah App. 1988)

The defendant was convicted of possession of a controlled substance with the intent to distribute it for value. The defendant appealed and the court held that: (1) the defendant did not fail to preserve objection to admissibility of evidence seized during the search of the van, and (2) even if the initial stop of the van was illegal, voluntary consent to search purged the taint of illegality and thus the evidence seized was admissible. The highway patrol officer that stopped Aquilar's car noted that it was travelling about 55 to 60 miles per hour and that it was about two car lengths behind a second vehicle, which the officer thought was too close.

State v. Arroyo, 796 P. 2d 684 (Utah 1990)

This case is on a writ of certiorari to review a decision of the court of appeals. This particular case presents important issues concerning the effect of consent searches and pretextual traffic stops under the Fourth Amendment of the United States Constitution. In summary, the decision of the court of appeals is reversed and the case is remanded to the trial court for an evidentiary hearing to determine the voluntariness of the consent, whether the consent was an exploitation of an illegal stop, and the scope of the consent. The decision of the court of appeals is, therefore, reversed and remanded. (The in-depth discussion of the above issue is much too long for use here. It is recommended that anyone wanting to explore the fine points of this should read the original court transcript.)

The facts on the case are that a highway patrol officer followed a pickup then pulled alongside it to observe its occupants and gauge its speed. The pickup's two occupants were Hispanic and the truck had out-of-state license plates. The trooper stopped the pickup and cited Arroyo, the driver, for following too closely and driving with an expired license. The trooper asked Arroyo's consent to search the truck and Arroyo agreed. The search uncovered approximately one kilogram of cocaine inside the passenger's side of the door panel of the pickup. Arroyo was arrested and charged with possession of a controlled substance with intent to distribute.

Arroyo moved to suppress the evidence on the grounds that the traffic stop was a pretext for searching the truck for evidence of a more serious crime. The trial court found that the testimony at the suppression hearing established the probability that no traffic violation occurred and that the alleged violation was only a pretext asserted by the trooper to justify his stop of a vehicle with an out-of-state license and with occupants of Latin origin. The trial court also ruled that the defendant consented to search of the vehicle. Nevertheless, the court granted Arroyo's motion and ordered suppression of the evidence. The state filed an interlocutory appeal in the court of appeals challenging the suppressed order. The court of appeals held that the traffic stop was an ''unconstitutional pretext.'' The appeals court also stated ''we are persuaded that a reasonable officer would not have stopped Arroyo and cited him for following too close except for some unarticulated suspicion of a more serious crime.''

Recommended Reading

Beattie, K. H. 1981. Warnings versus citations in wildlife law enforcement. *Wildlife Society Bull.* 9.

Davis, W. E. 1971. Outdoor recreation and law enforcement in the 1970s. In *Proceedings Western Assn. of Game and Fish Comms.*, Vol. 51.

Fansler, R. K. 1972. Water and boat safety-administered by a conservation agency. In *Proceedings Southeast Assn. of Game and Fish Comms.*, Vol. 26.

Gould, D. H. G. 1958. State conformance and enforcement of the new federal boating law (Public Law 85–911) on coastal waters. In *Proceedings Southeast Assn. of Game and Fish Comms.*, Vol. 12.

Jardine, A. O. 1971. Recreation-conservation law enforcement in the 70s. In *Proceedings Western Assn. of Game and Fish Comms.*, Vol. 51.

Lewis, O. W. 1969. The warden's role in recreational land use. In *Proceedings Western Assn. of Game and Fish Comms.*, Vol. 49.

Kennedy, J. L. 1977. An analysis of the effects of agency, officer, and user-related variables, on the provision of wildlife law enforcement. Ph.D. diss., Utah State Univ., Logan.

Steinhart, P. 1982. Killing for cactus. *Audubon* 84: 6.

Vance, D. C. 1982. Regulations and the wildlife resource. In *Proceedings Western Assn. of Fish and Wildlife Agencies,* Vol. 62.

White, J. 1964. A report on Florida's boating law after two years of operation. In *Proceedings Southeast Assn. of Game and Fish Comms.*, Vol. 18.

Multiple-Choice Questions

Mark only one answer.

1. Illegal activities on federal lands include:
   a. growing marijuana
   b. manufacturing amphetamines
   c. stealing Indian (Native American) artifacts
   d. robbing early American graves
   e. all of the above

2. Agricultural crimes include the theft of:
   a. grain
   b. farm machinery
   c. green oranges
   d. avocados
   e. all of the above

All but
B

3. The theft of heavy equipment from both farms and construction companies is a growing problem that is difficult to solve because: all but one
   a. the thieves are often professionals
   b. manufacturing numbering systems are universal
   c. the equipment has no standard permanently affixed identification numbers
   d. it is often located at out-of-the-way unsupervised places
   e. anti-theft devices are quite expensive

4. One of the following statements is not true:
   a. visitors seem to be loving some of our national treasures to death by simple overuse
   b. the damaging effects of overuse of national parks was never envisioned by early park supporters
   c. the mission of the Park Service has always been to preserve the park ''unimpaired for future generations''
   d. the Park Service believes that by the year 2010 an estimated ninety million people will visit their already over-stressed sanctuaries
   e. the easy way to solve the problem of park overuse is to arbitrarily limit the number of visitors on a day-by-day basis; everyone would understand and support this stand

5. Which of the following is the most destructive to a watershed?
   a. downhill racing by gasoline- or diesel-fueled vehicles
   b. riparian damage by domestic cattle
   c. illegal cutting of timber
   d. large area watershed fires
   e. the illegal creating of new roads by vehicles

6. What are some of the problems associated with illegal toxic waste disposal?
   a. a company defense is apt to try to equate cleanup costs with the loss of jobs
   b. public ignorance of the real dangers of toxic waste, perhaps because it is such a complex issue
   c. it is costly and needs highly trained people who may be in short supply, and the companies are apt to be slow to cooperate
   d. courts in the past have not always been sympathetic with toxic waste dumping problems, or perhaps it is that the investigators have not brought the proper legal proof before the court
   e. all of the above

Chapter 19

The Face of the Future

OVERVIEW

As a nation we are moving into some of the most turbulent years in history. Law enforcement will be more complex, dangerous, and different than ever before. The change in law enforcement is only a part of the overall, complex picture. We have moved through the agricultural and industrial ages and, now, we face the age of computers, satellites, space travel, fiber optics, fax machines, robots, bar coding and electronic data interchange, DNA genetic coding, and many others. That is probably only the beginning (Toffler and Toffler 1990).

Our national debt and continuing deficit will be more burdensome in the twenty-first century. Our educational processes will be more bogged down; water, air, and soil pollution will increase; and wildlife habitat will suffer from net loss of wetlands. Continued development for homes and industry will severely impact the animals' range. Hispanic and Asian immigrants will not understand our culture; nor we theirs. All of this will have an impact on wildlife resources.

We can expect the demographics to change noticeably in the next fifteen to twenty years. By the year 2000, one-third of all American children will be Black, Hispanic, or Asiatic. In some states this group will be a majority by the year 2010. In less than one hundred years the white race will be in the minority. The average age of whites will be older than other groups. These changes will have a profound influence on all aspects of law enforcement, including wildlife. This flood of immigrants of different races, ethnic groups, religions, and cultures will expect, or at least hope for, the material benefits of American society. Many will be untrained, uneducated, and only slightly conversant with the English language. Will they understand and how will they be expected to solve work force conditions in future America? Many will turn to welfare or menial jobs. Some will turn to crime. Some of the young will find an identity and self-respect simply by living outside the law. They probably will not understand American laws and ethics in renewable natural resource management (Trojanowicz and Carter 1990).

Historically, the role of law enforcement has been to maintain the status quo, but reliance on current practices will not prepare law enforcement agencies for the future. And to deal with change, enforcement personnel must understand the process of change. People may be viewed as the first hurdle in law enforcement. If you, as an officer, knock down that hurdle you might as well call it quits. For no matter how much law enforcement or law you know and understand, if you cannot react positively with people you will fail your law enforcement assignment. People view an officer either as friendly or as adversarial, that is, very few view the officer on neutral ground (Tafoya 1990). For example, the officer who is wearing a helmet, mirror glasses, and sidearm does not, at first glance, present a friendly or reassuring picture. Officers obviously should not part with their helmets and sidearms, but they certainly can take off their magnum-size mirror glasses when talking to someone. That can be the first step in establishing a pleasant relationship.

An area of enforcement that may have escaped public notice is the private security sector. It has expanded at a rapid rate and now employs close to two million people, twice the number of public law enforcement employees. In wildlife, private security is responsible in some areas for game ranches, private hunting preserves, salmon rearing, and private clubs on leased ground. Originally, enforcement agencies viewed private guards with cynicism or, at least, indifference. They are now viewed in a more positive light. They are accepted, if not as partners, at least as collaborators in many endeavors (Mangan and Shanahan 1990). It is obvious that research organizations doing confidential work cannot be adequately protected by public police. It is equally obvious that game ranches or large tracts of private land with an abundance of game cannot be protected by conservation officers. In most cases there is no conflict between conservation officers and private security agencies. There are potential exceptions: large licensee game farms may want to cut corners, as for example, capturing wild animals and claiming that they were reared on a farm. Toxic chemical plants may want to keep their disposal plans private, if they are not within the legal guidelines. Employers, with the aid of their private security agents, may be instrumental in influencing legislatures for their benefit. This very often will conflict with the best interests of the public. It may be that, sometime in the future, private security agents will be better paid, better educated, and more respected than the public law enforcement officers. If this happens it will be because the citizens defaulted in their duties of supporting and adequately financing the public law enforcement officers.

The first step in solving a problem is recognizing it. In looking toward the twenty-first century how do we recognize the multitude of problems? How do we adapt to the nearly irreversible changes that face us? These questions also imply the interplay of a multitude of governing sciences such as politics, economics, engineering, arts, history, social science, humanities, and perhaps above all others, ethics. It is understood that law enforcement is society's second line of defense against crime and sociopathic behavior. The first line is social disapproval of friends, family, neighbors, and peers. But, in change-racked America, people are less bonded to one another than they used to be and, therefore, social disapproval loses some, or all, of its power over some people. When this happens the law must step in until society is able to restore the "glue." And, until this is done, we can expect more, not less, crime. The technological revolution that is getting underway will bring new problems and also new tools to enforcement personnel and to criminals alike. Many of these will raise legal, political, and moral issues (Toffler and Toffler 1990). For example, James D. Watson, Director of the National Center for Human Genome Research, told a House Subcommittee in October 1991 that he fears a genetic fingerprinting in the hands of the malevolent. The Nobel laureate and one of the two people who first described the structure of DNA believes individuals should have the right to genetic privacy. Dr. Watson said the idea of a huge bank of genetic information on millions of people is repulsive, but he speculates that in ten years such forms of identification may be reduced to simple bar codes easily stored and transmitted. Witnesses at the hearing raised several questions about genetic testing, such as patient/doctor relationship and employer/prospective employee problems. David J. Galas of the Human Genome Research Program said that even for the more conventional medical information the ethical, legal, and social issues . . . have never been properly addressed. In the final analysis, politics will probably be the deciding factor. And it was admitted by Dr. Watson that ''fingerprinting'' individuals by genetic codes is in use in courts today (*The Washington Post* 1991).

The U.S. Fish and Wildlife Service established a national forensics laboratory at Ashland, Oregon, in 1987. Once they have an adequate baseline of data they may be able to identify animal species or subspecies through DNA analysis, and possibly even the general location from which the animal came.

We need to explore long-range options so that we can define the limits of government power and the rights of the individual. Both American society and its constitutional rights must be protected. We should try to understand the social changes resulting from fast-moving technology. Toffler and Toffler (1990) believe straight line trends are of little value. They are not recognized soon enough and they do not provide any explanation of why something is happening. Rather, the widening of our imagination is crucial to survival in a period of accelerating, destabilizing change. An early step is to ask questions. For example, what will the law enforcement budget be, how will its personnel be trained, what skills will they need, what tools will they have, what will be the structure of a department or division, what will be the racial or ethnic composition of the enforcement personnel, will recruits need to be bilingual or trilingual, how will pay scales compare to other professions (Toffler and Toffler 1990)? Will the power and wealth of the white male erode as their numbers decline? Will minorities band together or will they split apart and compete with each other? Are we entering an era of tolerance or one of hostility?

It is painfully obvious that society is changing and law enforcement agencies must change with it. To refuse to do so means partial or total failure. The military-type structure of rigid command and demand for unquestioning obedience is something that has worked in the past—and that is probably where it belongs. Ed Kozicky, on the other hand, firmly believes in the military structure of rigid command. He believes it will return (Personal communication 6–10–92). Law enforcement agencies need to objectively examine their policies and procedures and act accordingly. Service functions will be a very real part of enforcement in the future. It may be that many enforcement people will be more social engineers than crimebusters. It is obvious that there never will be enough enforcement people to catch all criminals. Therefore, enforcement needs help from the citizenry. They are on the same side and they should be a team. In the past, some administrators have felt that people, including their own personnel, needed to be coerced, controlled, and threatened. Many of the coming generation will respond negatively to this philosophy. Some will fight it, most will move away from it (Tafoya 1990). Tafoya also believes that it is vital that enforcement administrators understand that:

(1) there are powerful dynamics transfiguring virtually every facet of American society;

(2) the forces that are recasting social institutions will alter law enforcement;

(3) as society's values change so will that of law enforcement;

(4) to deal effectively with this diversity the process of change must be understood; and

(5) the role and goals of police must be clearly and concisely articulated.

In summary, it is believed that law enforcement can and will recognize the problems ahead and meet the challenge; there is too much talent and desire out there to do otherwise.

WILDLIFE LAW ENFORCEMENT

Not all, but many, of the problems that face law enforcement in general will affect wildlife law enforcement. The rapid advances in technology and the changes in racial composition in America in the twenty-first century will present a challenge to wildlife and environmental officers greater than they have ever seen before. There will be more people but relatively fewer will be white. There will be fewer hunters, trappers, and anglers, but even so they will have less space per capita than in earlier times. More homes are being built on what was once prime deer range. The result is fewer deer but more people-deer conflicts. In many states, deer-car collisions result in loss of game, damage to automobiles, and occasionally, injury or death to passengers. In the great cornfields of the Midwest there is no food or cover for pheasants or cottontail rabbits. The tendency toward a monoculture, the destruction of cover, plus the use of pesticides and herbicides, are devastating to wildlife. More lands are being closed to hunters and anglers because of damage to, or theft of, farm machinery and livestock, and because of potential liability claims or because the land has been leased to sports groups (Lindsley 1989).

Anti-kill individuals and groups will increase in the 1990s and the early part of the twenty-first century. One reason for this is that as America has become more urbanized there are fewer individuals with hunting, fishing, or trapping experience. They are not only not interested in these activities—they are less than sympathetic. The Asian and Hispanic immigrants will have little or no experience or cultural background to help them understand America's philosophy of wildlife management. If sports groups and managers continue to ridicule anti-kill people and shower them with what is considered irrefutable scientific data, the conflict will grow. An effort to find a common meeting ground should be made and hunters and anglers should try to understand, or at least appreciate, these people's often emotional approach. Lindsley (1989) points out that there are animals that can live in close proximity to people and those that cannot. In the former group there are coyote, fox, raccoon, skunk, deer, some species of fish, pheasant, several species of duck, cottontail rabbits, and in recent years, cougars, and many more. The latter group has wolf, grizzly bear, bison, wilderness animals, and animals of the far north. The people-compatible animals can cause serious problems when they become too abundant. In Cape Canaveral a couple of decades back, rabid raccoons bit several children. Coyotes live and reproduce inside the city limits of both Los Angeles and New York where they eat pet dogs and cats and occasionally bite children. Deer that are too abundant destroy shrubs, flowers, and seedling trees. Even cougars have recently become less afraid of people and, therefore, more dangerous.

The people who oppose killing not only should be informed but shown what happens when prolific animals become too abundant. Another way to counter the anti-kill movement is to form a strong coalition of constituents who favor scientific management of all wild animals, or to work with organizations such as the National Rifle Association (NRA) or the more recent Wilderness Impact Research Foundation (WIRF). WIRF is a movement designed primarily to offset extremist animal rights organizations such as PETA (People for the Ethical Treatment of Animals). The March 1991 WIRF meeting had more than two hundred thirty co-sponsoring organizations, totalling, it was claimed, twenty-five million members. They include the American Farm Bureau Federation, American Mining Congress, American Petroleum Institute, and the National Association of Manufacturers. They also include grass-roots coalitions of farmers, loggers, ranchers, off-road motorists, and hunters (Wood 1991).

The new activism includes regular conferences, impact studies, and documentaries. Through its network of phones, fax machines, newsletters, and grass-roots organizations it is learning how to muster defense for local fights. Groups such as the Sierra Club want the federal government to put much more land into wilderness, and WIRF says they have enough. Bruce Hamilton, Director of the Field Services for the Sierra Club, believes that at field hearings its opponents are turning out contingencies so sizeable they neutralize the activity under dispute (Wood 1991). Those who do not want to go that far to the right, or at least toward the center, should ask themselves, What are the options?

Education is an option that we fall back on when we don't know which way to go. It is very important at several levels. Young people who can be weaned at least partially from the TV, VCR, ATV, and other nonexercise recreation will be challenged mentally and physically and will have a chance to meditate on the wonders of nature if they go hunting or fishing.

News media are important tools that officers can use to educate the public. Captain Norman Galman, Chief Environmental Conservation Officer, Region Four, New York State Department of Environmental Conservation, proposes several ideas for law enforcement officers when dealing with the news media. When you get a call from the media find out what they want to talk about, their deadline, and then get the spokesperson they want. In an interview tell the truth, stay on the record, speak in understandable terms, and avoid playing favorites with the hard news. Giving scoops to a reporter on feature articles is okay. Understand the reporter's viewpoint and the pressures he or she faces in covering the news. Reporters are expected to know a little bit about everything, and they must constantly attempt to explain issues in understanding and interesting ways. There are several channels to the media: news conferences, informal briefings, media interviews, news releases, and media tours. When there is a crisis the cardinal rule is "tell it all, tell it fast." Before interviews, decide the points you want to make, then do it. Have a positive strategy for getting them across. Expect to be nervous but don't let it get to you. If possible, critique the interview by having someone videotape the newscast or send you a clip.

Lindsley (1989) has these comments for the future: use wildlife as an environmental barometer; develop species-specific methods of trapping; use examples of diseased wildlife to take the offensive; and point out the benefits of consumptive sport; and take advantage of environmental disasters which will cause the public to demand control over pollution. Lindsley believes that there is a future for people who want to be a part of it.

DEMOGRAPHIC CHANGES IN THE UNITED STATES IN THE 1990S AND THE TWENTY-FIRST CENTURY

Among the many demographic changes are four trends that will likely influence nearly all dimensions of American society:

(1) decreased rates of population growth;

(2) an aging population;

(3) an increase in minority populations; and

(4) a decrease in household composition (Murdock et al. 1992).

In the future fish and wildlife managers are more likely to find themselves forced to address the needs of these new groups in a more direct and visible way than previously. There may be increased service demands for elderly and minority residents at a time when they are exempt from license fees, or have lower license costs and, at the same time, an increase in management for the benefit of these groups.

ALLOCATION OF WATER BY MARKETS IN THE TWENTY-FIRST CENTURY

Essentially there are three ways to reallocate water. The courts can do it as the California Supreme Court did when it invoked the public trust to limit Los Angeles diversions from creeks that fed Mono Lake. Administrative agencies can shrink water rights by finding that a use is wasteful. Finally, water markets can reallocate water by transferring it from willing sellers to willing buyers. So states Dan Tarlock, Chairman of the National Academy of Sciences Panel on Western Water Policy. Tarlock points out, however, that there is a need for caution about the use of water markets to reallocate water to meet new needs. In some cases it can impose hardship on third parties. Markets alone cannot accurately reflect all the relevant values of water (Tarlock 1992).

Literature Cited

Lindsley, D. B. 1989. Fish and wildlife enforcement: the future. Paper presented at Northeast Fish and Wildlife Conference, Golden Gate Auditorium, Calif.

Mangan, T. J., and M. G. Shanahan. 1990. Public law enforcement/private security: A new partnership? *FBI Law Enforcement Bull.* (Federal Bureau of Investigation, U.S. Dept. of Justice, Washington, D.C.) 59 (1): 18–22.

Murdock, S. H., K. Backman, R. B. Ditton, N. Hoque, and D. Ellis. 1992. Demographic change in the United States in the 1990s and the twenty-first century: Implications for fisheries management. *Fisheries* 17 (2): 6–13.

Tafoya, W. L. 1990. The future of policing. *FBI Law Enforcement Bull.* (Federal Bureau of Investigation, U.S. Dept. of Justice, Washington, D.C.) 59 (1): 13–17.

Tarlock, D. 1992. Let markets allocate water-with brakes. *The Christian Science Monitor,* 6 Apr., 18.

Toffler, A., and H. Toffler. 1990. The future of law enforcement: Dangerous and Different. *FBI Law Enforcement Bull.* (Federal Bureau of Investigation, U.S. Dept. of Justice, Washington, D.C.) 59 (1): 2–5.

Trojanowicz, R. C., and D. L. Carter. 1990. The changing face of America. *FBI Law Enforcement Bull.* (Federal Bureau of Investigation, U.S. Dept. of Justice, Washington, D.C.) 59 (1): 6–12.

Washington Post. Experts want DNA data kept private. Salt Lake City, Utah, *The Salt Lake Tribune,* Vol. 243, No. 7, Page 1. 1991.

Wood, D. B. 1991. Land use advocates make gains. *The Christian Science Monitor,* 3 Oct.

Recommended Reading

Allen, J. P., and E. J. Turner. 1988. Where to find the new immigrants. *American Demographics.* Sept.

Dillin, J. 1985. Asian-American: Soaring minority. *The Christian Science Monitor.* 10 Oct.

Lumas, D. K., and R. B. Ditton. 1988. Techniques for projecting the future growth and distribution of marine recreational fishing demand. *North American Journal of Fishery Mgmt.* 8: 259–263.

Murdock, S. H., K. Backman, E. Colberg, M. Hoque, and R. Ham. 1990. Modeling demographic change and characteristics in the analysis of future demand for leisure services. *Leisure Science* 12: 75–102.

Robye, B. 1985. *The American people: A timely exploration of a changing America and the important new demographic trends around us.* New York: A. P. Dutton, Inc.

Schwartz, J., and T. Exter. 1989. All our children. *American Demographics.* May.

Toffler, A. 1970. *Future shock.* New York: Random House.
———. 1980. *The third wave.* New York: William Morrow.
———. 1985. *The adaptive corporation.* New York: McGraw Hill.

U.S. Bureau of the Census. 1986. Projection of the number of households and families: 1986–2000. *Current population reports,* ser. P–25, no. 986. Washington, D.C.: U.S. Govt. Printing Office.

Multiple-Choice Questions

Mark only one answer.

1. According to some forecasters the white race will be in the minority in the United States in less than how many years?
   a. ten
   b. twenty
   c. fifty
   d. seventy-five
   e. one hundred

2. One of the following statements is not true:
   a. we need to understand the social impacts of new technology
   b. there is a troublesome line that must be drawn between the limits of government power and the rights of the individual
   c. straight line trends of the past predict the future
   d. the stigma of peer disdain is not effective with a number of individuals today
   e. law enforcement personnel will need to be strongly people-oriented in the early twenty-first century

3. Which is not true of the private sector of law enforcement?
   a. originally public law enforcement personnel viewed the private sector with disdain
   b. there is strong indication that the numbers of people in the private sector have peaked and will drop in the twenty-first century
   c. the number of personnel in the private sector has expanded more rapidly than in the public sector
   d. the private sector employs a number twice that of the public sector
   e. there are areas such as wild game farms, private fish hatcheries, and so forth, where public law enforcement officers cannot be expected to function

4. One of the following statements is not true:
   a. people are the first hurdle an officer will meet on the job
   b. historically law enforcement personnel have maintained the status quo
   c. officers will have a much easier time in the early part of the twenty-first century than they have had in the late twentieth century

   d. people tend to put officers into two broad categories, friendly and adversarial
   e. Asiatics often have trouble understanding wildlife management ethics and rationale

5. Dr. James D. Watson and others have expressed deep concern about what they call genetic fingerprinting (DNA). Which is not true?
   a. people should have the right to genetic privacy
   b. in the hands of the irresponsible great damage could be done
   c. it is feared that in ten years a genetic signature may be a simple bar code
   d. fingerprinting by genetic code is not presently legal
   e. there are medical, legal, and ethical reasons why we should move cautiously with genetic fingerprinting

6. Norman Galman of New York State has several ideas for dealing with the news media. Which one is not his?
   a. keep the media at arms length and tell them as little as possible
   b. in an interview stay on the record and speak the truth
   c. find out when the media's deadline is and help them to meet it
   d. understand the reporters' points of view. They are working for their readers and not for you
   e. avoid playing favorites with hard news; it is different with feature articles

7. Lindsey suggests four ideas for the future. Which one is not his?
   a. the future for wildlife management is bleak and unpredictable
   b. take advantage of disasters to show the cost and cause of pollution
   c. use wildlife species and where they live as examples of environmental barometers
   d. develop species-specific methods of trapping
   e. use overcrowded, diseased wildlife populations as examples of why there is a need to reduce the population, preferably by hunting

Appendix A[1]

Glossary of Terms

ABANDONMENT. The intentional and voluntary relinquishment of the reasonable expectancy of privacy in premises. Note that proof of abandonment requires voluntary relinquishment of premises and intent to abandon. Mere absence is not abandonment, nor is involuntary absence due to arrest and incarceration or for other reasons.

ABET. To encourage or set another on to commit a crime. This word is always applied to aiding the commission of a crime. To abet is to assist; to cause; to hire; to command. Abetting imports a positive act in aid of the commission of an offense. The abettor must stand in the same relation to the crime as the criminal; approach it from the same direction; touch it at the same point. (See Aid and Abet.)

ABIDE. To accept the consequences of; to rest satisfied with. With reference to an order, judgment, or decree of a court, to perform, to execute.

ABROGATE. To repeal; to make void; to annul.

ABROGATION. The destruction of or annulling a former law, by an act of the legislative power, or by usage.

ACCESSORY. One who is not the chief actor in the perpetration of an offense or present at its performance, but in some way is concerned with it either before or after. Any thing which is joined to another thing as an ornament, or to render it more perfect. For example, the halter of a horse, the frame of a picture, the keys of a house, and the like, each belong to the principal thing.

ACCESSORY AFTER THE FACT. Normally any person who, knowing that a crime has been committed, thereafter receives, comforts, or assists the offender in order to hinder or prevent his apprehension, trial, or punishment is considered an accessory after the fact. These acts need not necessarily be designed to effect personal escape or detection but may include those acts which are performed to conceal the commission of the offense. Thus, a person is an accessory after the fact if he knowingly conceals illegally killed game in order to help the perpetrator escape detection by a conservation officer. It should be noted, however, that mere failure to report an offense does not make one an accessory after the fact. In a California wildlife case, an accessory after the fact is a principal in illegal possession of wild game.

ACCESSORY BEFORE THE FACT. One who, although absent at the time the crime is committed, yet procures, counsels, or commands another to commit it. In California, an accessory before the fact would be a principal to a fish and game violation.

ACCIDENT. An event which takes place without one's foresight, or expectation, an event that proceeds from an unknown cause, or an unusual effect of a known cause. An event happening without human agency, or if happening through human agency, an event which is unusual and not expected.

ACCOMPLICE. In criminal law. One who is concerned in the commission of a crime, though not as a principal. A person so connected with a crime that at common law he might himself have been convicted either as principal or as an accessory before the fact. An accomplice may be one of the principal actors, or an aider and abettor, or an accessory before the fact, and includes all persons who participate.

ACCUSATION. A charge made to a competent officer against one who has committed, or is believed to have committed, a crime, so that he may be brought to justice and punishment.

ACKNOWLEDGMENT. The act of one who has executed a deed, in going before some competent officer or court and declaring it to be his act or deed.

ACQUITTAL. Ordinarily, in criminal jurisprudence, the word means a discharge after a trial, or an attempt to have one, upon its merits; but under statutes it may refer to a discharge for other reasons. A release or discharge from an obligation or engagement. The absolution of a party charged with a crime or misdemeanor—a verdict of not guilty.

ACT. A thing done or established; a deed or other written instrument evidencing a contract or an obligation; a statute; a bill which has been enacted by the legislature into a law, as distinguished from a bill which is in the form of a law, presented to the legislature for enactment.

ACTION. An ordinary proceeding in a court of justice by which a party prosecutes another party for the enforcement or protection of a right, the redress or prevention of a wrong, or the punishment of a public offense; a judicial remedy for the enforcement or protection of a right.

1. In part from:
James A. Ballentine, *Law Dictionary with Pronunciations,* 2d ed. Lawyers' Cooperative Publishing Co., Rochester, New York, 1948.
Henry C. Black, *Black's Law Dictionary,* 4th ed., St. Paul, Minn.: West Publishing Co., 1951.
John Bouvier, *Bouvier's Law Dictionary and Concise Encyclopedia,* Third rev., Kansas City, Mo.: Vernon Law Book Co., and St. Paul, Minn.: West Publishing Co., 1914, 3 vol.
Department of Defense, *Manual for Courts-Martial,* Washington, D.C.: U.S. Govt. Printing Office, 1975.

ACT OF GOD. Any accident due to natural causes directly and exclusively without human intervention, such as could not have been prevented by any amount of foresight, pains, and care reasonably to have been expected. Any irresistible disaster, the result of natural causes, such as earthquakes, violent storms, lightning, and unprecedented floods, such a disaster arising from causes, and which could not have been reasonably anticipated, guarded against, or resisted.

ADMISSIBLE. Pertinent and proper to be considered in reaching a decision. Used with reference to the issues to be decided in any judicial proceeding.

ADULT. A person who is at least twenty-one years old. In the civil law, a boy of fourteen, a girl of twelve.

AFFIDAVIT. A statement or declaration reduced to writing, and sworn to or affirmed before some officer who has authority to administer an oath of affirmation.

AFFIRM. To declare solemnly instead of making a sworn statement. Persons having conscientious scruples against making oath are permitted to affirm.

AFFIRMATION OF JUDGMENT. A judgment of an appellate court which is a determination by that court that the proceedings under review are free from prejudicial error. If there is a decree partly in favor of and partly adverse to one who appeals from the adverse portion, an affirmance does not affirm the portion which was in his favor.

AGGRAVATED ARSON. See Arson.

AID AND ABET. To constitute oneself as an aider and abettor there must be an intent to aid or encourage the persons who commit the crime, and there must be a sharing of intent with the perpetrator of a crime. One is liable as a principal in a crime if the offense is committed as a common venture or as a natural or probable consequence of the offense directly intended. A person may be guilty of an act committed by those with whom he voluntarily associates himself if the execution of the unlawful design was planned by the accused; this, even though the consequences were more far-reaching and serious than was anticipated by the principal. A person who plans a crime which is actually carried out by his associates is a principal in the act and is guilty of whatever illegal acts his associates may carry on during the course of the illegal act or acts which he planned and which were carried out by the associates.

A person, on the other hand, who merely witnesses a crime without intervention does not become a party to the commission of the crime unless he has a duty to interfere and his noninterference was designed to and did operate to encourage and aid the perpetrator of the crime. A conservation officer, for example, is guilty as a principal in a game violation if he or she idly stands by and watches a violation being committed and the violator escaping without interference.

A person who counsels, commands, or procures another to commit an offense which is subsequently perpetrated in consequence of such counsel, is a principal whether present or absent at the commission of the offense. This is true even though the offense is effected by a different means other than those suggested by the counselor. For example, a person might suggest to another that he should, in a specified manner, proceed to a given area and shoot pheasants (out-of-season). If the perpetrator of the crime proceeds to an area and acquires possession of pheasants by illegal methods, even though it is other than by shooting as suggested by the counselor, then the counselor as well as the perpetrator is guilty of the crime of taking pheasants illegally.

The offense committed by those persons who, although not the direct perpetrators of a crime, act to render aid to the actual perpetrator thereof.

ALIAS SUMMONS. A new summons issued in the same form and to serve the same purpose as one previously issued, and usually issued where the original summons has been returned, and hence has become *functus officio,* without having been served on any or all of the defendants.

ALIAS WARRANT. A warrant to the issuing official unserved and one which may later be reissued.

ALIAS WRIT. A writ issued to take the place of a similar writ which has been lost or returned or for some other reason has not taken effect or has become *functus officio.*

ALIBI. Literally present in another place other than that described. When a person, charged with a crime, proves *(se eadem die fuisse alibi)* that he was, at the time alleged, in a different place from that in which it was committed, he is said to prove an alibi, the effect of which is to lay a foundation for the necessary inference that he could not have committed it.

An alibi is said to be proof that at the time a crime was committed the defendant was not at the scene of the crime and could not have participated in it. In order to be a complete alibi, it must be proved that it was impossible for the defendant to be at the place where the crime was committed.

ALLEGATION. The assertion, declaration, or statement of a party of what one can prove.

AMBIGUITY. Duplicity, indistinctness, or uncertainty of meaning of an expression used in a written instrument. The word "uncertainty" in a suit refers to the uncertainty defined in pleading and does not include ambiguity.

AMBUSH. The noun means the act of attacking an enemy unexpectedly from a concealed station; a concealed station where troops or enemies lie in wait to attack by surprise; an ambuscade. The verb "to ambush" means to lie in wait; to surprise; to place in ambush.

AMENABLE. To be amenable means to be liable to answer; responsible; answerable; liable to be called to account.

ANGER. A strong passion or emotion of displeasure or antagonism, excited by a real or supposed injury or insult to one's self or others by the intent to do such injury; resentment; wrath, rage, fury, passion; ire; gall, choler, indignation; displeasure; vexation; grudge; spleen.

ANIMAL. Any member of the group of living beings typically capable of spontaneous movement and rapid response to external stimulation (as distinguished from a plant). Animals, in contrast to plants, are able to move about and to take and ingest food. Strictly speaking, animals include human beings; however, in wildlife work the term frequently is used in a way which excludes the human race, and in many cases it refers only to vertebrates (animals with backbones).

Any animate being endowed with the power of voluntary motion.

ANIMALS *DOMITAE NATURAE.* Those animals which have been tamed by man; domestic. Those animals which are naturally tame and gentle or which by long continued association with man have become thoroughly domesticated and are now reduced to such a state of subjection to his will that they no longer possess the disposition or inclination to escape.

ANIMALS *FERAE NATURAE.* Those animals which still retain their wild nature. A man may have an absolute property in animals of a domestic nature, but not so in animals *ferae naturae,* which belong to him only after he has taken legal possession. Such animals as are of a wild nature or disposition and so require to be reclaimed and made tame by art, industry, or education, or else must be kept in confinement to be brought within the immediate power of the owner.

ANIMUS. Mind; intent; intention; disposition.

ANSWER. A plea interposed by a defendant to a declaration or complaint or any material allegation of fact therein, which, if untrue, would defeat the action. The traverse may deny all the facts alleged, or any particular material fact; but it is not the office of a plea or answer to raise an issue of law where such issue should be determined on demurrer.

A POSTERIORI. From the effect to the cause; from what comes after.

APPARENT. Webster defines the word as clear, or manifest to the understanding; plain; evident; obvious; appearing to the eye or mind.

APPEAL. Any complaint to a superior court of an injustice done by an inferior one. This is the general use of the word.

APPEAL AND ERROR. The methods of exercising appellate jurisdiction for the review by a superior court of the final judgment, order, or decree of some inferior court.

The most unusual modes of exercising appellant jurisdiction are by a writ of error, or by an appeal, or by some process of removal of a suit from an inferior tribunal. An appeal is a process of civil law origin, and removes a cause, entirely subjecting the facts as well as the law to a review and a retrial. A writ of error is a process of common law origin, and it removes nothing for reexamination but the law.

APPEARANCE. Coming into court as party to a suit, whether as plaintiff or defendant. It may be of the following kinds:

Compulsory—That which takes place in consequence of the service of process.

Conditional—One which is coupled with conditions as to its becoming general.

De bene esse—One which is to remain an appearance, except in a certain event.

General—A simple and absolute submission to the jurisdiction of the court.

Gratis—One made before the party has been legally notified to appear.

Optional—One made where the party is not under any obligation to appear, but does so to save his rights.

Special—That which is made for certain purposes only, and does not extend to all the purposes of the suit; as to contest and jurisdiction, or the sufficiency of the service.

Subsequent—An appearance by the defendant after one has already been entered for him by the plaintiff.

Voluntary—That which is made in answer to a subpoena or summons, without process.

APPELLANT. One who makes an appeal from one court to another.

APPELLATE JURISDICTION. The jurisdiction which a superior court has to rehear causes which have been tried in inferior courts.

APPREHENSION. The capture or arrest of a person on a criminal charge. The word strictly construed means the seizing or taking hold of a person and detaining him with a view to his ultimate surrender.

APPROPRIATE. To appropriate means to allot, assign, set apart, or apply to a particular use or purpose.

ARGUMENT. An effort to establish belief by a course of reasoning.

ARMED. Furnished or equipped with weapons of offense or defense. A person who has in hand a dangerous weapon with which he makes an assault, is certainly armed.

ARRAIGNMENT. Calling the defendant to the bar of the court, to answer the accusation contained in the indictment.

ARREST. The taking, seizing, or detaining of the person of another either by touching, or putting hands on, or by any act which indicates an intention to take the person into custody, and subjects the person arrested to the actual control and will of the person making the arrest. Touching is not necessary; it is sufficient if the person imprisoned understands that he is in the power of the officer, and submits. The taking of a person into custody by a legal authority. This can be accomplished either by arrest on sight or on a proper warrant. To deprive a person of liberty by legal authority. The taking, seizing, or detaining the person of another, touching or putting hands upon him in the execution of process, or any act indicating an intention to arrest. A restraint of the person, a restriction of the right of locomotion which cannot be implied in the mere notification, or summons on petition, or any other service of such process, by which no bail is required nor restraint of personal liberty.

ARRESTEE. The person arrested.

ARSON. To set fire to an inhabited dwelling or other structure of another person.

    Aggravated Arson—In this case the essential element is the danger to human life. It is immaterial that no one in fact is injured.

ASSAILANT. A person who assails, or who assaults; the aggressor. In a fight the person who commits the first assault is the assailant.

ASSAULT. An unlawful offer or attempt with force or violence to do a corporeal hurt to another. Force unlawfully directed or applied to the person of another under such circumstances as to cause a well-founded apprehension of immediate peril.

    Aggravated Assault—One committed with the intention of committing some additional crime. Simple assault is one committed with no intention to do any other injury.

AT LARGE. Within the comprehension of the statutes inflicting a penalty on one who suffers animals to be "running at large," the term means strolling without restraint or confinement, or wandering, roving, or rambling at will; unrestrained.

ATTEMPT. An endeavor to accomplish a crime carried beyond mere preparation, but falling short of execution of the ultimate design in any part of it.

    An attempt to commit an offense is an act or acts done with the specific intent to commit a particular crime which would have been completed except for some interference other than that which was voluntarily caused by the perpetrator. There must be a specific intent to commit some particular crime by an overt act which directly tends to accomplish the unlawful purpose. This act must be more than mere preparation. It must consist of devising or arranging means necessary for the commission of the offense, and goes beyond the preparatory stages in that it is a direct movement toward the commission of the offense. For example, the purchase of a gun with the intent of violating a game law is not an attempt to violate a game law; however, actual use of the weapon in attempting to kill game out-of-season is an illegal act even though the shot may go astray.

    It is not an attempt when every act by the accused is completed without committing any offense. However, an accused may be guilty of attempt even though the crime turns out to be impossible because of something beyond his power or knowledge. For example, the person knowingly raises a gun and levels it at a game animal, out-of-season, and pulls the trigger with obvious and admitted intention of killing the game animal. However, the gun misfired and the animal is not killed. Still there is an attempt to commit an illegal act. If, on the other hand, a person raises a gun, aims it at a game animal during a closed season, and pulls the trigger, he is not guilty of an attempt if he knew at the time he pulled the trigger that the gun would not fire because it was empty, or for some other reason.

ATTEST. A witness. To attest means to bear witness to; to affirm to be true or genuine.

ATTORNEY. The word, unless clearly indicated otherwise, is construed as meaning attorney-at-law. When used in connection with the proceedings of courts, it often has a fixed and universal significance, on which the technical and popular sense unite.

ATTORNEY GENERAL. The chief law officer of a state or nation, to whom is usually entrusted the duty of prosecuting all suits or proceedings wherein the state is concerned. The attorney general may also advise the chief executive and other administrative heads of government in legal matters.

AUTOPSY. A medical or surgical examination of a dead body made in order to determine the cause of death. Sometimes spoken of as a "postmortem."

BAIL. One who becomes surety for the appearance of the defendant in court. Money or other security furnished in behalf of a defendant to allow one's physical liberty, pending one's appearance for a hearing, a trial, or a sentence.

BAILIFF. A person to whom some authority, care, guardianship, or jurisdiction is delivered, committed, or entrusted. A sheriff's officer or deputy. A court attendant, sometimes called a tipstaff.

BAIL PIECE. A process authorizing the rearrest of a defendant who has defaulted an attendance at a hearing where bail has been provided for his liberty.

BATTERY. Any unlawful beating or other wrongful physical violence or constraint inflicted on a human being without his consent. An unlawful touching the person of another by the aggressor himself, or any other substance put in motion by him.

BEAST. Any four-footed animal which may be used for labor, food, or sport; as opposed to man; any irrational animal.

BELIEF. Conviction of the mind, arising not from actual perception or knowledge, but by way of inference, or from evidence received or information derived from others.

BENCH WARRANT. A warrant issued by proper authority of a court of record requiring the arrest of a person, and his appearance before the issuing authority. An order issued by or from a bench, for the attachment or arrest of a person. It may be issued either in case of a contempt or where an indictment has been found.

BEST EVIDENCE. The best evidence of which the nature of the case admits, not the highest or strongest evidence which the nature of the thing to be proved admits of: e.g., a copy of a deed is not the best evidence; the deed itself is better.

BIAS. A particular influential power which sways the judgment; the inclination or propensity of the mind towards a particular object.

BILL OF RIGHTS. That portion of the Federal Constitution and the constitution of each state which consists of the guarantees of such rights as are to a large extent declaratory of fundamental principles and the foundation of citizenship; those providing against excessive fines or cruel punishment.

BONA FIDES. Good faith, honesty, as distinguished from *mala fides* (bad faith).

BOND. An obligation in writing and under seal.

BRASS KNUCKLES. A weapon used for offense and defense, worn upon the hand to strike with as if striking with the fist. When first known and used, the weapon was originally made of brass, but it is now made of other heavy metal. It still retains the name of brass knuckles.

BREAKING AND ENTERING. What would be a breaking in burglary, it has been held, is equally a breaking by the sheriff to serve process. Even the right to lift a door latch has been denied. Other cases hold that the door may be opened in the ordinary manner such as by lifting the latch, turning the knob, or turning a key left in the lock.

BRIBE. A gift or promise, which if accepted, is of some advantage as an inducement for some illegal act or omission.

BRIBERY. The receiving or offering of any undue reward by or to any person in order to influence his behavior and to induce him to act contrary to his duties or to the known rules of honesty and integrity. Offering or receiving a present, with the intent and hope of causing dishonest behavior.

BRIEF. A detailed statement of a party's case. An abridgment of a plaintiff's or defendant's case. Tidd says a brief should contain an abstract of the pleadings; a statement of the facts of the case, with such observations as occur thereon. The great rule to be observed in drawing briefs consists in conciseness with perspicuity.

BURDEN OF PROOF. The duty of proving the facts in dispute on an issue raised between the parties in a case.

CANON LAW. A body of Roman ecclesiastical law, relative to such matters as the Church of Rome either had, or pretended to have jurisdiction over.

CARE. Charge or oversight; implying responsibility for safety and prosperity.

CASE. A question contested before a court of justice. An action or suit at law or in equity.

CAT. A domestic animal which if kept as a household pet is a thing of value. As such it is the proper subject of a civil action, it is property subject to taxation and is such an animal as is included in statutes against cruelty.

CAUSE. In Civil Law. The consideration or motive for making a contract.

CENSUS. A decennial official count by the government of the United States of the inhabitants and wealth of the country. In fish and game work it generally means an estimate derived from a presumably reliable sample.

CERTIFICATE. A writing made in any court, and properly authenticated, to give notice to another court of anything done therein. A writing by which testimony is given that a fact has or has not taken place.

CERTIFIED COPY. A copy of a document or record, signed and certified as a true copy by the officer to whose custody the original is entrusted.

CERVUS. A stag. Any member of the deer or elk family.

CHAIRMAN. The presiding officer of a deliberative body.

CHALLENGE FOR CAUSE. Those for which some reason is assigned.

CHALLENGE-PEREMPTORY. Those made without assigning any reason, and which court must allow.

CHALLENGE OF DOMICILE. A change of one's abiding place with no intention to live elsewhere than in the new abode. The residing permanently or indefinitely in the new abode.

CHANGE OF RESIDENCE. An actual removal to a residence outside of the state coupled with an actual intention to change one's residence from the state to residence out-of-the-state.

CHANGE OF VENUE. This is permission of a court having jurisdiction in a case which allows the case to be tried in some other court, usually in another county, where, for a number of reasons it appears that a fair trial cannot be had at the originally scheduled place.

CHARACTER. The possession by a person of certain qualities of mind or morals, distinguishing the person from others. In evidence—the opinion generally entertained of a person derived from the common report of the people who are acquainted with him; his reputation. The moral character and conduct of a person in society may be used in proof before a jury in three classes of cases: first, to afford a presumption that a particular person has not been guilty of a criminal act; second, to affect the damages in particular cases, where their amount depends on the reputation and conduct of any individual; and third, to impeach or confirm the veracity of a witness.

CHARGE. To charge a jury means that the court shall instruct the jury as to the essential law of the case, although, unless requested to do so by either party, the judge is not compelled to reduce the charge to writing, yet it is the better practice. A duty or obligation imposed upon some person. A lien, encumbrance, or claim which is to be satisfied out of the specific thing or proceeds thereof to which it applies. To impose such an obligation; to create such a claim.

CHARTA DE FORESTA. A collection of the laws of the forest, made in the reign of Henry III.

CIRCUIT COURT. A court presided over by a judge or by judges at different places in the same district.

CIRCUMSTANCES. The particulars which accompany an act. The surroundings at the commission of an act.

CIRCUMSTANTIAL EVIDENCE. Facts inferred from circumstances rather than those seen or otherwise established.

CITATION. A writ issued out of a court of competent jurisdiction, commanding a person therein named to appear on a day named and do something therein mentioned, or show cause why he should not. An order or summons by which a defendant is directed or notified to appear. The act by which a person is so summoned or cited.

CITE. To summon; to command the presence of a person; to notify a person of legal proceedings against him and require his appearance thereto.

CIVIL. In contradistinction to barbarous or savage, indicates a state of society reduced to order and regular government: thus, we speak of civil life, civil society, civil government, and civil liberty. In contradistinction to criminal, to indicate the private rights and remedies of men, as members of the community, in contrast to those which are public and relate to the government: thus, we speak of civil process and criminal process, civil jurisdiction and criminal jurisdiction. It is also used in contradistinction to military or ecclesiastical, to natural or foreign. The word is frequently used as an adjective to qualify a wrong or breach of duty and to import that the wrong or breach of duty may be the subject of a civil action, as distinguished from a criminal prosecution.

CIVIL ACTION. In Civil Law. A personal action which is instituted to compel payment, or the doing of some other thing which is purely civil.

CLAIM. The assertion of a demand, or the challenge of something, as a matter of right; a demand of some matter, as of right, made by one person upon another to do or to forbear to do some act or thing, as a matter of duty.

CLEAN HANDS. The maxim, ''He who comes into equity, must do so with clean hands,'' signifies that equity deals only with conscionable demands and usually the maxim has reference only to the plaintiff's rights against the defendant and only when one's equity has an immediate and necessary relation to the equity for which one sues.

CODE. A body of law established by the legislative authority of the state, and designed to regulate completely, so far as a statute may, the subject to which it relates. A system of law; a systematic and complete body of law.

COLD STORAGE. The term as used in the trade, means a storage space where the temperature is kept at a low degree, but above the freezing point; while a freezer is a place for the preservation of meat or poultry where the temperature is kept below the freezing point.

COLLUSION. An agreement to defraud a third party of one's rights by the forms of law, or to secure an unlawful object.

COMMITMENT. A writ issued by a judge, magistrate, or justice of the peace by which the body of a person is committed to prison, and requiring the jail custodian to retain the person until properly released by law.

COMMON FISHERY. A fishery which is distinguished from several or an exclusive fishery by the number of persons who have a right to resort to the place for fishing; it is one which is not exclusive to one person, but is open to a number of persons, generally to the public.

COMMON LAW. Laws which have gained their standing by usage and custom, sometimes dating back to the unwritten law of England.

That system of law or form of the science of jurisprudence which has prevailed in England and in the United States of America, in contradistinction to other great systems, such as the Roman or Civil Law. Those principles, usages, and rules of action applicable to the government and security of persons and of property, which do not rest for their authority upon any express and positive declaration of the will of the legislature. The body of rules and remedies administered by courts of law, technically so called, in contradistinction to those of equity and to the canon laws.

COMMON PLEAS. The name of a court having jurisdiction generally of civil actions. Such pleas or actions are brought by private persons against private persons, or by the government, when the cause of action is of a civil nature. In England, whence we derived this phrase, common pleas are called to distinguish them from pleas of the crown.

COMPETENCY OF A WITNESS. A competent witness is one who is legally qualified to be heard under oath before a judicial tribunal; one who has the requisite legal qualifications to give testimony in a court of justice. Competency is to be sharply distinguished from credibility. The former has to do with one's personal qualifications to testify and must be determined before one can give any testimony. On the other hand, the credibility of a witness relates to that quality in a witness which renders one's testimony worthy of belief. In earlier times it was held that in order to be a competent witness a person must ''possess a conscience alive with true accountability to a higher power than human law in case of falsehood.'' Otherwise, it was felt the oath was not binding upon him. It is not essential to the witness' competency that he be aware of God's existence or believe in a Supreme Being. It is generally said to be sufficient qualification if a witness understands and undertakes the obligation of an oath. As a general rule, a person offered as a witness is presumed to be competent to testify unless the contrary is shown. Competency is the rule and incompetency is the exception. Ordinarily the burden of showing incompetence rests upon the party asserting it. There is no arbitrary minimum age limit below which a child is automatically disqualified as a competent witness. The competency of a child of fourteen years is generally presumed, but below that age a judicial inquiry into the child's mental capacity usually is required, and becomes more searching in proportion to his chronological immaturity. A witness, be it child or adult, must be capable of distinguishing between truth and falsehood. It is not required that he be able to define the meaning of an oath, but rather that he appreciate the fact that as a witness he assumes a binding obligation to tell the truth and that, if he violates that obligation, he is subject to punishment by the court. An insane person, or one who is otherwise mentally incompetent, is not necessarily incompetent as a witness even though he may have been committed to an institution for the insane. Rather, it is necessary that he has sufficient mind to understand the nature and obligation of an oath and correctly to receive and impart his impressions of the matters of which he has seen or heard. A witness may be competent to testify although in some respects he is mentally unsound or has some mental impairment. However, where a witness is so impaired that he does not understand the obligation of an oath or has no respect for the truth he is not competent. Competency has two aspects: (1) the mental capacity to understand the nature of the questions, and the ability to form and communicate intelligent answers, and (2) the moral responsibility to speak the truth which is the essence and obligation of an oath.

COMPETENT. Able, fit, qualified; authorized or capable to act. (See competency of a witness.)

COMPETENT EVIDENCE. That evidence which the very nature of the thing to be proven requires, as the production of a writing where its contents are the subject of inquiry. Also evidence not excluded by a rule of evidence.

COMPLAINT. In Criminal Law. The allegation made to a proper officer that some person, whether known or unknown, has been guilty of a designated offense, with an offer to prove the fact, and a request that the offender may be punished. It is a technical term, descriptive of proceedings before a magistrate.

   To have a legal effect, the complaint must be supported by such evidence as shows that an offense has been committed and renders it certain or probable that it was committed by the person named or described in the complaint.

   A sworn legal statement usually made before a court which charges a person with the commission of a crime. The word complaint is sometimes used synonymously with the word information. (See Information.)

CONCLUSIVE. Shutting up a matter, shutting out all further evidence; not admitting of explanation or contradiction; putting an end to inquiry; final; irrefutable; decisive. Beyond question or beyond dispute; manifest; plain, clear; obvious; visible; apparent; indubitable; palpable; and ''notorious.''

CONCLUSIVE EVIDENCE. That which cannot be controlled or contradicted by any other evidence. Evidence which of itself, whether contradicted or uncontradicted, explained or unexplained, is sufficient to determine the matter at issue. Evidence upon the production of which the judgment is bound by law to regard some fact as proved, and to exclude evidence to exclude it.

CONFESSION. In Criminal Law. The voluntary admission or declaration made by a person who has committed a crime or misdemeanor, to another, of the agency or participation which he had in the same.

CONFIDENTIAL COMMUNICATIONS. Those statements with regard to any transaction made by one person to another during the continuance of some relation between them which calls for or warrants such communications. At law, certain classes of such communications are held not to be proper subjects of inquiry in courts of justice, and the persons receiving them are excluded from disclosing them when called upon as witnesses, upon grounds of public policy. Secrets of state and communications between the government and its officers are usually privileged.

CONFISCATE. To appropriate to the use of the state.

CONSPIRACY. A conspiracy is an act of two or more persons who have agreed by concerted action to accomplish an unlawful act or some purpose not in itself unlawful but by unlawful means, and the doing of some act by one or more of the conspirators to further the object of that agreement. This agreement need not be in any particular form, not even in formal words. It is sufficient that the parties arrive at an understanding as to what is to be accomplished and what part will be played by each of the conspirators. The overt act of conspiracy must be an independent act by one of the persons involved, but it need not be a crime in itself; but it must be a manifestation that the conspiracy is being executed. A telephone call to a conservation officer that something is taking place in some far corner of a county followed by an illegal act of hunting in the opposite by one of the conspirators is in itself an overt act and therefore becomes a part of a conspiracy. A person may be guilty of conspiracy although himself being incapable of committing the crime. Thus a man eighty years old who could not kill and carry away an illegally killed deer could plan for two youths to commit this act, and all three would, under the circumstances, be equally guilty.

CONSTITUTION. The fundamental law of a state, directing the principles upon which the government is founded, and regulating the exercise of the sovereign powers, directing to what bodies or persons those powers shall be confined and the manner of their exercise.

CONSTITUTION OF THE UNITED STATES OF AMERICA. The supreme law of the United States. It was framed by a convention of delegates from all the original thirteen states, except Rhode Island.

CONSTRUCTIVE ESCAPE. Such an escape as takes place when the prisoner obtains more liberty than the law allows, although he still remains in confinement.

CONTEMPT. A willful disregard or disobedience of a public authority.

CONTEMPT OF COURT. A despising of the authority, justice, or dignity of the court; such conduct as tends to bring the authority and administration of the law into disrespect or disregard, or to interfere with or prejudice parties litigant or their witnesses during litigations.

CONTINUANCE. The adjournment of a cause from one day to another of the same or subsequent term.

CONTRADICT. To prove a fact contrary to what has been asserted by a witness. A party cannot impeach the character of his witness, but may contradict him as to any particular fact.

CONTROL. To control, according to Webster, means to check, restrain, govern, have under command, and authority over. According to the Century dictionary, it means to hold in restraint or check, subject to authority, direct, regulate, govern, dominate.

CONVICTION. The confession of a person who is being prosecuted for crime, in open court, or a verdict returned against him by a jury, which ascertains and publishes the fact of his guilt.

The word applies simply and solely to the verdict of guilty.

In practice, that legal procedure which ascertains the guilt of the party and upon which the sentence or judgment is founded.

CORPUS DELICTI. From the Latin word, it literally means the body of the crime. In practice it may be a dead body of a person, horse, house burned, carcass of a deer, the elements of the crime, or anything in fact that establishes the essence of the crime. In all cases it is necessary to prove the *corpus delicti* to establish that a crime has actually been committed.

COURT. A place where justice is judicially administered; persons officially assembled under authority of law, at the appropriate time and place, for the administration of justice. A time when, a place where, and persons by whom judicial functions are to be exercised, are essential to complete a court, in contemplation of law.

A body, in the government, to which the administration of justice is delegated.

COURT CALENDAR. A court's list of the matters which are ready to be heard and the times which the court has assigned for their hearing.

COURTS OF THE FOREST. Courts held for the enforcement of the forest laws. The lowest of these was the Woodmote, or Court of Attachments. The next was the Swainimote. The highest was the Court of the Chief Justice. There was also a Survey of Dogs held by the Regarders of the Forest every three years for the lawing of dogs.

CRAZY. In ordinary language we speak of insane people as crazy, and vice versa; mad, demented.

CREDIBILITY. Capacity for being believed or credited. Worthiness of belief. The credibility of witnesses is a question for the jury to determine, as their competency is for the court.

CREDIBLE WITNESS. One who, being competent to give evidence, is worthy of belief. In deciding upon the credibility of a witness, it is always pertinent to consider whether one is capable of knowing thoroughly the thing about which one testifies; whether one was actually present at the transaction; whether one paid sufficient attention to qualify himself as a reporter; and whether one honestly relates the affair fully as one knows it, without any purpose or desire to deceive or to suppress or to add to the truth. Determining the credibility of a witness is the duty of the jury.

CRIME. An act committed or omitted in violation of a public law forbidding or commanding it. Crimes are public wrongs which a private individual has neither power to condone nor forgive. A crime is determined by the act the person did *in* committing a crime. A victim subsequently cannot forgive his wrongdoer and have that considered by the criminal as a defense.

CRIMINAL. Relating to or having the character of crime.

CRIMINAL CHARGE. A charge which, strictly speaking, exists only when a formal, written complaint has been made against the accused and prosecution initiated. Popular usage sometimes substitutes the word accusation, but in legal phraseology it is properly limited to such accusations as have taken shape in a prosecution. A person is legally charged with crime only when he is called upon in a legal proceeding to answer to such a charge.

CRIMINAL INTENT. The intent to commit a crime; malice, as evidenced by a criminal act.

CROSS-EXAMINATION. The examination of a witness by the party opposed to the party who called the witness, and who examined, or was entitled to examine the witness in chief. The purpose of the cross-examination is to test the truthfulness, intelligence, memory, bias, or interest of the witness; and any question to that end within reason is usually allowed.

CULPABLE NEGLIGENCE. This is defined as a degree of carelessness greater than simple negligence. It is an act or omission accompanied by a culpable disregard for foreseeable consequences to others. Leaving poison baits out where domestic animals or even children might reasonably be expected to pick them up is in this category.

CUMULATIVE PUNISHMENT. A punishment greater than a convicted person would suffer for a first offense. Such punishments of second and subsequent offenses are provided for by statute in England and in many states of the Union, and are held to be constitutional.

CURTILAGE. The enclosed space of ground and buildings immediately surrounding a dwelling-house. In its most comprehensive and proper legal signification, it includes all that space of ground and buildings thereon which is usually enclosed within the *general fence* immediately surrounding a principal messuage and outbuildings, and yard closely adjoining to a dwelling-house, but it may be large enough for cattle to be levant and couchant therein.

The curtilage of a dwelling-house is a space, necessary and convenient and habitually used for the family purposes, and the carrying on of domestic employments. It includes the garden, if there be one, and it need not be separated from other lands by fence.

CUSTODY. The bare control or care of a thing, as distinguished from the possession of it. The mere fact of putting one's property into the charge or custody of another person does not divest the possessions of the owner. So one having the mere custody of the property of another may commit larceny of it.

CUSTOM OR USAGE. The practice of some law enforcement officers of overlooking crimes does not establish a legal precedent or a legal defense. Even though the failure to enforce a law or erroneous interpretation of the law may have permitted violators to escape punishment, no right to violate the law is created. Furthermore, it does not provide a defense against criminal prosecution for a person committing a crime heretofore considered lightly.

DANGEROUS WEAPON. A weapon capable of producing death or great bodily harm. An unloaded gun, at some distance from the person alleged to have been assaulted, is not a dangerous weapon.

DECISION. A judgment given by a competent tribunal.

DECOY. To decoy is to entice; to tempt; to lure or allure. There can be no such thing as forcibly decoying a person, since fraud or deception is the moving element.

DE FACTO. Actually; in fact; in deed. A term used to denote a thing actually done. In fact, as distinguished from "*de jure*," by right.

DEFECT. A lack or absence of something essential to completeness. The want of something required by law.

DEFENDANT. The person or persons charged with the commission of a crime. A party sued in a personal action. The term does not in strictness apply to the person opposing or denying the allegations of the demandant in a real action, who is properly called the tenant. The distinction, however, is very commonly disregarded; and the term is further frequently applied to denote the person called upon to answer, either at law or in equity, and as well in criminal as civil suits.

DELINQUENT CHILDREN. Those who have committed offenses, or who are falling into bad habits, or are incorrigible.

DEMURRER. In Pleading. An allegation, that, admitting the facts of the preceding pleading to be true, as stated by the party making it, he has yet shown no cause why the party demurring should be compelled by the court to proceed further. A declaration that the party demurring will go no further, because the other has shown nothing against him.

DE NOVO PROCEEDINGS. A case in which the appeal is argued anew upon the merit of the matter, rather than an attack upon the record.

DEPUTY. A person subordinate to a public officer whose business and object is to perform the duties of the principal.

DIRECT EVIDENCE. Straightforward; not collateral. Evidence is termed direct which applies immediately to the fact to be proved, without any intervening process as distinguished from circumstantial, which applies immediately to collateral facts supposed to have a connection, near to remote, with the fact in controversy.

DISMISS. To remove. To send out of court. Formerly used in chancery of the removal of a cause out of court without any further hearing. The term is now used in courts of law also.

DISQUALIFIED WITNESS. No party to or person interested in the event of, any action or proceeding in which the opposite party has succeeded to the interest of a deceased person, shall be examined as a witness in his own behalf, in regard to any personal communication or transaction with the deceased.

DISTRICT ATTORNEY. Public officers elected or appointed, as provided in several state constitutions or by statute, to conduct suits, generally criminal, on behalf of the state in their respective districts. They are sworn ministers of justice, quasi-judicial officers representing the commonwealth.

District attorneys of the United States are appointed for a term of four years in each judicial district, whose duty it is to prosecute, in such district, all delinquents, for crimes and offenses cognizable under the authority of the United States, and all civil actions in which the United States shall be concerned, except in the Supreme Court, in the district in which the court shall be holden. The District Attorney must appear upon the record for the United States plaintiff, in order that the United States should be recognized as such on the record.

DOCKET. A formal record of judicial proceedings; a brief writing. A small piece of paper or parchment having the effect of a large one. An abstract.

DOCUMENTARY EVIDENCE. Written evidence includes such things as confessions, licenses, permits, tags, checks, registration cards, and statements by the officer, or by observers. It is frequently necessary to identify these statements by oral testimony in court.

Under physical evidence are such tangible materials as weapons, bullets, fingerprints, clothing, confiscated game, and fishing gear. It should be emphasized that it is very important to properly identify all physical evidence by marking it at once, so that it may be later properly identified when introduced at the trial.

DOCUMENTS. The deeds, agreements, title-papers, letters, receipts, and other written instruments used to prove a fact.

DOMESTIC ANIMALS. Those animals which by habit or training live in association with man. This class of animals includes cattle, horses, sheep, goats, pigs, poultry, cats, and dogs. Such animals, like other personal and movable chattels, are the subject of absolute property. The owner retains his property in them even if they stray or are lost.

DOMICILE. That place where an individual has his true, fixed, and permanent home and principal establishment, and to which whenever he is absent he has the intention of returning.

DOUBLE JEOPARDY. The second jeopardy of a person who has been previously in jeopardy for the same offense.

The test is not whether the defendant has already been tried for the same act, but whether he has been put in jeopardy for the same offense.

DOUBT. The uncertainty which exists in relation to a fact, a proposition, or other thing; an equipoise of the mind arising from an equality of contrary reasons. Some rules, not always infallible, have been adopted in doubtful cases, in order to arrive at the truth.

(1) In civil cases, the doubt ought to operate against one who, having it in one's power to prove facts to remove the doubt, has neglected to do so. In cases of fraud, when there is a doubt, the presumption of innocence ought usually to remove it.

(2) In criminal cases, whenever a reasonable doubt exists as to the guilt of the accused, that doubt ought to operate in one's favor. In such cases, particularly when the liberty, honor, or life of an individual is at stake, the evidence to convict ought to be clear and devoid of all reasonable doubt. The term reasonable doubt is often used, but not easily defined. Failure to explain reasonable doubt in a charge is not error.

DURESS. Personal restraint, or fear of personal injury or imprisonment. One is considered guilty of illegal acts even when one commits them under the *direction* of one's superior. For example, a superior officer orders another officer to make an illegal arrest; the latter is in violation.

EMINENT DOMAIN. The superior right of property subsisting in a sovereignty, by which private property may in certain cases be taken or its use controlled for the public benefit, without regard to the wishes of the owner. The power to take private property for public use.

EVIDENCE. That which tends to prove or disprove any matter in question, or to influence the belief respecting it. Belief is produced by the consideration of something presented to the mind. The matter thus presented, in whatever shape it may come, and through whatever material organ it is derived, is evidence.

The instruments of evidence, in the legal acceptance of the term, are:

1. *Judicial notice or recognition:* there are diverse things of which courts take judicial notice, without the introduction of proof by the parties, such as: the territorial extent of their jurisdiction, local division of their own countries, seats of courts, all public matters directly concerning the general government, the ordinary course of nature, divisions of time, the meanings of words, and generally, of whatever ought to be generally known in the jurisdiction. If the judge needs information on subjects, the judge will seek it from such sources as deemed authentic.
2. *Public records:* the registers of official transactions made by officers appointed for the purpose; as, the public statutes, the judgments and proceedings of courts, etc.
3. *Judicial writings:* such as inquisitions, depositions, etc.
4. *Public documents* having a semiofficial character: as, the statute-books published under the authority of the government, documents printed by the authority of Congress, etc.
5. *Private writings:* as, deeds, contracts, wills.
6. *Testimony of witness.*
7. *Personal inspection,* by the jury or tribunal whose duty it is to determine the matter in controversy: as, a view of the locality by the jury, to enable them to determine the disputed fact, or the better to understand the testimony, or inspection of any machine or weapon which is produced in the cause.

Real evidence is evidence of the thing or object which is produced in court. When, for instance, the condition or appearance of any thing or object is material to the issue, and the thing or object itself is produced in court for the inspection of the tribunal, with proper testimony as to its identity, and, if necessary, to show that it has existed since the time at which the issue in question arose, this object or thing becomes itself "real evidence" of its condition or appearance at the time in question.

Extrinsic evidence is external evidence, or that which is not contained in the body of an agreement, contract, and the like.

Presumptive evidence is that which shows the existence of one fact, by proof of the existence of another or others, from which the first may be inferred; because the fact or facts shown have a legitimate tendency to lead the mind to the conclusion that the fact exists which is sought to be proved.

Presumptions of fact are not the subject of fixed rules, but are merely natural presumptions, such as appear, from common experience, to arise from the particular circumstances of any case. Some of these are founded upon a knowledge of the human character, and of the motives, passions, and feelings by which the mind is usually influenced.

Circumstantial evidence is the proof of facts which usually attend other facts sought to be proved; that which is not direct evidence. For example, when a witness testifies that a man was stabbed with a knife, and that a piece of the blade was found in the wound, and it is found to fit exactly with another part of the blade found in the possession of the prisoner, the facts are directly attested, but they only prove circumstances; and hence this is called circumstantial evidence.

Circumstantial evidence is of two kinds, namely, certain and uncertain. It is certain when the conclusion in question necessarily follows: as, where a man had received a mortal wound, and it was found that the impression of a bloody left hand had been made on the left arm of the deceased, it was certain some person other than the deceased must have made such mark; but it is uncertain whether the death was caused by suicide or by murder, and whether the mark of the bloody hand was made by the assassin, or by a friendly hand that came too late to the relief of the deceased.

Circumstantial evidence warrants a conviction in a criminal case, provided it is such as to exclude every reasonable hypothesis but that of guilt of the offense charged to the defendant, but it must always rise to that degree of convincing power which satisfies the mind beyond reasonable doubt of guilt. This can never be the case when the evidence, as produced, is entirely consistent with innocence in a given transaction.

Secondary evidence is that kind of proof which is admissible when the primary evidence cannot be produced, and which becomes by that event the best evidence that can be adduced.

Prima facie evidence is that which appears to be sufficient proof respecting the matter in question, until something appears to controvert it, but which may be contradicted or controlled.

Conclusive evidence is that which, while uncontradicted, establishes the fact: as in the instance of conclusive presumptions; it is also that which cannot be contradicted.

Res gestae. But where evidence of an act done by a party is admissible, his declaration made at the time, having a tendency to elucidate or give a character to the act, and which may derive a degree of credit from the act itself, are also admissible, as part of the *res gestae.*

Dying declarations are an exception to the rule excluding hearsay evidence, and are admitted, under certain limitations in cases of homicide, so far as the circumstances attending the death and its cause are the subject of them.

Opinions of persons of skill and experience, called experts, are also admissible in certain cases, when, in order for a better understanding of the evidence or to the solution of the question, a certain skill and experience are required which are not ordinarily possessed by jurors. A nonexpert witness on the question of the sanity of one accused of crime after stating such particulars as he can remember—generally only the more striking facts—is permitted to sum up the total remembered and unremembered impressions of the senses by stating the opinion which they produced.

EXAMINATION. In criminal law. The investigation by an authorized magistrate of the circumstances which constitute the grounds for an accusation against a person arrested on a criminal charge, with a view to discharging the person so arrested, or to securing his appearance for trial by the proper court, and to preserving the evidence relating to the matter.

EXCLUSION. Denial of entry.

EXECUTE. To complete; to sign, seal, and deliver.

EXEMPLIFIED. Exemplifications are copies verified by the great seal or by the seal of a court. Also defined as: an official transcript of a document from public records, made in form to be used as evidence, and authenticated as a true copy.

EXHIBIT. To produce a thing publicly, so that it may be taken possession of and seized.

EXPERT TESTIMONY. There are several things on which laypersons may not testify because of their lack of experience or special training. A ballistics expert, for example, is qualified to testify as to whether or not a bullet was or was not fired from some particular weapon. A doctor is better qualified to form an opinion as to the cause of death. A ditch digger may be better qualified to testify as to the cause of a bank cave-in. An expert witness must be qualified before the court; the witness may then state facts, and opinions as based on the facts presented. It is always up to the jury to weigh the evidence and draw their own conclusions.

EXPERT WITNESS. One who has acquired such special knowledge of the subject matter about which one is to testify, either by study of the recognized authorities on the subject or by practical experience, so that one can give the jury assistance and guidance in solving a problem for which their equipment, good judgment, and average knowledge is inadequate.

One who is skilled in some art, trade, profession, or science, or a person who has knowledge and experience in relation to matters at hand which are generally not within the knowledge of persons of common education and experience. An expert witness may express an opinion on a state of facts which is within one's specialty. An expert witness may be qualified by the court or one may be accepted without being qualified if there is no objection from either side. An expert witness may state one's relevant opinions based on personal experience or observation or study without specifying data on which it is based. However, on cross-examination, one may be required to bring forth data upon which one's opinion is based. An expert witness may also be asked to testify on a hypothetical question if this question is based on facts and evidence at the time the question is asked. When, at a later date, no such facts are actually introduced, the opinion based on them should be excluded from admissible evidence. An admissible expert opinion is regarded as evidence when it pertains to the matter on which the opinion relates.

EXTORTION. The communication of threats to another with the intention of obtaining something of value. For example, to gain an acquittance, or an immunity. The offense is complete when the communication has been completed. The success or failure of the venture is immaterial in determining guilt. The threat may be communicated by word of mouth or in writing. It may be in the form of a threat of unlawful injury to the person, or his property or any member of his family, or any other person held dear to him, or a threat to expose or impute any deformity or disgrace to the individual threatened, or to any member of his family or other person held dear to him, or a threat to expose any secret affecting the individual or others previously mentioned.

EXTRADITION. The surrender by one sovereign state to another, on its demand, of persons charged with the commission of crime within its jurisdiction, that they may be dealt with according to its laws. The surrender of persons by one federal state to another, on its demand, pursuant to their federal constitution and laws.

A process of requisition by which a defendant charged with a crime in one state, and apprehended in another state, is returned to the place of jurisdiction. Violations of fish and game laws in many states are considered as nonextraditable offenses.

EYE-WITNESS. One who saw the act or fact to which he testifies. When an eye-witness testifies, and is a person of intelligence and integrity, much reliance must be placed on the witness' testimony.

FALSE ARREST. Any unlawful, physical restraint by one person of the liberty of another, whether in prison or elsewhere.

FALSE PRETENSE. An intentional, false statement concerning a material matter of fact, in reliance on which the title or possession of property is parted with. If possession is obtained by fraud, the offense is that of obtaining by false pretenses, provided the means by which it is acquired are false. A false pretense must be false when made and it must be knowingly false in that it is made without an honest belief in its truth. A false pretense is a false representation of past or existing facts. For example, one who pretends that he is about to perform a certain act but who at the time of the representation has no honest intention of doing so makes a false representation. A person who states that he will appear in a particular court at a given time but at that time has no intention of doing so, is guilty of false pretense. False representations and statements, made with a fraudulent design to obtain ''money, goods, wares, and merchandise,'' with intent to cheat.

FALSE STATEMENT. The term false statement refers to statements which are wilfully false or fraudulent. Wilfully false is generally, though not invariably, the determining mark of crime.

FEAR. In robbery, the fear which moves the victim to part with one's goods. If a transaction be attended with such circumstances of terror as in common experience are likely to create an apprehension of danger and to induce a person to part with one's property for the safety of one's person, one is put in fear.

FEDERAL COURTS. The courts of the United States comprise the following: The Senate of the United States, sitting as a court of impeachment; the Supreme Court; the courts of appeals; the district courts; the court of claims; the court of customs and patent appeals; the customs court; the tax court of the United States; and provisional courts; courts of territories and outlying possessions. For the purpose of enforcing federal law applicable to the whole country and therefore applicable to the District of Columbia, the courts of the District are held to be federal courts.

FELONY. Whether a criminal act is a felony or a misdemeanor is usually made to depend upon the character of the punishment provided by statute. An offense which occasions a total forfeiture of either lands or goods, or both, at common law, to which capital or other punishment may be superadded, according to the degree of guilt. A more serious crime than a misdemeanor; sometimes separated from the latter on the basis of the monetary values involved.

This is the most serious grade of crime and includes murder, manslaughter, treason, burglary, rape, robbery, arson, kidnapping, and most of the other more serious illegal acts.

FERAE NATURAE. A term used to designate animals not usually tamed, or not regarded as reclaimed so as to become the subject of property. Such animals belong to the person who has captured them only while they are in his power. Wild animals; wild beasts. Animals denominated *ferae naturae* include deer in the forest, birds in the air, and fish in the public waters or in the sea.

FERAL. Untamed, unbroken, undomesticated, or uncultivated; hence, wild; savage; bestial. Having escaped from domestication or cultivation and become wild.

FINE. Pecuniary punishment imposed by a lawful tribunal upon a person convicted of crime or misdemeanor.

FIREARMS. As sometimes used in statutes prohibiting parades of unauthorized bodies with firearms, any gun which to the ordinary observer appears to be an efficient weapon capable of being discharged is held to be a firearm whether it has in fact been rendered harmless or not. A weapon that propels a projectile by firing a powder charge. Whether or not a weapon that fires a projectile by spring or compressed air force is a firearm is something that must be determined by state or local statutes.

FISH. An animal which inhabits the water, breathes by means of gills, and swims by aid of fins. Fish may be either egg-laying or live-bearing. Fish themselves are normally *ferae naturae*—the common property of the public, or of the state—in the United States. Any of the numerous cold-blooded, strictly aquatic, water-breathing animals with backbones. When limbs are present they are developed as fins and the body is typically scaled or may be partially scaled or naked. Under certain terminology, any of the class Pisces of zoological nomenclature.

FISHERY. The right to employ lawful means for the taking of fish. It differs from a fishing place or the right to use a particular shore or beach as the basis for carrying on the business; the latter being vested in the shore owner and entirely distinct from the right to take fish from the water.

FISHING. Taking or attempting to take fish.

FISHING EXPEDITION OR FISHING EXAMINATION. The improper calling for and examination of documents of an adversary, or the questioning of an adverse witness, on bare suspicion and with no reason to believe that evidence pertinent to the issues will be disclosed.

FISH, ROYAL. Fish which when brought ashore were the property of the king; such fish were confined to sturgeons, whales, and porpoises. (The latter two are mammals.)

FLIGHT. The evading the course of justice by a person's voluntarily withdrawing oneself.

FOREST LAW. The old law relating to the forest, under which the most horrid tyrannies were exercised, in the confiscation of lands for the royal forests.

FORGERY. Forgery may be committed either by falsely writing or by knowingly uttering a falsehood which was made in writing. It is the altering, with intent to defraud, of a signature or any part thereof or of any writing which would, if genuine, apparently impose a legal liability on another, or change one's legal right of liability or prejudice one. It is the uttering, offering, issuing, or transferring, with intent to defraud, of a writing known by the offender to have been fraudulently made or to have been altered. It should be noted that it is not forgery to make a false statement or to alter a letter if it does not, by this alteration, impose a legal liability on another or change a legal right or liability to one's prejudice. For example, a letter of introduction. Forgery is not committed by the genuine making of a false statement for the purpose of defrauding another. For example, a check bearing the signature of the maker, although drawn on a bank in which the maker has no money or credit, and even with intent to defraud the payee or the bank, is not a forgery, for the instrument, although false, is not falsely made. Likewise, if a person makes a false signature to an instrument, and signs another name but puts the word "by" with his own signature after the original signature, thus indicating authority to sign, the offense is not forgery even if no such party exists. False statements of facts in a genuine document do not constitute forgery, as, for example, when a conservation officer states that he or she was in the field eight hours on a particular day when in fact actually playing billiards at "Joe's Place." The primary aspects of forgery involve altering a signature or writing and composing a legal liability as a result thereof. The altering of an instrument already invalid is not forgery because the instrument already had no legal standing. However, the making of another's signature on a statement with intent to defraud, even though there is little or no resemblance to the genuine signature, is forgery.

In order to constitute a forgery by altering a writing, the alteration must effect a material change in the legal tender of the writing. An alteration in the charge on a complaint which reduces it to a lesser offense is an example. As regards to the intent to defraud, it need not be directed toward anyone in particular nor need it be to the advantage of the offender. Furthermore, it is actually immaterial whether or not anyone is actually defrauded.

FRAUD. An endeavor to alter rights, by deception touching motives, or by circumvention not touching motives.

FUGITIVE FROM JUSTICE. One who, having committed a crime, flees from the jurisdiction within which one was committed, to escape punishment.

FUR. The hairy covering or coat of certain mammals which is normally fine, soft, and thick, and has an undercoat as well as an outer coat. This is distinguished from the ordinarily thin, coarse hair of certain animals such as the mountain lion.

Skins valuable chiefly on account of the fur. Skins is a term applied to those valuable chiefly for the skin. The word hides is inapplicable to fur skins.

GAME. Wild animals, excluding fish. Generally only mammals and birds. Any animals so defined by law or proclamation having the effect of law.

This term, as it is presently used, generally means all wild animals excluding fish. However, more specifically, it frequently applies to only birds and mammals. In many states all wild animals (wildlife) are either listed as fish or as game. This is broadly interpreted to mean that all cold-blooded (Poikilothermic) animals fall under the category of fish; and all warm-blooded animals fall under the category of game. Strictly speaking, a game animal or a game fish is any animal so defined by the legal code of any particular state or territory. All others are excluded. The word includes all animals of both land and sea which are hunted for sport. Fish, deer, pheasants, partridges, and even wild bees, are held to be within the meaning of it.

GAME LAWS. Laws regulating the killing, taking, or possession of game. The English game laws are founded on the idea of restricting the right of taking game to certain privileged classes, generally landholders, and are said to be directly descended from the old forest laws.

GILL NET. A fish net so constructed that the fish is usually caught by its gills when it attempts to escape from the meshes of the net. Such nets are regulated as to their use in some jurisdictions and forbidden in others.

GUILT. That which renders one criminal and liable to punishment. In general, everyone is presumed innocent until guilt has been proved; but in some cases the presumption of guilt overthrows that of innocence; as, for example, where a party destroys evidence to which the opposite party is entitled.

GUILTY. The state or condition of a person who has committed a crime, misdemeanor, or offense.

GUN. In the usual sense, a weapon which throws a projectile or missile to a distance; a firearm, for throwing a projectile with gunpowder. A firearm is sometimes defined as a weapon which acts by the force of gunpowder.

HABEAS CORPUS. Meaning literally that "you have a body." A writ of *habeas corpus* is an order of the court directed to the jailer or other police officer detaining a certain person to produce the detained person in court and to otherwise justify his being kept in custody.

A writ directed to the person detaining another and commanding him to produce the body of the prisoner at a certain time and place, with the day and cause of his capture and detention, to do, submit to, and receive whatsoever the court or judge awarding the writ shall consider in that behalf. A writ which has for centuries been esteemed the best and only sufficient defense of personal freedom, having for its object the speedy release by judicial decree of persons who are illegally restrained of their liberty, or illegally detained from the control of those who are entitled to the custody of them.

HEARING. The examination of a prisoner charged with a crime or misdemeanor, and of the witnesses for the accused.

The action brought before court or other official bodies having jurisdiction in a case, in which the testimony of the prosecutor, the defendant, and their witnesses are heard.

HEARSAY EVIDENCE. That kind of evidence which does not derive its value solely from the credit to be given to the witness himself, but rests also, in part, on the veracity and competency of some other person. Evidence not open to cross-examination.

A statement offered as evidence to prove the truth of a matter, but which is not made by the author as a witness to the act. This applies to written or oral statements, by the use of symbols, or any other substitute for words which is offered as the equivalent of a statement. Hearsay does not become competent evidence even though it is received by court without objection. This simply means that a fact cannot be proven by showing that someone stated it as a fact. Hearsay is someone else's statement repeated.

HOSTILE WITNESS. A witness who is subject to cross-examination by the party who called the witness, because of his evident antagonism toward the party as exhibited in direct examination.

HUE AND CRY. See: Statute of Hue and Cry.

HUNTING. The act of pursuing, taking, or attempting to take wild game.

HUNTING LICENSE. A license required in most, if not all of the states, under statutes which usually prescribe the fee to be paid therefore and make the procuring of such a license a prerequisite to hunting wild game in the state.

IGNORANCE OF FACT. If a person technically commits a crime but is ignorant of a *fact* essential to the crime or makes a mistake regarding the facts, he may *have* a good defense. For example, a person who receives a game animal from another in the belief that the animal was legally taken and therefore may be legally accepted.

IGNORANCE OF LAW. The courts in the United States have supported the principle that every person is presumed to know the law of his country and ignorance of the law is no defense.

ILLEGAL. The word is synonymous with the word "unlawful."

IMPEACH. To impeach is defined by *Webster's New International Dictionary* as to bring or throw discredit on; to call in question to challenge; to impute some fault or defect to.

IMPLICATION. An inference of something not directly declared, but arising from what is admitted or expressed.

IMPORT. To import goods is to bring them into a country from a foreign country. In a constitutional sense, articles brought from one state of the Union into another state of the Union are not imported.

INDICTMENT. An accusation made by the grand jury, recommending that the defendant or defendants be tried in court.

INFORMATION. Strictly speaking, an information is an accusation made in writing against an individual by the prosecuting attorney, and direct to the court without the matter having been passed upon by the grand jury. (See Complaint.)

IN PERSON. Without counsel in the conduct of one's action or defense.

INSANITY. A diseased or disordered condition or malformation of the physical organs through which the mind receives impressions or manifests its operations, by which the will and judgment are impaired and the conduct rendered irrational.

INSTIGATE. According to the *Standard Dictionary,* to instigate means to stimulate or goad to an action, especially a bad action. One of the synonyms of the word is "abet," which in law means to aid, promote, or encourage in the commission of an offense.

INTENT. In a legal sense, the word is quite distinct from motive and may be defined as the purpose to use a particular means to effect a certain result, while motive is the reason which leads the mind to desire that result.

INTENTION, INTENT. A design, resolve, or determination of the mind.

INVALID. The word is defined by Webster as having no force or effect or efficacy; void, null.

INVESTIGATE. To make inquiry, judicially or otherwise, for the discovery and collection of facts concerning the matter or matters involved. Authority to a legislative committee to make an investigation includes the power to call witnesses and to compel them to testify under oath.

JEOPARDY. The situation of a prisoner when a trial jury is sworn and impanelled to try his case upon a valid indictment, and such jury has been charged with his deliverance.

JOHN DOE WARRANT. A warrant for the arrest of a person who is described by a fictitious name because person's real name is unknown.

JUDGE. One who publicly is charged with and performs judicial functions; one who presides at the trial of causes involving justiciable matters in which the public at large is interested. A public officer lawfully appointed to decide litigated questions according to laws. An officer so named in his commission who presides in some court. In its most extensive sense the term includes all officers appointed to decide litigated questions while acting in that capacity, including justices of the peace.

JUDGMENT. The final consideration and determination of a court of competent jurisdiction upon the matter submitted. The sentence of the law upon the record; the application of the law to the facts and pleadings. The last word in a judicial controversy; the final consideration and determination of a court of competent jurisdiction upon matters submitted to it in an action or proceeding.

JUDICIAL NOTICE. A term used to express the doctrine of the acceptance by a court for the purposes of the case, of the truth of certain notorious facts without requiring proof. It is the process whereby proof by parol evidence is dispensed with, where the court is justified by general considerations in assuming the truth of a proposition without requiring evidence from the party setting it up.

JURISDICTION. The authority by which judicial officers take cognizance of and decide causes.

JURISPRUDENCE. The science of the law. The practical science of giving a wise interpretation to the laws and making a just application of them to all cases as they arise.

JURY. A body of lay people, selected by lot, or by some other fair and impartial means, to ascertain, under the guidance of a judge, the truth in questions of fact arising either in civil litigation or a criminal process. At common law, a jury must consist of twelve people. A body of people sworn to declare the facts of a case as they are proven from the evidence placed before it.

JUSTICE OF THE PEACE. A public officer invested with judicial powers for the purpose of preventing breaches of the peace and bringing to punishment those who have violated the laws. Usually elected by popular ballot and not necessarily a person with formal legal training.

JUSTICE'S DOCKET. A book required to be kept by a justice of the peace wherein it is a duty to enter the titles of all causes commenced before the justice, the time when the first and subsequent process was issued against the defendant, and the particular process issued, the judgment rendered by the justice and the time of rendering the same, the time of issuing execution and the officer to whom delivered.

JUSTIFIABLE TRESPASS. An intentional trespass which the law has authorized; as, an entry into a house through an open door to serve a civil process.

JUVENILE COURTS. Courts which have recently been created by statute, as a product of the solicitude of the law for the welfare of infants, to deal with dependent, neglected, and delinquent children.

LAW. A law is a rule governing human conduct and prescribed by a governing authority and enforced by the courts. A statute; a bill; a legislative enactment; a constitutional provision; the whole body or system of rules of conduct, including both decisions of courts and legislative acts. That which is laid down; that which is established. A rule or method of action, or order of sequences.

The rules and methods by which society compels or restrains the action of its members. The aggregate of those rules and principles of conduct which the governing power in a community recognizes as those which it will enforce or sanction, and according to which it will regulate, limit, or protect the conduct of its members. The aggregate of rules set by people as politically superior or sovereign, to people as politically subject.

In its relation to human affairs there is a broad use of the term, in which it denotes any of those rules and methods by which a society compels or restrains the action of its members. Here the idea of a command is more generally obvious, and has usually been thought an essential element in the notion of human law.

LAWFUL. Legal. That which is not contrary to law. That which is sanctioned or permitted by law. That which is in accordance with law. The term ''lawful,'' ''unlawful,'' and ''illegal'' are used with reference to that which is in its substance sanctioned or prohibited by the law. The term ''legal'' is occasionally used with reference to matter of form alone; thus, an oral agreement to convey land, though void by law, is not properly said to be unlawful, because there is no violation of law in making or in performing such an agreement; but it is said to be not legal, or not in lawful form, because the law will not enforce it, for want of that written evidence required in such cases.

LAWYER. An attorney or counselor at law; a barrister; a solicitor; a person licensed by law to practice the profession of law.

LEADING QUESTION. A question which puts into the witness' mouth the words to be echoed back, or plainly suggests the answer which the party wishes to get from the witness. In that case the examiner is said to lead the witness to the answer. It is not always easy to determine what is or is not a leading question. Certain questions cannot be put to a witness unless he is a hostile witness. But, questions may be put to lead the mind of the witness to the subject of inquiry; and they are allowed when it appears that the witness wishes to conceal the truth or to favor the opposite party, or where from the nature of the case the mind of the witness cannot be directed to the subject of inquiry without a particular specification of such subject. In cross-examination, the examiner generally has the right to put leading questions; but not perhaps when the witness has a bias in his favor.

LEGAL. According to the principles of law; according to the method required by statute; by means of judicial proceedings; not equitable.

LETHAL. Capable of producing death or great bodily harm.

LETHAL WEAPON. A gun, sword, knife, pistol, or the like, is a lethal weapon, as a matter of law, when used within striking distance of the person assaulted; and all other weapons are lethal or not, according to their capacity to produce death or great bodily harm in the manner in which they are used.

LICENSE. A right, given by some competent authority, to do an act, which without such authority would be illegal. A permission to do some act or series of acts on the land of the licensor, without having a permanent interest in it; it is founded on personal confidence, and not assignable.

MAGISTRATE. A public civil officer, invested with some part of the legislative, executive, or judicial power given by the constitution. A word commonly applied to the lower judicial officers, such as justices of the peace, police judges, town recorders, and other local judicial functionaries.

MAGNA CHARTA. The greater charter of liberties which was wrung from King John by the barons in the year 1215, and later confirmed in Parliament by Henry the Third, son of John. It contained some new grants, but was principally a declaration of the principal grounds of the fundamental laws of England.

MALA FIDE. In bad faith.

MALICE. In Criminal Law. The doing of a wrongful act intentionally without just cause or excuse. A wicked and mischievous purpose which characterizes the perpetration of an injurious act without lawful excuse.

MALICIOUS ARREST. A wanton arrest made without probable cause. The term is applied when the arrest on which an action for malicious prosecution is based was under civil, and not under criminal process.

MALICIOUS PROSECUTION. A wanton prosecution made by a prosecutor in a criminal proceeding, or a plaintiff in a civil suit, without probable cause, by a regular process and proceeding, which the facts did not warrant, as appears by the result.

MALINGERING. Feigning illness, disability, mental derangement, or intentionally inflicting self-injury on oneself; all for the purpose of avoiding an act or duty as requested by an officer or by the court.

MALTREATMENT. In order for a charge of maltreatment of a prisoner to be brought against an officer the act must be real and without justifiable cause although not necessarily of a physical nature. Abuse may be by derogatory words, assault, striking, or depriving the prisoner of certain benefits and the necessities of life within a reasonable time.

MAMMAL. The highest class of animals with backbones, including humans and all other animals that nourish their young with milk. Mammals are "warm-blooded" and, except in one group, the young are born alive.

MANSLAUGHTER. The unlawful killing of another, but without malice. It may be voluntary, as when the act is committed with a real design and purpose to kill, but through the violence of sudden passion which in tenderness for the frailty of human nature the law considers sufficient to palliate the offense. It is involuntary when the death of another is caused by some unlawful act not accompanied with any intention to take life.

MARTIAL LAW. That government and control which military commanders may lawfully exercise over the persons and property of citizens and individuals not engaged in the land and naval service. Military law applies to those rules enacted by the legislative power for the government and regulation of the United States Armed Forces.

MATERIAL EVIDENCE. Such as is relevant and goes to the substantial matters in dispute, or has a legitimate and effective influence or bearing on the decision of the case.

MATTER OF LAW. Something that can be decided by the application of statutory rules or the principles and determination of the law, not investigative facts.

MID-CHANNEL. If there be more than one channel of a river, the deepest channel is regarded as the navigable mid-channel for the purpose of territorial demarcation; and the boundary line will be the line drawn along the surface of the stream corresponding to the line of deepest depression of its bed.

MIDDLE OF THE STREAM. In international law, and by the usage of European nations, the term, as applied to a navigable river, is the same as the middle of the channel of the stream.

MISCONDUCT. A transgression of some established and definite rule of action, where no discretion is left, except what necessity may demand; it is a violation of definite law; a forbidden act. It differs from carelessness.

MISDEMEANOR. A term used to express every offense inferior to felony, punishable by indictment, or by particular prescribed proceedings. In its usual acceptation, it is applied to all those crimes and offenses for which the law has not provided a particular name.

A grade of crime less serious than a felony, but more serious than a summary conviction offense in some states.

MITIGATION. The reduction of damages or punishment by reason of extenuating facts or circumstances.

MODUS OPERANDI. Method of operation. The techniques used by a criminal in the commission of a crime. The type of operation that is peculiar to an individual.

MOOT QUESTION. A mooted or undecided point of law.

MORAL CERTAINTY. That degree of certainty which will justify a jury in grounding on it their verdict. The phrase is one the use of which is not likely to assist a jury in charging upon the question of reasonable doubt. It is an artificial form of words having no precise and definite meaning.

A probability sufficiently strong to justify action upon it.

MORAL CONSIDERATION. A consideration which is good only in conscience. The idea that in every case where a person is under a moral obligation to do an act, as to relieve one in distress by personal exertions, or to spend money, a promise to that effect would bind one in law, is not supported by principle or precedent.

MORAL EVIDENCE. Evidence not only of that kind which is employed on subjects connected with moral conduct, but all the evidence which is not obtained either from intuition, or from demonstration. In the ordinary affairs of life, we do not require demonstrative evidence.

MORAL INSANITY. A term sometimes employed to denote such mental disease as destroys the ability to distinguish between right and wrong as to a particular act, and sometimes to denote a mere perversion of the moral sense. It is sometimes used as being synonymous with irresistible impulse.

MORAL OBLIGATION. Where courts generally recognize the existence of a moral obligation as a sufficient consideration, it refers to promises that are enforceable because they are based on obligations, which were formerly valid but have since become barred, as by limitations or bankruptcy.

MOTION. An application to a court by one of the parties in a cause, or one's counsel, in order to obtain some rule or order of court which one thinks becomes necessary in the progress of the cause, or to get relieved in a summary manner from some matter which would work an injustice. It is said to be a written application for an order. An application made to a judge, chancellor, or court for the purpose of obtaining a rule or order directing some act to be done in favor of the applicant. It is usually a proceeding incidental to an action.

MOTION FOR DIRECTED VERDICT. A motion made by either plaintiff or defendant, after all the evidence is in, requesting the court to instruct the jury to return their verdict in favor of the moving party. A motion for a directed verdict in favor of the defendant is tantamount to a demurrer to the evidence.

MOTION FOR JUDGMENT ON PLEADINGS. A motion made by either party to an action in the nature of a general demurrer to the pleadings of the adverse party, and which, for all purposes of the motion admits the truth of all the allegations contained in those pleadings.

MOTION FOR NEW TRIAL. A motion the office of which is primarily to afford the court an opportunity to correct errors in the proceedings before it without subjecting the parties to the expense and inconvenience of an appeal or petition in error.

MOTIVE. The inducement, cause, or reason why a thing is done. It is an inducement, or that which leads or tempts the mind to indulge the criminal act; it is resorted to as a means of arriving at an ultimate fact, not for the purpose of explaining the reason of a criminal act which has been clearly proved, but from the important aid it may render in completing the proof of the commission of the act when it might otherwise remain in doubt. It is not indispensable to conviction for murder that the particular motive for taking the life of a human being shall be established by proof to the satisfaction of the jury.

MURDER. The unlawful killing of a human being with malice aforethought.

MUTILATE. To deprive of some essential part, as to tear off a portion of a railroad ticket.

NAVIGABILITY. The test of navigability is whether the river, in its natural state, is used, or capable of being used, as a highway for commerce, over which trade and travel are or may be conducted. Navigability is not destroyed because the watercourse is interrupted by occasional natural obstruction or portages; nor need the navigation be open at all seasons of the year, or at all stages of the water.

NAVIGABLE. Rivers must be regarded as public navigable rivers in law which are navigable in fact, and they are navigable in fact when they are used, or are susceptible of being used, in their ordinary condition, as highways for commerce. And it has been held that a stream of sufficient depth and width, in its natural state, to be used for the transportation of timbers or logs, is subject to the public right of user.

NAVIGABLE IN FACT. Navigability in fact is the test of navigability in law, and whether a river is navigable in fact is to be determined by inquiring whether it is used, in its natural and ordinary condition, as a highway for commerce.

NAVIGABLE STREAM. A river capable of floating to market the products of the country, and upon which boats, barges, rafts, or logs may be borne, is a navigable stream both in fact and in law. The criterion of navigability is the use to which the stream may be put.

NAVIGABLE WATERS. Formerly the term was confined to those waters which were affected by the ebb and flow of the tides, but in the development of this country the meaning has been enlarged so as to apply to all waters which are in fact navigable.

NEGATIVE TESTIMONY. Testimony which either denies that certain alleged facts are true, or that certain things occurred, or which denies all knowledge of the matter. Testimony which is positive in form may amount merely to negative testimony.

NEGLIGENCE. The word ordinarily used to express the foundation of civil liability at common law for injury to person or property, when such injury is not the result of premeditation and formed intention. The question arises as to whether the basis of an action for negligence is the negligent act of the defendant or the injury resulting therefrom. It would seem that it is the injury and not alone the negligent act which gives rise to the right of action.

NEW TRIAL. A re-examination of an issue of fact in the same court after a trial. When there has been no trial upon issues of fact, a new trial will not be granted, as where the case has been submitted on the pleadings by stipulation leaving the court to decide under the law and upon the facts stated.

NO AWARD. A name which was given to a plea which denied that an award which was sued upon was ever made.

NO BILL. An indictment which was not found by the grand jury was endorsed ''no bill'' or ''ignoramus'' (we do not know).

NOLO CONTENDERE. Literally means ''I will not contest it.'' In practice the term does not mean that no defense will be given to the charge nor does it mean that the defendant is pleading guilty. A plea usually made only in federal courts.

A plea sometimes accepted in criminal cases, not capital, whereby the defendant does not directly admit oneself to be guilty, but tactly admits it by throwing oneself upon the mercy of the court and desiring to submit to a small fine, which plea the court may either accept or decline. The difference in effect is that, after the latter, not guilty cannot be pleaded in an action of trespass for the same injury, whereas it may be pleaded at any time after the former.

NON EST INVENTUS. From the Latin meaning, ''He is not to be found.'' A warrant which has been placed in the hands of an officer and cannot be served, because of the inability of the officer to locate the defendant, is usually returned *non est inventus,* which is sometimes abbreviated to NEI.

NONRESIDENT. One who is not residing or living in the state, or one who has no home or abode within the state, or one who has no intention, when leaving, of returning to a home or abode in the state.

NOTARY, NOTARY PUBLIC. An officer appointed by the executive or other appointing power, under the laws of different states. Notaries are of ancient origin; they existed in Rome during the republic.

Their duties differ somewhat in the different states, and are prescribed by statute. They are generally as follows: to protest bills of exchange and draw up acts of honor; to authenticate and certify copies of documents; to receive the affidavits of mariners and draw up protests relating to the same; to attest and take acknowledgments of deeds and other instruments; and to administer oaths. Ordinarily notaries have no jurisdiction outside the county or district for which they are appointed; but in several states they may act throughout the state.

NOT GUILTY. The name given to a plea of the general issue in trespass and certain other civil actions. The plea of a defendant in a criminal prosecution, the effect of which is to deny the truth of the allegations of the indictment or information.

OATH. Any form of attestation by which a person signifies that one is bound in conscience to perform an act faithfully and truthfully. It invokes the idea of calling on God to witness what is averred as truth, and it is supposed to be accompanied with an invocation of His vengeance, or a renunciation of His favor, in the event of falsehood. An outward pledge given by the person taking it that his attestation or promise is made under an immediate sense of his responsibility to God.

OATH OF OFFICE. An oath, the form of which is prescribed by law and which in most of the states is required as a qualification for a public office. The constitutions or laws of the states frequently provide that every state officer shall, before entering on the discharge of duties, take a prescribed oath.

OBJECTION. Where evidence is objected to at a trial, the nature of the objections must be distinctly stated, whether an exception be entered on the record or not, and, on either moving for a new trial on account of its improper admission, or on arguing the exception, the counsel will not be permitted to rely on any other objections than those taken at *nisi prius.*

OFFENSE. The doing that which a penal law forbids to be done, or omitting to do what it commands. In this sense, it is nearly synonymous with crime. In a more confined sense, it may be considered as having the same meaning with misdemeanor; but it differs from it in this, that it is not indictable, but punishable summarily by the forfeiture of a penalty.

OFFICER. A person who holds an office, either public or private.

OPEN FIELD. Areas beyond the curtilage.

ORDER. A written order of a court or judge embodying a determination of some preliminary matter pertaining to the proceedings, which is not implicated with the essential rights of the litigants.

OUT OF THE STATE. As used in a proviso tolling the statute of limitations while the defendant is ''out of the state,'' the expression has been held to apply not only to a resident who is absent for a time from the state, but also to a person who resides altogether out of the state.

PENAL STATUTE. A statute which imposes a penalty for transgressing its provisions. Such a statute is one that imposes a penalty or creates a forfeiture as the punishment for the neglect of some duty, or the commission of some wrong, that concerns the good of the public, and is commanded or prohibited by law.

PEREMPTORY. Final; positive; conclusive.

PEREMPTORY CHALLENGE. A challenge to proposed jurors which a defendant in a criminal case may make as an absolute right, and which cannot be questioned by either opposing counsel or the court. The motive which may influence a defendant or his attorney in the exercise of such right is not the subject of inquiry, nor comment in the presence of the jury.

PEREMPTORY DEFENSE. A defense which denies the right of the plaintiff to sue.

PEREMPTORY EXCEPTION. An answer which merely raises an issue of law, the legal effect of which is the same as that of a demurrer.

PEREMPTORY PLEA. A plea which sets up the defense that the plaintiff has no right to sue.

PERJURY. The wilful assertion as to a matter of fact opinion, belief, or knowledge, made by a witness in a judicial proceeding as part of witness' evidence, either upon oath or in any form allowed by law to be substituted for an oath, whether such evidence is given in open court, or in an affidavit, or otherwise, such assertion being known to such witness to be false, and being intended by witness to mislead the court, jury, or person holding the proceeding. The wilful and corrupt false swearing or affirming, after an oath lawfully administered, in the course of a judicial or quasi-judicial proceeding as to some matter material to the issue or point in question. Some cases add the requirement that the false statement must be made after deliberation, others say that it must be knowingly false.

PERMANENT ABODE. A home, which a party may leave as interest or whim may dictate, but which one has no present intention to abandon.

PERMIT. The word is derived from the Latin word *permittere,* which means to concede, to give leave, to grant. To permit is to grant leave or liberty to by express consent; to allow expressly; to give leave, liberty, or license to; to allow to be done by consent or by not prohibiting.

PHOTOGRAPH. A photograph of a document is but a copy of that document, and its admission in evidence as a document is governed by the rules pertaining to copies.

PHYSICAL IMPOSSIBILITY. In the law of contracts, the term means practical impossibility according to the state of knowledge of the day; as for example, a promise to go from New York to London in an hour (1955), or to discover treasure by ''magic.''

PLAINTIFF. The person or persons making the charges against another in court. One who complains. One who, in personal action, seeks a remedy for an injury to one's rights.

PLEA. In Equity. A special answer showing or relying upon one or more things as a cause why the suit should be either dismissed, or delayed, or barred.

At Law. The defendant's answer by matter of fact to the plaintiff's declaration, as distinguished from a demurrer, which is an answer by matter of law. It includes as well the denial of the truth of the allegations on which the plaintiff relies, as the statement of facts on which the defendant relies.

PLEA OF *NOLO CONTENDERE.* The so-called plea of *nolo contendere* (I will not contest it) raises no issue of law or fact under the indictment, is not one of the pleas, general or special, open to the accused in all criminal prosecutions, and is allowable under leave and acceptance by the court. It is in reality a formal declaration that the accused will not contend with the prosecuting authority under the charge. *Nolo contendere* is a common defense plea in California fish and game cases.

POSSESSION. The detention or enjoyment of a thing which a person holds or exercises by oneself, or by another who keeps or exercises it in one's name.

POST MORTEM. From the Latin meaning ''after death,'' the term in legal circles is synonymously used with the word ''autopsy.''

POWER OF ATTORNEY. An instrument authorizing a person to act as the agent or attorney of the person granting it. It is often called letter of attorney. A general power authorizes the agent to act generally in behalf of the principal. A special power is one limited to particular acts.

PRECEDENTS. Legal decisions which are deemed worthy to serve as rules or models for subsequent cases.

PRELIMINARY EXAMINATION. The hearing given to a person of crime, by a magistrate or judge, exercising the functions of a committing magistrate or judge, to ascertain whether there is evidence to warrant and require the commitment and holding to bail of the person accused. Coroners generally have the powers of a committing magistrate as also have the mayors of cities in many of the states.

PREMEDITATION. An act is not premeditated unless the thought of completing it was consciously conceived and the action was taken with the completion of the act intended. It is the formation of a specific intent and the consideration of the intended act. It is not necessary that the intent shall have been entertained for any particular length of time. For example, a person legally hunting rabbits may suddenly decide to attempt to kill the next quail (which is not in season) that flies up. On the other hand, a person hunting rabbits that has no intention of killing quail, fires at one on the spur of the moment, is not guilty of a premeditated act, although it is an illegal one. Most officers and judges prefer to deal more harshly with a premeditated crime than with a spontaneous one, even though the acts involved may be identical.

PRESUME. To presume is derived from the Latin word *praesumere,* and signifies to take or assume a matter beforehand, without proof—to take for granted.

PRESUMPTION. An inference affirmative or disaffirmative of the truth or falsehood of any proposition or fact drawn by a process of probable reasoning in the absence of actual certainty of its truth or falsehood, or until such certainty can be ascertained.

PRIMA FACIE. Literally meaning ''at first sight.'' It means that there is sufficient evidence to justify or strongly infer the facts stated. For example, when buildings are fired by sparks from a locomotive, the fire has been held to be *prima facie* evidence of neglect on the part of those in charge of the locomotive. The proof of the mailing of a letter, duly stamped, is *prima facie* evidence of its receipt by the person to whom it is addressed. That is *prima facie* just, reasonable, or correct which is presumed to be just, reasonable, or correct until the presumption has been overcome by evidence which clearly rebuts it. Evidence which if unexplained or uncontradicted would of itself establish the fact alleged.

At first view, on first appearance. For example, the holder of a bill of exchange, endorsed in blank, is *prima facie* its owner.

PRIVATE POND. To fall within the meaning of the term, the pond must be essentially private. It must be a sheet of water covering exclusively the land of its owner, and must be such as no one could forbid the owner its use, any more than the cultivation of the soil underneath, if it was free from water.

PROCEDURE. The methods of conducting litigation and judicial proceedings.

PROOF. The conviction or persuasion of the mind of a judge or jury, by the exhibition of evidence, of the reality of a fact alleged. Thus, to prove is to determine or persuade that a thing does or does not exist.

PROSECUTING ATTORNEY. The attorney who conducts proceedings in a court. A public prosecutor. Public officers elected or appointed, as provided by the constitution or statutes of the various states, to conduct suits, generally criminal, on behalf of the state in their respective districts. They are sworn ministers of justice.

PROSECUTION. In Criminal Law. The means adopted to bring a supposed offender to justice and punishment by due course of law. The well-understood, legal signification of the word is, a criminal proceeding at the suit of the government.

PUNISHMENT. In Criminal Law. Some pain or penalty warranted by law, inflicted on a person for the commission of a crime or misdemeanor, or for the omission of the performance of an act required by law, by the judgment and command of some lawful court. The penalty for the transgression of the law.

PUNITIVE. That which inflicts or awards punishment. Whatever is concerned with punishment or penalties such as punitive laws or punitive justice. Also, whatever is involved in or aimed at punishment; such as, a punitive expedition or a punitive section of a code book.

QUALIFIED. Possessing fitness or capacity; a person is said to have qualified, or to be qualified for an office to which one has been elected or appointed, when one has complied with the requirements of the law; as by taking the oath of office and giving an official bond.

RATIONAL. Capable of reasoning; sane.

REASONABLE DOUBT. Such a doubt as will leave the juror's mind, after a careful examination of all the evidence, in such a condition that one cannot say that one has an abiding conviction, to a moral certainty, of the defendant's guilt. A doubt that, arising from a candid and impartial investigation of all the evidence, would cause a reasonable and prudent person to hesitate and pause.

REASONABLE OR PROBABLE CAUSE. As the expression is used in malicious prosecution, it is defined as a reasonable amount of suspicion, supported by circumstances sufficiently strong to warrant a cautious person in believing that the accused is guilty; but mere suspicion alone is not sufficient.

REBU HABLE PRESUMPTION. In the law of evidence. A presumption which may be rebutted by evidence. Otherwise called a ''disputable'' presumption. A species of legal presumption which holds good until disproved. (It shifts burden of proof. And which standing alone will support a finding against contradictory evidence.)

REBUT. To deny; to contradict; to avoid.

REBUTTAL. Testimony addressed to evidence produced by the opposite party; rebutting evidence.

REBUTTING EVIDENCE. Rebuttal evidence is that produced for the purpose of contradicting something previously testified.

That evidence which is given by a party in the cause to explain, repel, counteract, or disprove facts given in evidence on the other side. The term rebutting evidence is more particularly applied to that evidence given by the plaintiff to explain or repel the evidence given by the defendant. It is a general rule that anything may be given as rebutting evidence which is a direct reply to that produced on the other side.

RECLAIMED ANIMALS. Animals wild by nature which have been domesticated.

RECOGNIZE. To try a question of fact; to ratify; to adopt; to become bound by a recognizance.

RE-EXAMINATION. A second examination of a thing. A witness may be re-examined, in a trial at law, in the discretion of the court; and this is seldom refused. In equity, it is a general rule that there can be no re-examination of a witness after witness has once signed his name to the deposition and turned his back upon the commissioner or examiner. The reason for this is that the witness may be tampered with or induced to retract or qualify what he has sworn to.

REHEARING. A second consideration which the court gives to a cause on a second argument. A rehearing cannot be granted by the Supreme Court after the record has been remitted to the court below.

RELEVANT EVIDENCE. See Evidence. Any species of proof, or probative matter, legally presented at the trial of an issue, by the act of the parties and through the medium of witnesses, records, documents, concrete objects, etc., for the purpose of inducing belief in the minds of the court or jury as to their contention.

REMIT. To pardon; to remand for a new trial or for further proceedings; to transmit.

REPEAL. The abrogation or destruction of a law by a legislative act. A repeal is expressed, as when it is literally declared by a subsequent law, or implied, when the new law contains provisions contrary to or irreconcilable with those of the former law. The power to revoke or annul a statute or ordinance is equivalent to the power to repeal it; and in either case the power is legislative and not judicial in its character.

RES GESTAE. The circumstances, facts, and declarations which grow out of the main fact and are contemporaneous with it, and serve to illustrate its character. The facts of the transaction.

RESIDENCE. Personal presence in a fixed and permanent abode. A residence is different from a domicile, although it is a matter of great importance in determining the place of domicile. The essential distinction between residence and domicile is that the first involves the intent to leave when the purpose for which one has taken up one's abode ceases. The other has no such intent.

RESIST. To resist is to oppose by direct, active, and quasi-forcible means to stand against, to withstand. Refusal to obey an officer is not resistance to the officer.

ROTENONE. A certain substance that poisons fish by suffocating them.

RUSTLING. Stealing cattle, or in some cases other domestic livestock. In some states it is a felony.

SANCTUARY. A sacred place where a person who has committed a crime is immune from arrest. A place where wildlife may not be hunted or molested.

SATISFACTORY EVIDENCE. That which is sufficient to induce a belief that the thing is true; in other words, it is credible evidence.

SEARCH WARRANT. A legal writ, executed by competent authority, authorizing the search of the premises named therein and for the express purpose of determining if such unlawful article or articles as are named within the warrant are being secluded or held within the described premises. The goods may be either stolen or illegally possessed.

A warrant requiring the officer to whom it is addressed to search a house, or other place, therein specified, for property therein alleged to have been stolen, and, if the same shall be found upon such search, to bring the goods so found, together with the person occupying the same, who is named, before the justice, or other officer granting the warrant, or some other justice of the peace, or other lawfully authorized officer. It should be given under the hand and seal of the justice, and dated.

SELF-DEFENSE. In Criminal Law. The protection of one's person and property from injury. A person may defend oneself, and even commit a homicide, for the prevention of any forcible and atrocious crime which, if completed, would amount to a felony.

SENTENCE. A judgment, or judicial declaration made by a judge in a cause. The term ''judgment'' is more usually applied to civil, and ''sentence'' to criminal, proceedings. Sentences are final, when they put an end to the case; or interlocutory, when they settle only some incidental matter which has arisen in the course of its progress.

SEVERAL FISHERY. An exclusive right to fish in a given place, either with or without the property in the soil at such place. No person other than the owner of the fishery can lawfully take fish at such place.

SHERIFF. The office of sheriff is one of the oldest known to the common law. It is inseparably associated with the county. The name itself signifies keeper of the shire or county. The office is said to have been created by Alfred when he divided England into shires, but Coke believed it to have been of Roman origin.

SOUND MIND. That state of a person's mind which is adequate to reason and comes to a judgment upon ordinary subjects like other rational persons. The law presumes that every person who has acquired full age is of sound mind and consequently, competent to make contracts and perform all civil duties; and one who asserts to the contrary must prove the affirmation of one's position by explicit evidence, and not by conjectural proof.

SPECIAL AGENT. An agent who is only authorized to do specific acts in pursuance of particular instructions, or with restrictions necessarily implied from the acts to be done.

SPECIAL JUDGE. A judge appointed to act in a particular case because of the disqualification of the regular judge. The authority of such a judge is not limited to the term during which one is appointed, but extends to subsequent terms until the disability of the regular judge is removed.

STATE. A body politic, or society of people united together for the purpose of promoting their mutual safety and advantage, by the joint efforts of their combined strength.

STATEMENT. The act of stating, reciting, or presenting verbally or on paper.

STATE POLICE POWER. That power under which the states or their municipalities may enact statutes and ordinances to protect the public health, the public morals, the public safety, and the public convenience.

STATE'S EVIDENCE. The evidence of an accomplice who testifies for the prosecution in the hope of being released or punished more lightly.

When an accomplice becomes a witness for the prosecution and discloses his guilt and that of his associates the law says there is an implied promise that he will not be prosecuted. It is within the discretion of the public prosecutor to determine whether to use an accomplice as *state's evidence.*

STATUS QUO. The existing state of things at any given date.

STATUTE. A law properly enacted by the authorized law-making body of a state or nation. A law established by the act of the legislative power. An act of the legislature. The written will of the legislature, solemnly expressed according to the forms necessary to constitute it the law of the state. This word is used to designate the written law in contradistinction to the unwritten law.

STATUTE OF HUE AND CRY. According to Webster, a loud outcry with which felons were anciently pursued, and which all who heard it were obligated to take up and join in pursuit. Later a written proclamation for the capture of a felon or stolen goods.

A statute or act by which the inhabitants of an area were liable for the loss of the goods taken unless they produced the robber.

STATUTE OF LIMITATIONS. The time within which an action may be brought against a person for a crime. The time varies with the crime and the laws of the state where the crime is committed.

SUBPOENA. A legal instrument issued by competent authority, requiring the presence of a person to testify of one's own knowledge concerning certain acts of information.

A process to cause a witness to appear and give testimony, commanding witness to lay aside all pretenses and excuses, and appear before a court or magistrate therein named, at a time therein mentioned, to testify for the party named, under a penalty therein mentioned. This is called distinctively a *subpoena ad testificandum*. On proof of service of a subpoena upon the witness and that he is material, an attachment may be issued against him for a contempt if he neglects to attend, as commanded.

SUMMONS. The name of a writ commanding the sheriff, or other authorized officer, to notify a party to appear in court to answer a complaint made against party and in the said writ specified, on a day therein mentioned.

SUPREME COURT. A court of superior jurisdiction in many of the states of the United States. The name is properly applied to the court of last resort, and is so used in most of the states.

SWEAR. To take an oath administered by some officer duly empowered. One may swear who is not duly sworn; and in such case the oath is not administered, but self-imposed, and the swearer incurs no legal liability thereabout.

TESTIMONY. The words of a living witness speaking under oath. Oral evidence spoken by a human witness, as contrasted with evidence presented in documentary form by innate objects.

TRESPASS ACCIDENTAL. A situation may be judged accidental trespass when the act is completely accidental and nonintentional on the land of another. This is different from an intentional though mistaken trespass and ordinarily does not result in legal liability.

TRESPASS BY MISTAKE. One who intentionally enters the land of another under the mistaken belief that one is entitled to possession or ownership of the land or that one has consent or privilege to enter the land may be liable as a trespasser.

TRIAL. In Practice. The examination before a competent tribunal, according to the laws of the land, of the facts put in issue in a cause, for the purpose of determining such issue.

VENUE. The county in which the facts are alleged to have occurred, and from which the jury are to come to try the issue. Some certain place must be alleged as the place of occurrence for each traversable fact.

VERDICT. The decision of the proper authority hearing the action. The decision made by a jury and reported to the court on the matters lawfully submitted to them in the course of a trial of a cause.

VESTED INTEREST. An interest when vested, whether it entitles the owner to the possession now or at a future period, is fixed and present; so that the right of ownership, to the extent of the estate, may be aliened.

VIOLATION. The result of an act done unlawfully and often with force. This word has also been construed under this statute to mean carnal knowledge.

VOLUNTARY STATEMENT. In criminal proceedings, a defendant's statement, to have been voluntarily made, must have proceeded from the spontaneous suggestion of his own mind, free from the influence of any extraneous disturbing cause.

WANTONLY. Done in a licentious spirit, perversely, recklessly, without regard to propriety or the right of others; careless of consequences, and yet without settled malice.

WARRANT. A legal instrument, properly executed by competent authority, requiring the arrest and apprehension of the person named therein. A writ issued by a justice of the peace or other authorized officer, directed to a constable or other proper person, requiring him to arrest a person therein named, charged with committing some offense, and to bring him before that or some other justice of the peace. An order authorizing a payment of money by another person to a third person.

WEIGHT OF EVIDENCE. This phrase is used to signify that the proof on one side of a cause or issue is greater than on the other. When a verdict has been rendered against the weight of the evidence, the court may, on this ground, grant a new trial; but the court will exercise this power not merely with a cautious but a strict and sure judgment, before they send the case to a second jury.

WILD ANIMALS. Animals wild by nature, such as deer in the forest, birds in the air, and fish if in public waters or the ocean.

WILDLIFE. Any wild animal. Either fish or game, or both.

WILFULLY. Intentionally as distinguished from accidentally or involuntarily.

WITNESS. A person testifying concerning matters of fact or expert opinion and sworn to speak the truth, the whole truth, and nothing but the truth.

One who testifies to what one knows. One who testifies under oath to something which one knows of firsthand.

WRIT. A mandatory precept, issued by the authority and in the name of the sovereign or the state, for the purpose of compelling the defendant to do something therein mentioned.

WRITTEN INSTRUMENT. A Missouri statute defines the term as including every instrument, partly printed and partly written, or wholly printed with a written signature thereto, and every writing purporting to be a signature.

X-RAY PHOTOGRAPH. A photograph made by the aid of a particular electrical ray whereby there may be secured reliable representations of the bones of a flesh-and-bones body, although they are hidden from direct view by the surrounding flesh, and also of metallic or other solid substances which may be imbedded in the flesh.

Appendix B

Syllabus

WILDLIFE LAW ENFORCEMENT

Credit Four Hours: Four hours of lecture or three hours of lecture and one three-hour lab.

Texts: Wildlife Law Enforcement, the code book of a state wildlife agency or a regional office of the U.S. Fish and Wildlife Service.

Content: The course material is from lectures, telelectures, mock arrests and seizure of evidence, mock trials, reading assignments, training films, and guest speakers.

Mock Arrests and Seizure of Evidence: Fish and game departments are almost invariably willing to cooperate with university students in practical aspects of their training. In a field operation one or more students may be the investigating and arresting officer while the fish and game personnel play the part of the bad guys. This should not be an impromptu performance but one that is planned ahead and put in writing so that it may be critiqued by the class and by the fish and game personnel. Since it may be a case of one-on-one or two- or three-on-one there may be several ''incidents'' going on at the same time. In all cases the actions should have a post mortem critique and discussion.

Mock Trials: One of the most trying and nerve wracking experiences that an officer, particularly a recruit, has is that of testifying in court. The officer knows that if he or she is not careful the defense attorney may attempt to put the officer rather than the suspect on trial. The Houston, Texas, police department, with the aid of representatives from a district attorney's office, have developed a viable mock trial program (Toettmeier 1982). Their objectives are: (1) to develop specific courtroom experience by testifying under realistic conditions; (2) to experience the stress of having to testify in a court before an active judge, a resourceful prosecutor, and a very determined and sometimes ruthless defense attorney; (3) to magnify errors on an offense report that may jeopardize the successful prosecution of a case; (4) to illustrate the importance of prepping oneself prior to going to court; and (5) to identify common mistakes made by officers which may contribute to the failure of a prosecution. This can be very embarrassing for the officer. The Houston model follows the crime scene program. The students must first respond to and resolve several crime-in-progress calls, prepare necessary paperwork generated by these calls, and then be prepared to testify in one of the cases that go to court some weeks later. Houston believes this type of sequencing is important to the recruit because it prepares him or her in a manner that is consistent with the demands placed on officers. It also facilitates the learning process. The students are able to participate actively in procedures that are similar to their expectations about what a police officer may do on the job. An important part of the program is the successful recruitment of veteran prosecutors and judges as well as defense attorneys. The students learn early on that when they look to the judge for help when the defense attorney embarrasses them, they are usually on their own. In fact it may appear that the judge all too often will let the defense attorney badger and abuse them. Given time they discover that this discomfort can be attributed largely to their inexperience and to the fact that they are letting it be taken personally. Testifying is obviously the most exciting aspect of a mock trial for the students. What they may fail to realize is that successful prosecution depends on preparation. To aid students in this aspect of the program the training staff critiques their offense reports at least a week before the trial, and a copy of the critique is given to the cadet while other copies are sent to the prosecutors participating in the mock trial program. Students have ample time to analyze their mistakes and seek answers to other questions. The prosecutors also review the offense reports prior to the start of a program. The purpose of this is: (1) to gauge the progress of the student's writing capabilities; (2) to analyze reports in terms of inconsistencies; and (3) to evaluate the content of the narrative portion of the report in terms of completeness and accuracy. Since only one student can testify at a time the remaining recruits may play the roles of jurors and spectators. It soon becomes apparent to a student testifying in a mock trial that one must be firm in one's statements, accurate, patient, and courteous at all times. Above all, one must not lose one's cool.

Training Films: These can generally be obtained through one of the local or federal law enforcement officers. They may cover such things as handling of guns, how to conduct a search, how to make an arrest, how to gather and record evidence and crime scene search.

Guest Speakers: Local conservation officer, sheriffs, judges, or prosecuting attorneys are generally willing to appear before a law enforcement class.

Telelectures: People of national prominence from several geographic areas.

Literature Cited

Toettmeier, T. 1982. The Houston mock trial program. *FBI Law Enforcement Bull.* (Federal Bureau of Investigation, U.S. Dept. of Justice, Washington, D.C.) 51: 2.

Appendix C

A Survey of Wildlife Law Enforcement Activities 1991

The States

In 1991, a law enforcement survey was sent to fifty-three state agencies (Washington, Florida, and Pennsylvania each have two natural resource agencies) and Guam, Puerto Rico, and Washington, D.C. (latter returned as undeliverable). Fifty-three of the fifty-five delivered requests (96.4 percent) responded. The questions covered educational requirements for employment, duties, sidearms, and enforcement problems.

Forty-five percent of the states require a high school diploma as a prerequisite for employment. Two percent have no educational requirements for employment but state that the civil service exam is so written that most or all of the acceptable applicants are graduates of a four-year college or university. However, several of these states note that while it is not a legal requirement, in practice they hire mostly people with four-year degrees. They also point out there are inservice training schools after employment, lasting as long as nine months. Thirteen percent of the respondents require two years of college, but they actually hire mostly graduates of four-year schools. Forty percent of the respondents require a four-year college or university degree. Four percent require a liberal college education only, ten percent a bachelor of science degree in science, and eighty-six percent a bachelor of science degree in natural resources. A few states indicate that wildlife law enforcement experience can be substituted for one year of college training. Nine percent of the states require experience in addition to formal education.

Nine percent of the respondents indicate they enforce wildlife laws only. Of the ninety-one percent that enforce more than wildlife laws the breakdown for all activities is as follows: management, fourteen percent; public relations, sixteen percent; and law enforcement, seventy percent. Some states enforce all state laws including environmental pollution. Others enforce other than wildlife laws when called upon by another agency. Most conservation officers, ninety-two percent, are also peace officers. Seventy-four percent are under civil service. The others note they are protected by a different set of standards. Women represented 3.5 percent of the field officer force.

Only one type and caliber of sidearm per department was listed by seventy-three percent of the states. The other states listed two or three guns indicating it is an officer's choice. For the states listing only one weapon the .357 Magnum revolver is by far the most popular, forty-seven percent. The second choice is the 9-millimeter automatic pistol, twenty-six percent; the .40-caliber automatic pistol, fourteen percent; the Colt .45 automatic pistol, eight percent; and the .38 special revolver, five percent. The guns listed by the states with more than one choice were the same sidearms as in the one choice states. However, the order was slightly different. The .357 Magnum revolver was first, the 9-millimeter automatic pistol was second, and the .38 special revolver was third. Eighty-three percent of field officers, other than covert operation people, wear uniforms all of the time. Of the remaining seventeen percent most of them wear uniforms at least eighty-five percent of the time.

Volunteer officers are not used by seventy-seven percent of the states. The number of volunteers in the other twenty-three percent of the states varied widely. One state listed 2, another 5, another 66, still another 330, and one 950.

The states list their law enforcement problems in decreasing order of importance by percent. The number one problem is lack of funds to hire additional personnel, forty-two percent; negative attitude of the public, twenty-three percent; buying or selling of game animals or parts, nineteen percent; loss of habitat, six percent; irresponsible hunters and anglers, four percent; biopolitics in wildlife management, two percent; the fair labor act, two percent; and the necessity of enforcing other than wildlife laws, two percent. In summary, most state wildlife law enforcement administrators believe that they are overloaded with work and underpaid, equipment is inadequate and in poor repair, and the situation will get worse before it gets better.

Canada

A total of twelve survey forms were sent to Canada, nine responded. Twenty-two percent indicate the high school diploma is all that is required, but one province requires one year of related experience and graduation from a forest techniques course. Another indicates that the applicant must be a grade V specialist serving as a CO or have two years of wildlife experience. Fifty-six percent require two years of college and twenty-two percent a bachelor of science degree. One respondent requires five to six years of experience in addition to formal education and another two years experience. All indicate the law enforcement duties are much broader than just wildlife laws.

The officer's time in various duties is broken down as follows: management, nineteen percent; public relations, sixteen percent; and law enforcement, sixty-five percent. One province notes that seventy-eight percent of the officers are peace officers. In all other cases all officers are peace officers. In only one province are the officers not protected by civil service. Women in field officer positions make up

2.5 percent. In six of the nine provinces officers carry side-arms. The most popular gun is the .38 special revolver, sixty-eight percent; the .357 Magnum revolver and the 9-millimeter automatic pistol, sixteen percent each. Only one percent of the field force do not wear uniforms on duty. This excludes those in covert action. With regard to volun-teers one province listed two percent and another has "separate units of five hundred plus (two hundred percent)." Commercialization listed at fifty percent is the number one problem; professionalism and lack of training, ten percent; lack of access to private land, ten percent; special interest groups, ten percent; funds, ten percent; and court decisions, ten percent.

Appendix D

General Multiple-Choice Questions

Mark only one answer.

1. The first federal "game" law to be enacted was the:
   a. Migratory Bird Treaty Act
   b. Migratory Bird Hunting and Conservation Act
   c. Lacey Act
   d. Bass Act
   e. Bald Eagle Act

2. Taking the body temperature of a dead animal may be useful as evidence:
   a. when the officer is checking on opening day of the hunting season and believes the animal was taken before the season opened
   b. when the officer thinks the animal may have been sick
   c. when the officer finds illegal game in a cold storage locker
   d. if someone sees the officer take the animal's temperature
   e. only in the summertime is this technique of value

3. Fingerprint identification:
   a. is based on the fact that no two prints are identical
   b. is never useful in wildlife law enforcement
   c. must always be made with consent of the suspect
   d. is useless unless there also is a confession
   e. must always include all ten prints

4. Double jeopardy is:
   a. the charging of a defendant with two violations at the same time
   b. the uncertainty which exists in relation to a fact
   c. the second jeopardy of a person who has previously been in jeopardy for the same offense and with the same evidence
   d. the second item introduced as evidence in a court trial
   e. a poacher who has taken two limits of game

5. You, as a conservation officer, come across a father and two boys, nine and eleven years of age, fishing on a reservoir. The law states that the person attending the fishing gear must be no farther than ten feet from it. As you watch the youngsters playing, many times they were not within ten feet of their poles. Your action as a conscientious officer is as follows:
   a. do nothing
   b. point out to the father and the boys that the law requires them to be within ten feet of their poles, while at the same time realizing that active boys find this difficult at times; possibly issue a warning citation
   c. arrest the father
   d. arrest the older boy
   e. arrest both boys

6. An expert witness is:
   a. one with special knowledge of the subject matter about which one is to testify
   b. one who has testified in court many times
   c. one who makes a good impression on the judge and jury
   d. one who will bend the truth skillfully to get a conviction
   e. one who refuses to answer questions which one does not believe to be true

7. Elwood Johnson illegally shoots three Canada geese in Idaho and carries them across the state line into Wyoming. He has violated the:
   a. Black Bass Act
   b. Lacey Act
   c. Migratory Bird Treaty Act
   d. Dyer Act
   e. none of these

8. Documentary evidence is also called demonstrative or visual evidence. Mark the answer which is not documentary:
   a. confessions
   b. photographs
   c. fingerprints
   d. empty cartridge cases
   e. field notes

9. The surrender, by one state to another state, on its demand, of a person or persons charged with the commission of a crime within the borders of the second state is termed:
   a. extradition
   b. *mala fide*
   c. *modus operandi*
   d. *corpus delicti*
   e. abrogation

10. A person planning professional wildlife training as a future occupation should possess or develop the following traits. Mark the one that is not correct:
    a. have no physical handicaps
    b. suspect every hunter as guilty until proven otherwise
    c. be able to take care of oneself in the out-of-doors
    d. be intelligent, resourceful and possess a great capacity for detail
    e. have at least an elementary training in automotive repairs such as may be needed in the field

11. The management of natural resources in a national forest is provided for by:
    a. the governor of the state
    b. the regional forester
    c. the National Forest Administration Act
    d. the Forest Game Act
    e. the Chief of the U.S. Forest Service

12. Basically wildlife legislation is passed:
    a. either as a result of pressure from special interest groups or because of professional recommendations or a combination of the two
    b. because of public pressure alone
    c. because one state is trying to keep up with another state
    d. to educate the public to national wildlife problems
    e. because of the need for it or the inefficiency of existing laws

13. The punitive section of a code book deals with:
    a. punishment
    b. states rights
    c. interstate commerce
    d. abrogation
    e. court trials

14. Who was authorized in 1916 to establish preserves for the protection of game animals, birds, and fish?
    a. the President of the United States
    b. the Congress of the United States
    c. the Secretary of the Interior
    d. the U.S. Fish and Wildlife Services
    e. the Secretary of the Treasury

15. Fingerprints are a form of:
    a. circumstantial evidence
    b. direct evidence
    c. indirect evidence
    d. real evidence
    e. documentary evidence

16. Which of the following may be a *corpus delicti?*
    a. a dead person
    b. a forest fire
    c. a string of fish
    d. a dead deer
    e. all of the above

17. What act gives jurisdiction of waterfowl and other migratory birds to the federal government?
    a. no specific act gives this jurisdiction
    b. Lacey Act
    c. Migratory Bird Treaty Act
    d. Migratory Bird Hunting and Conservation Act
    e. Enabling Act of Congress

18. Jurisprudence is best defined as:
    a. the leader of the jury
    b. the alertness of a lawyer
    c. the act of presiding over the court by the judge
    d. the charge to be brought before the jury
    e. the science of the law

19. A wildlife law enforcement officer should have:
    a. many close associates and few friends
    b. a good life insurance policy
    c. many friends but few close associates
    d. many close friends
    e. few enemies

20. The *Charta De Foresta* was:
    a. a body of Roman ecclesiastical law ·
    b. a collection of the laws of the forest, made in the reign of Henry the III
    c. a question contested before a court of justice
    d. a book required to be kept by a justice of the peace
    e. the great charter of liberties which was introduced by King John

21. If a person fires a 12-gauge shotgun using number six shot, at a conservation officer four hundred yards away, the officer can charge the man with:
    a. aggravated assault
    b. assault
    c. nothing
    d. careless handling of firearms
    e. two of the above

22. In court the officer should be all of the following except:
    a. neatly dressed
    b. alert
    c. witty
    d. dignified
    e. at ease

23. The FBI was established in:
    a. 1900
    b. 1908
    c. 1915
    d. 1926
    e. 1895

24. The Black Bass Act when valid protected only:
    a. black bass
    b. black bass and trout
    c. black bass, trout, and salmon
    d. game fish
    e. all fish

25. The term CRIME refers to:
    a. a criminal
    b. *primae facie*
    c. something always punishable by fines and/or imprisonment
    d. an act committed against a public law
    e. *corpus delicti*

26. A crime is defined as:
    a. a forbidden act or omission
    b. the physical accomplishment of something the law forbids
    c. failure to do an act that is required by law
    d. doing something overt
    e. all of the above

27. Federal arrest warrants are usually issued by a:
    a. justice of the peace
    b. U.S. commissioner
    c. city magistrate
    d. U.S. marshal
    e. U.S. judge

28. The United States can reserve water rights in a national forest for what purpose?
    a. recreation
    b. aesthetics
    c. wildlife preservation
    d. preserving timber
    e. cattle grazing

29. Waterfowl market hunters often use which of the following as a defense?
    a. entrapment
    b. encarpment
    c. illegal procedure on the part of the officers
    d. illegal search and seizure
    e. unlawful use of the Fifth Amendment

30. Which of the following are game wardens not authorized to do?
    a. demand to see a valid hunting license
    b. search a vehicle when they know a violation has been committed
    c. search a private abode in a routine check
    d. search a licensed wholesale fish dealer without a warrant
    e. search a licensed food fish canner without a warrant

31. In regard to early English background as described by Nelson (1762) one of the following statements is not true:
    a. during the war between the Britons and Saxons, so many of the Britons were killed that the cultivated lands were more than sufficient to maintain the conquerors
    b. The Woods, Waste, and Bruery lands that were not appropriated to any particular person, remained the property of the King
    c. William the Conqueror laid waste to thirty-six townships in Hampshire to make a forest
    d. in Edgar's time he, having an eloquent taste but being very tolerant of his subjects, let the peasants hunt and fish as they saw fit
    e. Rufus, the son of William the Conqueror, is recorded for the severity of his proceedings against all that hunted in his forest

32. The public domain has been and will continue to be of considerable interest to most citizens of the United States because of its size and for many other reasons. One of the following statements is not true:
    a. in spite of the size of the public domain it does not include great numbers of song birds and nongame animals such as are found on private land of the eastern United States

b. in 1763, England acquired all land in North America east of the Mississippi River by defeating France in the Seven-Years War
    c. many major domestic and foreign problems that have confronted this nation have been directly concerned with matters of governing, acquiring, and disposing of public land
    d. most of the public land in the United States is in the West
    e. the percent of the United States that is public land is roughly one-third

33. Evidence which includes physical objects such as bodies of game animals, blood, weapons, empty shell cases, projectiles, glass fragments, and articles of clothing is called:
    a. direct evidence
    b. circumstantial evidence
    c. real evidence
    d. documentary evidence
    e. indirect evidence

34. One of the much publicized and real conflicts in the western United States is that of use of public lands by conflicting interest groups. Which of the following statements do you regard as misleading or untrue?
    a. most of the early and even present day conflicts are caused by wealthy cattle ranchers trying to control large areas of public land
    b. probably the first human conflict in the public domain was that of the Native Americans v. whites
    c. over the years, mineral interest and timber operators have in general been in conflict with conservationists
    d. recreationists, at first, considered the public domain a vast, unexploited free resource area, but they soon discovered this was not true
    e. early western livestock ranchers sought and won the use of unappropriated public lands and they believed they had a real, if not a vested interest in the system

35. Concepts of wildlife management vary somewhat between regions of North America and between people with different backgrounds. In general, however, there is good agreement. One of the following statements is not true:
    a. it is recognized that most species of big game animals prosper in climax, rather than subclimax, vegetation
    b. there were relatively few deer in North America when our forbearers first landed on this continent
    c. the cutting of vast tracts of timber in early North America left large areas of land which were ideal for browsing animals, notably deer and elk
    d. regulated timber cutting, controlled grazing by domestic animals, and limited big game populations can live in harmony if acceptable management practices are properly carried out

e. mule deer prospered because of heavy grazing of domestic animals around the turn of the century

36. One of the following statements is not true in regard to nongame wildlife:
   a. they may not be killed, harassed, or harmed for any reason
   b. one of the reasons for their classification as nongame animals may be they are endangered
   c. any animals so defined by state or federal law are nongame
   d. some nongame animals, such as the snail darter, are protected by the federal Endangered Species Act and also by state law
   e. only certain shore birds and song birds are protected by the federal Migratory Bird Treaty Act and by state law

37. Amenable means:
   a. to be easy-going
   b. one who is not guilty of duplicity
   c. not antisocial
   d. responsible; answerable
   e. dependable

38. Which of the following animals best qualify as farm wildlife?
   a. coyotes
   b. feral horses
   c. muskrats
   d. black bear
   e. cottontail rabbits

39. U.S. Supreme Court Justice Holmes, speaking on a suit filed by the U.S. Attorney General against the Sanitary District of Chicago, made the now famous statement which is:
   a. the Tenth Amendment at the time the Constitution was adopted was intended to confirm the understanding of the people
   b. the custodial power of the state may be exercised in some ways that contravene the provisions of the Federal Constitution
   c. this is not a controversy among equals. United States is asserting its sovereign power. . . .
   d. today it is apparent that the Tenth Amendment does shield the states and their political subdivisions from the impact of the federal government
   e. all legislative power must be vested in the state or national government. Legislative powers other than these belong to the people

40. One of the following statements is not true concerning special federal acts:
   a. the Eagle Act protects both the bald and the golden eagles
   b. the Wild, Free-roaming Horses and Burros Act was passed in 1971
   c. the Marine Mammal Protection Act was passed in 1972

d. originally the Bass Act covered only black basses
   e. there are legally no wild, free-roaming horses or burros in any national park

41. Permission of a court having jurisdiction in a case which allows the case to be tried in some other court, when it appears that a fair trial cannot be had at the originally scheduled place, is the definition of:
   a. change of domicile
   b. change of civil action
   c. change of residence
   d. *de novo* proceedings
   e. change of venue

42. One of the following statements is not true:
   a. horses and burros protected under the Wild, Free-roaming Horses and Burros Act are not protected if they stray onto private land
   b. owners of private land that find wild burros on their land can call a federal marshal or some other agent of the federal government to have the animals removed
   c. the Wild, Free-roaming Horses and Burros Act provides that the animals may not be captured, branded, or harassed
   d. bald eagles were protected earlier than golden eagles by the Eagle Act
   e. the former Bass Act covered all fish, both game and nongame

43. One of the following is the correct chronological order of enactment of federal laws:
   a. Lacey Act, Bass Act, Eagle Act, Migratory Bird Hunting and Conservation Act, Marine Mammal Protection Act
   b. Lacey Act, Migratory Bird Treaty Act, Bass Act, Eagle Act, Endangered Species Act
   c. Lacey Act, Eagle Act, Wild Free-roaming Horses and Burros Act, Marine Mammal Protection Act, Bass Act
   d. Lacey Act, Endangered Species Act, Eagle Act, Bass Act, Marine Mammal Protection Act
   e. Bass Act, Eagle Act, Wild Free-roaming Horses and Burros Act, Marine Mammal Protection Act, Lacey Act

44. The Marine Mammal Protection Act does not cover which of the following?
   a. polar bears
   b. order Sirenia
   c. order Pinnipedia
   d. brown bears which forage in the ocean off the coast of Kodiak Island
   e. sea otters

45. Preponderance means:
   a. the greater weight of evidence
   b. taking several days to think about it
   c. a juror's mind is made up before evidence is presented

d. consultation with the judge

e. evidence against

46. One of the following statements is inaccurate:

a. the term "migratory waterfowl" as used in the act, means ducks, geese, brant, and swans

b. there is no expiration date on a "lifetime" migratory waterfowl stamp

c. any person violating the Eagle Act, if convicted, may be fined

d. the two secretaries that administer the Wild Free-roaming Horses and Burros Act are those of Agriculture and Interior

e. free-roaming wild horses are not easily distinguished from horses on the same range that are under private ownership

47. *Non est inventus* means literally:

a. a nonresident

b. a notary public not in residence

c. he is not to be found

d. a case in which the appeal is argued anew

e. a person subordinate to a public officer

48. One of the following statements is true regarding crime:

a. courts held for the enforcement of early forest laws were not different than those of today

b. the conviction of a person, following a confession, is not proof of a crime

c. criminal intent is necessary in order to prove that a crime was committed

d. a lawyer is a person licensed to defend people who have allegedly committed crimes

e. crimes are public wrongs which a private individual does not have the power to either condone or forgive

49. Collusion is:

a. a preventable accident

b. an agreement to defraud a third person of rights or to obtain an unlawful object

c. a writ issued by a judge committing a person to prison

d. an indirect reference to a subject or object

e. an investigation by an unauthorized person

50. A search warrant is:

a. a legal writ, executed by competent authority, authorizing the search of a premises

b. in criminal law a writ entitling an officer to enter another person's property

c. a writ giving any officer an exclusive right to search a given place

d. any officer with such a writ may search as he or she pleases within the broad confines of the writ

e. similar in substance to a subpoena, but having some rather drastic differences

51. "The right of the people to be secure in their persons, houses, papers, and effects, against unreasonable searches and seizures, . . ." is from which Amendment?

a. First

b. Second

c. Third

d. Fourth

e. Fifth

52. "This signifies that the proof on one side of a cause or issue is greater than on the other." The preceding statement is part of a definition of:

a. warrant

b. weight of evidence

c. violation

d. venue

e. sanctuary

53. Forensic science is based upon the belief that:

a. every criminal has his own method of operation

b. the criminal always returns to the scene of the crime

c. all crimes can be solved scientifically

d. no two objects are exactly alike, and the difference can always be measured

e. every criminal leaves behind or takes away some evidence of the crime

54. The willful assertion as to a matter of fact, opinion, belief, or knowledge made by a witness in a judicial proceeding as part of his evidence, generally under oath; such assertion being known to such witness to be false is:

a. a plea

b. a plea of *nolo contendere*

c. a penal statute

d. peremptory

e. perjury

55. "Indistinctness or uncertainty of meaning of an expression used in a written instrument." This is the definition of:

a. ambiguity

b. animus

c. counter argument

d. verbal assault

e. abet

56. Criminals may approach their "work" in a particular set or routine way each time they perpetrate a crime. For example, they may break match sticks and drop them while waiting or may always approach a dwelling by a rear window and use a glass cutter to gain access to the door or window lock. This is generally known as:

a. operational procedure

b. *modus operandi*

c. general procedure and habits for a particular group of criminals

d. a way of life that has nothing to do with crime

e. routine and habitual behavior

57. If due to extenuating circumstances a justice of the peace reduces the penalty normally given for a certain violation, this could be termed:
    a. with good cause
    b. malingering
    c. legislative
    d. licentious
    e. mitigation

58. When an officer is approaching an unknown suspect, what should be his or her frame of mind?
    a. this person is innocent until proved guilty
    b. this person may be armed and dangerous and willing to do me great bodily harm
    c. this person has the appearance of being innocent and harmless
    d. this person has all the "earmarks" of being guilty
    e. there is no particular frame of mind that an officer should have when approaching a suspect. The best attitude is to leave one's mind free to make decisions

59. Which of the following forms of evidence is very often used, but frequently or at least occasionally, is unreliable?
    a. written
    b. dying declarations
    c. demonstrative
    d. photographic
    e. oral

60. To execute, in legal terms, means:
    a. to follow to a completion
    b. to experience
    c. to impeach without motive
    d. to motivate beyond ordinary circumstance
    e. to indict

61. To prove is:
    a. to determine or persuade that a thing does or does not exist
    b. to present evidence that disproves a statement
    c. to make certain a statement is understood as stated
    d. to substantiate circumstantial evidence
    e. to show any kind of evidence is relevant

62. The examination of a person before a competent tribunal, according to the laws of the land, and which puts the facts at issue in a case before the court for the purpose of determining such issue, is defined as a:
    a. verdict
    b. trial
    c. penal statute
    d. direct examination
    e. prehearing

63. In *Geer v. Connecticut* the United States Supreme Court held that:
    a. the killing of game is a privilege and a right
    b. the killing of game is a privilege but not a right
    c. the ownership of wildlife does not differ from that of other property
    d. after an animal becomes possessed by an individual, it is subject to no further restrictions
    e. under this court ruling fish and game were given still further protection and could not be illegally shipped from one state to another

64. One of the following statements is not true:
    a. at the Constitutional Convention of 1787, the state delegates invested the federal government with certain powers
    b. at the Constitutional Convention of 1787, the state delegates specifically mentioned the control of wildlife
    c. it was not until the Supreme Court upheld the constitutionality of the Migratory Bird Treaty Act that federal government control over migratory wildlife was decided
    d. the Tenth Amendment provides that all powers not delegated to the federal government and not prohibited to the states, are reserved to the states
    e. the right of states to govern the taking of fish and game was established in the case of *Geer v. Connecticut*

65. During cross-examination when the defense attorney demands a yes or no answer but the officer doesn't believe it can be answered simply by yes or no, what should the officer do?
    a. answer as the defense attorney asks even though it means trouble
    b. refuse to answer the question at all
    c. use the Fifth Amendment
    d. state that the questions cannot be answered by either yes or no and that he or she is willing to give the facts
    e. get up and walk out of the court room

66. Assume that the state of Minnesota has just completed a new state fish hatchery on the upper Mississippi River. To the consternation of the state administrators, they find the federal government is planning a power dam below the fish hatchery which will impound water far enough upstream to cover the hatchery. The federal government initiates condemnation procedures and the state files suit against the federal government to prevent it from building the dam. What will likely be the outcome of this suit?
    a. inasmuch as the federal government did not warn the state when they were building the fish hatchery, that there was a dam on the drawing boards, the state will win this suit
    b. the state will win because of its right of eminent domain
    c. because of the federal government's inability to appropriate funds to compensate the state of

Minnesota, for the destruction of the fish hatchery, the plans for the dam will be canceled or deferred
  d. since this is a controversy among equals the case will go to the United States Supreme Court
  e. the federal government acting under the Federal Power Act, will be authorized to proceed and build the dam, thus destroying the Minnesota state fish hatchery

67. Joe Smith illegally spears sixteen frogs in Iowa and takes them to his home in Nebraska. He has violated the:
  a. Mann Act
  b. Lacey Act
  c. Amphibian and Reptile Act
  d. Bass Act
  e. Migratory Bird and Amphibian Act

68. A landowner cannot:
  a. prohibit a hunter from retrieving birds which have been shot over someone else's property but have fallen on the landowner's property
  b. prohibit a hunter from shooting over one's land although the hunter is standing on another person's property
  c. post one's land in such a way that one retains title to the game therein
  d. exercise absolute and exclusive domain over one's property as regarding hunting
  e. maintain for oneself the exclusive right to hunt and fish on one's land

69. Which of the following is not one of the four Ss?
  a. superiority
  b. simplicity
  c. sound mind
  d. speed
  e. surprise

70. A conservation officer meets an irate sportsman in the woods who proceeds to curse the fish and game department and uses vile language in describing various employees. You, as a conservation officer, would immediately:
  a. arrest him for threats to an officer
  b. arrest him for vile language unbecoming a gentleman
  c. do nothing, for he was within his rights even though his judgment may have been poor and his statements incorrect
  d. arrest him for using ''deadly words''
  e. arrest him for aggravated assault

71. Which of the following does not constitute battery?
  a. causing a person to take poison
  b. pushing a second person into a third
  c. cutting a person's clothes
  d. spitting on a person
  e. accidentally hitting a person during an argument

72. John Hansen, convicted of killing a deer out-of-season, appealed to a higher court. In the latter case he:
  a. was still guilty
  b. could not be a defense witness
  c. was the appellant
  d. was the defendant
  e. could not appeal a misdemeanor conviction

73. A legal instrument properly executed by proper authority, generally a magistrate, which commands the arrest and apprehension of the person listed therein is:
  a. a citation
  b. a warrant arrest
  c. a misdemeanor or complaint
  d. a citizen's arrest
  e. a writ

74. The common house cat is:
  a. a predator when it is in the wild and therefore considered a wild animal
  b. *domitae naturae* and may be taxable
  c. never taxable
  d. never a feral animal
  e. never subject to protection when more than one hundred yards away from its home

75. An officer's field notes should not contain:
  a. authorized abbreviations
  b. date and time incident to arrest
  c. place of arrest
  d. personal remarks which indicate the officer's belief of guilt or innocence of the accused party
  e. specific location where the incident occurred

76. Who is authorized and directed by Congress to establish seasons and the method of hunting, shipping, and transportation of migratory birds?
  a. the President of the United States
  b. the fish and game directors of the respective states
  c. the Congress of the United States
  d. the Federal Game Commission
  e. the Secretary of the Interior

77. A deposition is a form of:
  a. evidence
  b. complaint
  c. attesting to the validity of the suspect's so-called unlawful act
  d. the result of the trial as delivered by the judge
  e. a special kind of brief

78. A hunter gains legal title to game by:
  a. becoming a landowner and posting his property against hunting and fishing, or trespassing
  b. seeing and claiming it first even though it may have been killed by another hunter
  c. capturing, killing, or reclaiming it even though not permitted by law

d. taking it under legal regulations and reducing it to possession

e. buying a game license and properly using it

79. Circumstantial evidence:
    a. is of no value whatsoever in a trial
    b. is competent evidence, is often as good as real evidence, and may often be the only type that exists
    c. will not be accepted by a jury under any circumstance
    d. is usually evidence enough in itself to convict a person
    e. does not prove or help to prove any particular thing under any circumstances

80. A person driving a car which accidentally hits and kills a deer:
    a. should leave the deer where it is and drive on in order to avoid involving oneself in complications with the law
    b. should claim the deer inasmuch as it was killed accidentally and this may, in part, offset the damage to the car
    c. should report the deer kill to the proper state authorities; in that no person has a right of possession under such circumstance
    d. should report the accident to the game department and then sue them for damages
    e. should report to the nearest city police station

81. The circumstances, facts, and declarations which grow out of the main fact and are contemporaneous with it, and serve to illustrate its character, are known as:
    a. *a posteriori*
    b. circumstantial evidence
    c. *res gestae*
    d. a confession
    e. collusion

82. Wildlife law provides that, except as permitted by regulations, it is unlawful to pursue, hunt, take, catch, kill, possess, offer for sale, barter, purchase, deliver for shipment, exchange, cause to be shipped or exported, any migratory bird, any part thereof, or any egg or nest of such bird. This law is known as:
    a. the Migratory Bird Hunting and Conservation Act
    b. the Attorney General's opinion
    c. the Lacey Act
    d. the International Bird Treaty Act
    e. the Migratory Bird Treaty Act

83. Post-mortem is a term used synonymously in legal circles with:
    a. preliminary examination
    b. illegal evidence
    c. autopsy
    d. murder
    e. pre-examination

84. Flem Snopes, a poacher, was caught in the act of pointing a loaded shotgun at Warden Slim Jim by Sheriff Tom Alert. Snopes was nominally guilty of:
    a. attempted murder
    b. resisting arrest
    c. disturbing the peace
    d. upsetting Warden Slim Jim
    e. assault

85. Aggravated assault is:
    a. an assault that an officer incurs
    b. an assault that a violator incurs
    c. necessarily premeditated
    d. any assault other than in defense
    e. assault with a dangerous weapon

86. Grog Hunter was convicted of illegally shooting an animal *domitae naturae*. He most likely had inadvertently killed a:
    a. muskrat
    b. golden eagle
    c. Hereford cow
    d. mule deer
    e. domestic lion gone wild

87. Munk Odelthorpe, when tried for poaching carp, told Judge Grinnup that he would hunt him down and kill him if he fined him so much as one dollar. Munk should be guilty of:
    a. extradition
    b. extortion
    c. obstructing justice
    d. threatening
    e. murder

88. Warden James who was sent to serve a warrant on Munk Odelthorpe's accomplice, returned to the judge and announced the warrant *non est inventus*. Warden James had:
    a. shot the accomplice
    b. been shot at by the accomplice
    c. not found the accomplice
    d. forgotten the name of the accomplice
    e. needed assistance to capture the accomplice

89. Under ordinary circumstances a witness may testify only:
    a. as to what one has learned through one's own senses
    b. as to what one believes has taken place
    c. as to what others have told one has taken place
    d. as to what one has learned about the crime through news media and other general sources
    e. as to what one has learned by talking with the officers in charge of the case

90. Zeek Beak was caught with two live bald eagles in his possession in Cowtown, Arizona. He was within the law if he had:
    a. bought the birds legally in Mexico and brought them to the United States

b. lawfully raised them from eggs

c. taken them lawfully on or before 1938

d. captured them in Wyoming where they were preying upon sheep

e. traded a legally taken whooping crane for them

91. Conservation officer, Jim Smith, observes Clyde Ankle shoot eight blue-winged teals on his first shot on opening morning of waterfowl season. The legal limit is four:

a. Conservation Officer Smith cannot prosecute Ankle because it was an accident

b. Officer Smith should pick up four of the eight ducks and cite Ankle for being over his limit

c. Officer Smith should take all eight ducks and cite Ankle

d. Officer Smith should pick up four of the eight ducks and caution Ankle to be more careful on future duck hunting trips

e. Officer Smith should drop the whole incident because it was an unenforceable situation

92. A John Doe warrant is a warrant for the arrest of:

a. John Doe

b. a person whose real name is not known

c. a person who uses an alias

d. a person who is out of the country

e. a person who has never been arrested before but is believed to be an alien

93. A leading question is:

a. one which is followed by more questions

b. the first question asked by the prosecuting attorney

c. one which is asked by the presiding judge

d. one which suggests that a confession is imminent

e. one which plainly suggests the answer which is desired

94. One of the fastest methods of reversing direction when travelling in a patrol car on a narrow road is known as the "bootlegger's turn." It consists of:

a. pulling off on the shoulder, gunning the motor, popping the clutch, and spinning around

b. pulling off on the righthand shoulder, backing into a U-turn, and then proceeding in the opposite direction

c. pulling off on the shoulder, flashing red lights to stop traffic, and jockeying the car around

d. there is no fast way to turn around on a narrow road

e. a "bootlegger's turn" consists of three maneuvers—two forward and one reverse

95. The first science used in criminal investigation was:

a. study of microscopic materials

b. forensic medicine

c. the use of challenge and debate

d. the use of plaster of paris casts

e. the cataloging of criminals by facial characteristics

96. The court should always be addressed as:

a. your honor

b. the court

c. sir

d. your majesty

e. judge

97. *Habeas corpus* means:

a. the corpse of an executed criminal

b. a person who refuses to present "his body" in court

c. the body of the crime

d. unlawful possession of a corpse

e. an influence of something directly declared

98. Which of the following may be a *corpus delicti?*

a. a piece of paper

b. a forest fire

c. a string of fish

d. a stolen gun

e. all of the above

99. Jurisprudence is best defined as:

a. the leader of the jury's verdict to the court

b. the alertness of a lawyer

c. the act of presiding over the court by the judge

d. the charge to be brought before the jury

e. the science of the law

100. The term "migratory waterfowl" as referred to in the Migratory Bird Hunting and Conservation Act refers specifically to:

a. all migratory birds

b. ducks and geese only

c. geese only

d. ducks only

e. ducks, geese, brant, and swans

101. Extortion is:

a. investigation by an authorized magistrate

b. communication of a threat to another person with the intention of obtaining something of value

c. an intentional false statement concerning the material at hand

d. the surrender of persons by one state to another

e. an ex-convict's testimony in court

102. Which one of the following federal agencies has direct and absolute control over waterfowl?

a. U.S. Fish and Wildlife Service

b. U.S. Forest Service

c. Bureau of Land Management

d. Corps of Engineers

e. Bureau of Indian Affairs

103. Which one of the following federal agencies has complete control over the habitat, but not the animals, of a number of big game species?

a. U.S. Forest Service

b. U.S. Fish and Wildlife Service

c. Corps of Engineers

d. National Park Service

e. Bureau of Indian Affairs on Indian Lands

104. Which federal agency, appointed by the president, has been given general powers to issue licenses and regulate construction, operation, and maintenance of dams and reservoirs, for the development of navigation and power?
    a. National Power Commission
    b. Bureau of Outdoor Recreation
    c. Federal Energy Regulatory Commission
    d. U.S. Fish and Wildlife Service
    e. President's Power Commission

105. In order to enforce the law, the first step is to:
    a. be suspicious of everyone
    b. determine that a crime has been committed
    c. search the person
    d. take the person's name and other information
    e. arrest the people concerned

106. The right of states to govern the taking of fish and game was established in the famous court case of:
    a. *Whitaker v. Stangvick*
    b. *State v. Wilson*
    c. *State v. Lessard*
    d. *Geer v. Connecticut*
    e. *Silz v. Hesterberg*

107. The final authority in any judicial case is:
    a. the conservation officer at the trial
    b. the judge or jury that sits in trial
    c. the administrative officer at the regional level
    d. the U.S. Attorney General
    e. the state's Attorney General

108. Which one of the following statements is incorrect?
    a. the first step in solving a crime is to establish the fact that a crime has been committed
    b. there are certain cases where suspicion rather than probable cause is adequate
    c. probable cause for a warrant also requires facts sufficient to support a belief that instrumentalities of the crime are located in the place to be searched
    d. warrants should be read in the common everyday language and not in a technical manner
    e. warrants are generally drafted by nonlawyers and in the haste of criminal investigation

109. A man draws an antelope permit in Wyoming. He injures himself before the hunt so he gives the permit to a friend. Would the wildlife law enforcement officer have a case and if so why?
    a. no, because the man with the permit stayed home
    b. no, because the permit allows one animal to be shot
    c. no, because the man that killed the animal was just doing the injured man a favor
    d. yes, licenses and permits are nontransferable
    e. no, this type of violation is overlooked

110. Following a hard day of hunting Joe Smith was walking back to his car. It was late afternoon. He was in a hurry and tired, and therefore cut across a section of a Federal Bird Refuge which was closed to hunting. Although he did not unload his gun, no shots were fired. He was guilty of:
    a. hunting without proper authorization
    b. hunting in an illegal place
    c. hunting with improper gear
    d. taking or attempting to take game illegally
    e. he is not guilty

111. Bill Jones shoots an elk out-of-season and asks Bob Smith to hide it in his barn for a few days. Technically Smith is classified as:
    a. a principal
    b. an accessory before the fact
    c. an accessory after the fact
    d. a principal in the first degree
    e. equally guilty

112. To be used in court, evidence must be:
    a. material
    b. legally adequate
    c. competent
    d. relevant
    e. all of the above

113. Abrogate means:
    a. annul
    b. give relief to
    c. apprehend
    d. decentralize authority of
    e. delegate to

114. The Tenth Amendment to the U.S. Constitution provides that:
    a. the Federal Government has the power to make treaties to control migratory waterfowl
    b. a defendant may refuse to testify on the grounds that he may incriminate himself
    c. the powers not granted or delegated to the federal government by the Constitution are reserved for the states
    d. an individual has the right to bear arms
    e. the states have limited police power

115. Legal acts of instruments which are deemed worthy to serve as rules or models for subsequent cases are termed:
    a. antecedents
    b. precedents
    c. amendments
    d. summons
    e. citations

116. When a court finds that a person was in fact entrapped:
    a. the defendant is acquitted
    b. the guilty officer is censured and the defendant is acquitted
    c. the officer explains why entrapment was necessary
    d. the jury decides if the trial goes on
    e. the judge decides if the defendant is guilty under entrapment

117. Jane Bohan, on hearing the verdict of guilty, shouted to the court, ''I think you are crazy.'' The judge promptly said, ''You are in contempt of court,'' which means:
   a. I am not insane
   b. you are wrong
   c. you are being disrespectful to the court
   d. you have had it
   e. the court is not incompetent

118. Jones is hunting ducks, during the season, he has only three shells in his gun; however, there is no plug in the gun and two more shells can be placed into the gun. He can be arrested for which of the following?
   a. improper equipment
   b. illegal gear
   c. improper license
   d. illegal place if he fires the gun
   e. illegal possession of gun

119. The planning of an offense by an officer and the procurement by improper inducement of the commission of the offense by a person who would not have perpetrated it, except for the trickery or fraud on the part of the officer is:
   a. trickery
   b. fraud
   c. unlawful action
   d. illegal procurement
   e. entrapment

120. A brief means:
   a. a short trial
   b. a short hearing
   c. a detailed statement of a case
   d. a long exhaustive dialogue of a case
   e. a brief statement of a case

121. Which is the best definition of probable cause?
   a. an apparent set of facts or circumstances which would lead a reasonably cautious person to believe a conviction would come about
   b. proof that a crime has been committed
   c. facts relating to the reason a crime was committed
   d. circumstances that justify the crime
   e. evidence beyond suspicion

122. A feral animal is:
   a. one full of fear
   b. one which has escaped from domestication and has become wild
   c. one which usually cannot be tamed
   d. a wild animal in a zoo
   e. any animal which is legally taken by a hunter from the wild

123. Proof of battery will support a conviction of:
   a. felony
   b. misdemeanor
   c. violation of a federal law
   d. assault
   e. illegal possession of firearms

Appendix E

Answers to Multiple-Choice Questions

Chapter 1—Historic to Present Role of Wildlife Law Enforcement
1. c   **2.** d   **3.** d   **4.** e   **5.** c   **6.** e   **7.** b
8. b   **9.** d   **10.** b   **11.** e   **12.** e   **13.** e

Chapter 2—Administration
1. e   **2.** d   **3.** e   **4.** b   **5.** e   **6.** e
7. e   **8.** e

Chapter 3—The Rights of the Individual
1. b   **2.** a   **3.** a   **4.** e   **5.** c   **6.** d   **7.** c
8. d   **9.** d   **10.** e   **11.** e   **12.** b

Chapter 4—State Jurisdiction over Wildlife
1. e   **2.** b   **3.** a   **4.** b   **5.** d   **6.** d
7. c   **8.** e   **9.** d

Chapter 5—Federal Jurisdiction over Wildlife
1. a   **2.** d   **3.** b   **4.** e   **5.** b   **6.** e
7. a   **8.** c   **9.** c   **10.** a   **11.** c   **12.** c

Chapter 6—The Wildlife Law Enforcement Officer
1. b   **2.** e   **3.** e   **4.** c   **5.** d   **6.** a   **7.** a
8. a   **9.** e   **10.** c   **11.** e   **12.** a

Chapter 7—The Wildlife Officer in Court
1. e   **2.** d   **3.** a   **4.** e   **5.** d   **6.** c
7. a   **8.** b   **9.** a   **10.** a

Chapter 8—Violation of Wildlife Law
1. c   **2.** c   **3.** e   **4.** e   **5.** e   **6.** a
7. c   **8.** d   **9.** c   **10.** e   **11.** a   **12.** b

Chapter 9—Arrests
1. e   **2.** d   **3.** e   **4.** e   **5.** c   **6.** a
7. e   **8.** c   **9.** e   **10.** a

Chapter 10—The Stopping and Search of Motor Vehicles
1. a   **2.** c   **3.** e   **4.** e   **5.** c   **6.** e
7. b   **8.** d   **9.** a   **10.** d

Chapter 11—Evidence
1. d   **2.** e   **3.** a   **4.** b   **5.** e   **6.** c
7. c   **8.** e   **9.** b   **10.** a   **11.** c   **12.** e

Chapter 12—Undercover Investigations in Wildlife Law Enforcement
1. a   **2.** d   **3.** c   **4.** e   **5.** d   **6.** a
7. b   **8.** e   **9.** c   **10.** a

Chapter 13—Forensics
1. e   **2.** c   **3.** d   **4.** d   **5.** a   **6.** e
7. a   **8.** e

Chapter 14—Conservation Officer Law Enforcement Training

Chapter 15—An Historical Perspective on the European System of Wildlife Management

Chapter 16—Indian Hunting and Fishing Rights

Chapter 17—Aircraft in Wildlife Law Enforcement
1. e   **2.** d   **3.** a   **4.** e   **5.** e

Chapter 18—Recreational Law Enforcement
1. e   **2.** e   **3.** b   **4.** e   **5.** d   **6.** e

Chapter 19—The Face of the Future
1. e   **2.** c   **3.** b   **4.** c   **5.** d   **6.** a   **7.** a

Answers to General Multiple-Choice Questions

1. c	**2.** a	**3.** a	**4.** c	**5.** b	**6.** a
7. c	**8.** d	**9.** a	**10.** b	**11.** c	**12.** a
13. a	**14.** a	**15.** d	**16.** e	**17.** c	**18.** e
19. c	**20.** b	**21.** d	**22.** c	**23.** b	**24.** e
25. d	**26.** e	**27.** b	**28.** d	**29.** a	**30.** c
31. d	**32.** a	**33.** c	**34.** a	**35.** a	**36.** e
37. d	**38.** e	**39.** c	**40.** e	**41.** e	**42.** a
43. b	**44.** d	**45.** a	**46.** b	**47.** c	**48.** e
49. b	**50.** a	**51.** d	**52.** b	**53.** e	**54.** e
55. a	**56.** b	**57.** e	**58.** b	**59.** e	**60.** a
61. a	**62.** b	**63.** b	**64.** b	**65.** d	**66.** e
67. b	**68.** c	**69.** c	**70.** c	**71.** e	**72.** c
73. b	**74.** b	**75.** d	**76.** e	**77.** a	**78.** d
79. b	**80.** c	**81.** c	**82.** e	**83.** c	**84.** e
85. e	**86.** c	**87.** b	**88.** c	**89.** a	**90.** c
91. c	**92.** b	**93.** e	**94.** b	**95.** b	**96.** a
97. c	**98.** e	**99.** e	**100.** e	**101.** b	**102.** a
103. a	**104.** c	**105.** b	**106.** d	**107.** b	**108.** b
109. d	**110.** b	**111.** c	**112.** e	**113.** a	**114.** c
115. b	**116.** a	**117.** c	**118.** b	**119.** e	**120.** c
121. a	**122.** b	**123.** d			

Index